Face Analysis, Modeling and Recognition Systems

Face Analysis, Modeling and Recognition Systems

Editor

Brygida Cullen

Face Analysis, Modeling and Recognition Systems
Edited by **Brygida Cullen**

ISBN: 978-1-68117-201-9
Library of Congress Control Number: 2016934748

© 2017 by
SCITUS Academics LLC,
www.scitusacademics.com
Box No. 4766, 616 Corporate Way,
Suite 2, Valley Cottage,
NY 10989

Notice

Preface

Face recognition is a type of biometric software application that can identify a specific individual in a digital image by analysing and comparing patterns. A facial recognition system is a computer application capable of identifying or verifying a person from a digital image or a video frame from a video source. One of the ways to do this is by comparing selected facial features from the image and a facial database. Facial recognition systems are commonly used for security purposes but are increasingly being used in a variety of other applications. In addition to being used for security systems, establishments have found a number of other applications for facial recognition systems. Some facial recognition algorithms identify facial features by extracting landmarks, or features, from an image of the subject's face. For instance, an algorithm may analyse the relative position, size, and/or shape of the eyes, nose, cheekbones, and jaw. These features are then used to search for other images with matching features. Other algorithms normalize a gallery of face images and then compress the face data, only saving the data in the image that is useful for face recognition. A probe image is then compared with the face data. One of the earliest successful systems is based on template matching techniques applied to a set of salient facial features, providing a sort of compressed face representation.

The book named Face Analysis, Modeling and Recognition Systems offers a concise and inclusive analysis of artificial face recognition domain through major areas of interest: biometrics, robotics, image databases and cognitive models. The topics in this book describe numerous novel face analysis techniques and approach many unsolved issues. It will be of special interest to academics and researchers interested in computer vision, biometrics, image processing, pattern recognition and medical diagnosis.

Table of Contents

CHAPTER 1

New Principles in Algorithm Design for Problems of Face Recognition

Vitaliy Tayanov

Vyacheslav Chornovil State Institute of Modern Technologies and Management of Lviv, Ukraine

1. INTRODUCTION

This chapter is devoted to two main problems in pattern recognition. First problem concerns the methodology of classification quality and stability estimation that is also known as classification reliability estimation. We consider general problem statement in classification algorithms design, classification reliability estimation and the modern methods solution of the problem. On the other hand we propose our methods for solution of such kind of a problem. In general this could be made by using of different kind of indicators of classification (classifier) quality and stability. All this should summarize everything made before with our latest results of solution of the problem. Second part of the chapter is devoted to new approach that gives the possibility to solve the problem of classifier design that is not sensitive to the learning set but belongs to some kind of learning algorithms.

Let us consider recognition process, using the next main algorithms, as some parts of the some complicated recognition system: algorithm for feature generating, algorithms for feature selection and classification algorithms realizing procedure of decision making. It is really important to have good algorithms for feature generating and selection. Methods for feature generating and selection are developed a lot for many objects to be recognized. For facial images the most popular algorithms use 3D graph models, morphological models, the selection some geometrical features using special nets, etc. From other hand very popular algorithms for feature generating and selection are those which use Principle Component Analysis (PCA) or Independent Component Analysis (ICA). PCA and ICA are good enough for a number of practical cases. However there are a lot of different deficiencies in classifier building. For classifiers using learning the most essential gap is that all such classifiers can work pretty well on learning stage and really bad for some test cases. Also they can work well enough if classes are linearly separated e.g. Support Vector Machines (SVM) in linear case. For non-linear case they have a number of disadvantages. That is why it is important to develop some approaches for algorithms building that are not sensitive for the kind of sample or complexity in classification task.

All classification algorithms built for the present could be divided almost into 5 groups: algorithms built on statistical principles, algorithms, built on the basis of potential functions and non-parametrical estimation, algorithms, using similarity functions (all kinds of metrical classifiers like 1NN and kNN classifiers) algorithms, using logical principles like decision lists, trees, etc, hierarchical and combined algorithms. A lot of recognition systems use a number of algorithms or algorithm compositions. For their optimization and tuning one uses special algorithms like boosting. These compositions can be linear or non-linear as well.

To build any effective classifier we need to use some algorithms that allow us to measure the reliability of classification. Having such algorithms we can find estimates of optimal values of classifier parameters. Accuracy of these estimates allows us to build reliable and effective classification algorithms. They perform the role of indicators of measurement of different characteristics and parameters of the classifier.

We propose the new approach for object classification that is independent of the learning set but belongs to some kind of learning algorithms. Methods, using for new classification approach concerns the field of results combination for the classification algorithms design. One of the most progressive directions in this area assumes using of the consensus in recognition results, produced by different classifiers.

Idea of usage of consensus approach is the following. One divides all objects to be recognized into three groups: objects, located near separation hyperplane (ambiguity objects), objects, located deeply inside the second class and belong to the first one (misclassified objects) and objects that are recognized correctly with large enough index of reliability (easy objects). The group of ambiguity objects is the largest one that can cause errors during recognition due to their instability. Because of that it is extremely important to detect such kind of objects. Next step will be detecting of the true class for every of ambiguity objects. For this it is planned to use apparatus of cellular automata and Markov models. It is important to mark that such an approach allows us to reduce the effect of overestimation for different recognition tasks. This is one of the most principle reasons for using such kind of algorithms.

The practical application of consensus approach for the task of face recognition could be realized by the following way. If we use one of the following classifiers e.g. 1NN, kNN, classifier, built on the basis of potential functions, SVM, etc. we can have some fiducial interval of the most likely candidates. Fiducial interval in general is the list of the candidates that are the most similar to the target object (person). If we use decision making support system controlled by operator result of the system work could be one or two candidates, given to the operator for the final decision or expertise. If we use an autonomous system the final decision should be made by the system that has to select one of the most likely candidates using some special verification algorithms that analyse the dynamics of behaviour of the object in the fiducial interval under the verifying external conditions and parameters of the algorithm.

2. ESTIMATIONS BUILDING IN PATTERN RECOGNITION

2.1. Some Important Tasks of Machine Learning

The modern theory of machine learning has two vital problems: to obtain precise upper bound estimates of the overtraining (overfitting) and ways of it's overcoming. Now the most precise familiar estimates are still very overrated. So the problem is open for now. It is experimentally determined the main reasons of the overestimation. By the influence reducing they are as follows [Vorontsov, 2004]:

1. The neglect of the stratification effect or the effect of localization of the algorithms composition. The problem is conditioned by the fact that really works not all the composition but only part of it subject to the task. The overestimation coefficient is from several tens to hundreds of thousands;
2. The neglect of the algorithms similarity. The overestimation coefficient for this factor is from several hundreds to tens of thousands. This factor is always essential and less dependent from the task than first one;
3. The exponential approximation of the distribution tail area. In this case the overestimation coefficient can be several tens;
4. The upper bound estimation of the variety profile has been presented by the one scalar variety coefficient. The overestimation coefficient is often can be taken as one but sometimes it can be several tens.

The reason of overtraining effect has been conditioned by the usage of an algorithm with minimal number of errors on the training set. This means that we realize the one-sided algorithms tuning. The more algorithms are going to be used the more overtraining will be. It is true for the algorithms, taken from the distribution randomly and independently. In case of algorithm dependence (as rule in reality they are dependent) it is suggested that the overtraining will be reduced. The overtraining can be in situation if we use only one algorithm from composition of two algorithms. Stratification of the algorithms by the error number and their similarity increasing reduces the overtraining probability.

Let us consider a duplet algorithm-set. Every algorithm can cover a definite number of the objects from the training set. If one uses internal criteria [Kapustii et al., 2007; Kapustii et al., 2008] (for example in case of metrical classifiers) there is the possibility to estimate the stability of such coverage. Also we can reduce the number of covered objects according to the stability level. To cover more objects we need more algorithms. These algorithms should be similar and have different error rate.

There is also interesting task of redundant information decrease. For this task it is important to find the average class size guaranteeing the minimal error rate. The reason in such procedure conditioned also by the class size decrease for the objects interfering of the recognition on the training phase.

The estimation of the training set reduction gives the possibility to define the data structure (the relationship between etalon objects and objects that are the spikes or non-informative ones). Also the less class size the less time needed for the decision making procedure. But the most important of such approach consists in possibility to learn precisely and to understand much deeper the algorithms overtraining phenomenon.

In this paper we are going to consider the metrical classifiers. Among all metrical classifiers the most applied and simple are the *k*NN classifiers. These classifiers have been used to build practical target recognition systems in different areas of human's activity and the results of such classification can be easily interpreted. One of the most appropriate applications of metrical classifiers (or classifiers using the distance function) concerns the biometrical recognition systems and face recognition systems as well.

2.2. Probabilistic Approach to Parametrical Optimization of the Knn Classifiers

The most advanced methods for optimization composition algorithm, informative training set selection and feature selection are bagging, boosting and random space method (RSM). These methods try to use the information containing in the learning sample as much as they can. Let us consider the metrical classifier optimization in feature space, using different metrics. The most general presentation of the measure between feature vectors **x** and **y** has been realized through Manhatten measure as the simple linear measure with weighted coefficients a_i [Moon & Stirling, 2000]:

$$d(x,y) = \sum_{i=1}^{n} a_i \mid x_i - y_i \mid,$$

(1)

where $d(x,y) = \sum_{i=1}^{n} a_i \mid x_i - y_i \mid$ is the arbitrary measure between vectors **x** and **y**.

Minkovski measure as the most generalized measure in pattern recognition theory can be presented in form of

$$d(x,y) = \left(\sum_{i=1}^{n} \mid x_i - y_i \mid^p\right)^{\frac{1}{p}} = \left(\sum_{i=1}^{n} a_i \mid x_i - y_i \mid\right)^{\frac{1}{p}} = C(p)\sum_{i=1}^{n} a_i \mid x_i - y_i \mid,$$

(2)

where parametrical multiplier $C(p)$ have been presented in form of

$$C(p) = \left(\sum_{i=1}^{n} a_i \mid x_i - y_i \mid\right)^{\frac{1-p}{p}} \; ; a_i = \mid x_i - y_i \mid^{p-1} ; p > 0$$

(3)

One can make the following conclusions. An arbitrary measure is the filter in feature space. It determines the weights on features. The weight must be proportional to the increase of one of indexes when it has been added to general feature set used for class discrimination procedure. Such indexes are: correct recognition probability, average class size, divergence between classes, Fisher discriminant [Bishop, 2006]. One can use another indexes, but the way of their usage should be similar. If one of the features does not provide the index increase (or worsen it) the value of such feature weight should be taken as zero. So by force of supplementary decrease of feature number one can accelerate the recognition process retaining the qualitative characteristics. The feature optimization problem and measure selection has been solved uniquely. This procedure has been realized using weighted features and linear measure with weighted coefficients. Feature selection task at the same time has been solved partially. First the feature subset from general set is determined. Such set has been determined by some algorithm (for example by the number of orthogonal transforms). Such algorithm should satisfy the definite conditions like follows: class entropy minimization or divergence maximization between different classes. These conditions have been provided by the Principle Component Analysis [Moon & Stirling, 2000]. The last parameter using in the model is the decision function or decision rule. Number of decision functions can be divided into functions working in feature space and the functions based on distance calculation. For example the Bayes classifier, linear Fisher discriminant, support vector machine etc. work in feature space. The decision making procedure is rather complex in multidimensional feature space when one uses such decision rules. Such circumstance is especially harmful for continuous recognition process with pattern series that have been recognized. Thus realizing the recognition system with large databases in practice one uses classifiers based on distance function. The simplest classifier is 1NN. But this classifier has been characterized by the smallest probability indexes. Therefore one should use kNN one. So the task consists in selection of k value that is optimal for decision making procedure in bounds of fiducial interval. This interval corresponds to the list of possible candidates. Unlike the classical approach k value has upper bound by class size. In classical approach the nearest neighbor value should be taken rather large, approximating Bayes classifier.

Let us consider RS with training. The calculation and analysis of the parameters of such systems is carried out on the basis of learning set. Let there exists the feature distribution in linear multidimensional space or unidimensional distribution of distances. We are going to analyse the type of such distribution. The recognition error probability for $\mu = 0$ could be presented as $\int_{|x|\geq\theta} p(x)dx$, where θ is the threshold. According to the Chebyshev inequality [Moon & Stirling, 2000] we obtain $\int_{|x|\geq\theta} p(x)dx \leq \frac{\sigma^2}{\theta^2}$.

Let us consider the case of mean and variance equality of $p(x)$ distribution. The upper bound for single mode distributions with mode $\mu = 0$ with help of Gauss inequality [Weinstein, 2011] is:

$$P(|x - \mu| \geq \lambda\tau) \leq \frac{4}{9\lambda^2}$$

(4)

where $\tau^2 \equiv \sigma^2 + (\mu - \mu_0)^2$.

Let $\mu = \mu_0 = 0$ and $\tau \equiv \sigma$. Then the threshold θ is $\theta = \lambda\tau = \lambda\sigma$ and $\lambda = \dfrac{\theta}{\sigma}$. Thus the Gauss inequality for the threshold θ could be presented in form of:

$$\int_{|x|\geq\theta} p(x)dx \leq \frac{4\sigma^2}{9\theta^2}.$$

(5)

As seen from (Eq. 5), the Gauss upper bound estimate for the single-mode distribution is better in 2.25 times then for the arbitrary distribution. So the influence of the distribution type on the error probability is significant. The normal distribution has equal values of mode, mean and median. Also this distribution is the most popular in practice. On the other hand the normal distribution has been characterized by the maximum entropy value for the equal values of variance. This means that we obtain the minimal value of classification error probability for the normally distributed classes. For the algorithm optimization one should realize the following steps:

- to calculate the distance vector between objects for the given metric;
- to carry out the non-parametrical estimation of the distance distribution in this vector by the Parzen window method or by the support vector machines;
- to estimate the mean and variance of the distribution;
- on the basis of estimated values to carry out the standardization of the distribution ($\mu = 0$, $\sigma = 1$);
- to build the distributions both for the theoretical case and estimated one by the non-parametrical methods;
- to calculate the mean square deviation between the distributions;
- to find out the parameter space, when deviation between the distributions less then given δ
- level.

2.2.1. Probability Estimation for Some Types of Probability Density Functions

Let us consider some probability density functions (pdfs) that have a certain type of the form (presence of the extremum, right or left symmetry). If pdf have not one of such types of structure one can use the non-parametrical estimation. As the result of such estimation we get the uninterrupted curve describing pdf. This function can be differentiated and integrated by the definition. Because the Gaussians have been characterized by the minimal error of the classification for the given threshold θ and does not exceed $\dfrac{4\sigma^2}{9\theta^2}$ (see Eq. 5) for the unimodal and symmetric pdf or pdf with right asymmetry, the double-sided inequality for the given value of recognition error can be presented in form of :

$$0.5(1 - erf(\frac{\theta}{\sigma})) \leq \varepsilon \leq \frac{4\sigma^2}{9\theta^2}.$$

(6)

where $\mu = 0$.

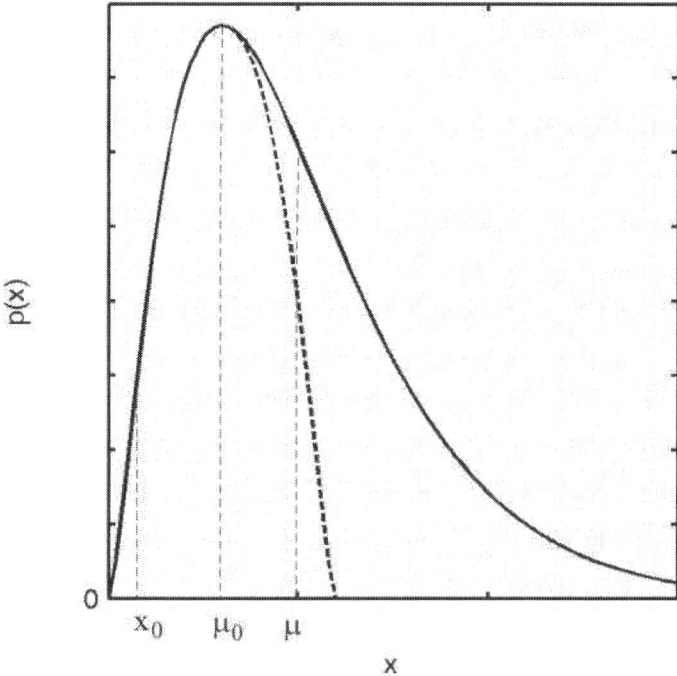

Figure 1. Right asymmetry of pdf

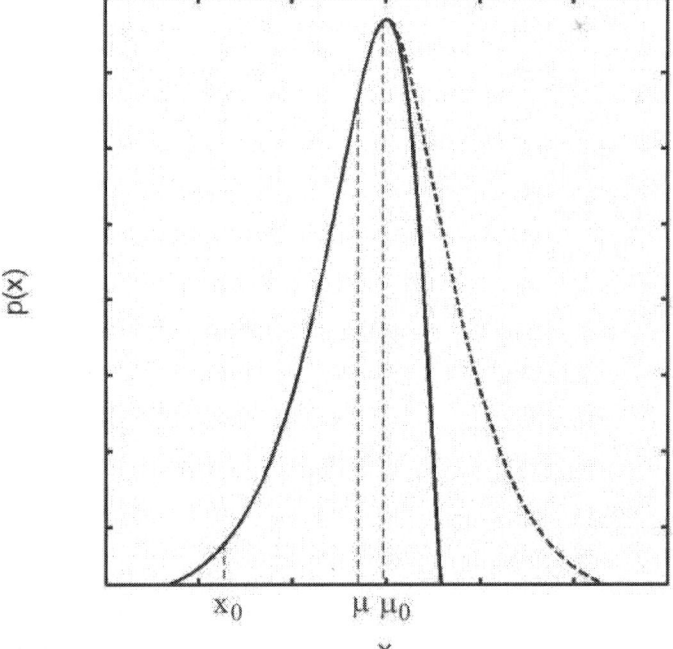

Figure 2. Left asymmetry of pdf

Let us analyse the form of potentially generated pdfs of distances between objects. All of the distributions will have extremum. This will be conditioned by following facts. All of the pdfs have been determined on the interval $[0,\infty)$ and the density near zero and for the large distances is not high because these values are mostly unlikely. The right asymmetry is much more likely because pdf of distances is limited by zero and from the other side it has no strictly determined limitations.

Let's consider a widespread problem of classification in the conditions of two classes. We will denote the size of classes as s_1 and s_2 correspondingly. Then if the probability of replacement of object of a class having size s_1 within a fiducial interval is equal to ε_1 the probability of no replacement of objects from the same class by objects from a class s_2 in this interval is equal to $(1 - \varepsilon_1)^{s_2}$ under the condition of independence of objects [Kapustii et al, 2008; Kyrgyzov, 2008; Tayanov & Lutsyk, 2009]. For other class at corresponding changes this probability is equal to $(1 - \varepsilon_2)^{s_1}$. If now one selects some virtual class and admits that replacement of any object of this class by objects from the mentioned two classes is authentic event it is possible to write down a following equation:

$$\gamma\left((1-\varepsilon_1)^{s_2} + (1-\varepsilon_2)^{s_1}\right) = 1,$$

(7)

where the proportionality multiplier is calculated trivially.

Sometimes there are situations when distances between objects are equal to 0. Thus non-parametrically estimated distribution of one of the classes can have a maximum in a point corresponding to zero distance. Let density of distributions are equal $p_1(0)$ and $p_2(0)$ in a zero point. The estimation of relation between probabilities can be set in a form of $\frac{p_1(0)^{s_2}}{p_2(0)^{s_1}}$ $\ln\frac{p_1(0)^{s_2}}{p_2(0)^{s_1}}$. Thus it is necessary to make boundary transition from cumulative density function (cdf) to pdf as they are connected among themselves by differentiation operation. The relation $\ln\frac{p_1(0)^{s_2}}{p_2(0)^{s_1}}$ or generally $\left(\ln\frac{p_2(0)^{s_1}}{p_1(0)^{s_2}}\right)$ can be used for construction of the following classifier

$$\ln\frac{p_1(\theta)^{s_2}}{p_2(\theta)^{s_1}} > \gamma_1; \qquad \ln\frac{p_2(\theta)^{s_1}}{p_1(\theta)^{s_2}} > \gamma_2;$$

or

$$\ln\frac{p_1(\theta)^{s_2}}{p_2(\theta)^{s_1}} < \gamma_1, \qquad \ln\frac{p_2(\theta)^{s_1}}{p_1(\theta)^{s_2}} < \gamma_2,$$

(8)

where values $\ln\frac{p_1(\theta)^{s_2}}{p_2(\theta)^{s_1}} = 0$ or $\ln\frac{p_2(\theta)^{s_1}}{p_1(\theta)^{s_2}} = 0$ have no influence on classification results and the decision can be accepted for benefit of any class. In case of non-parametric estimation the probability of such value is almost equal to 0. This approach is especially useful for the recognition tasks with similar objects i.e. objects that are week separated in the feature space. It should be noted that such type of algorithms have been oriented on the tasks with high level of class overlapping. Face recognition belongs to the tasks that have sufficiently a lot of objects that could not be separated so easy.

2.3. Combinatorial Approach

Let us present the recognition results for kNN classifier in form of binary sequence:

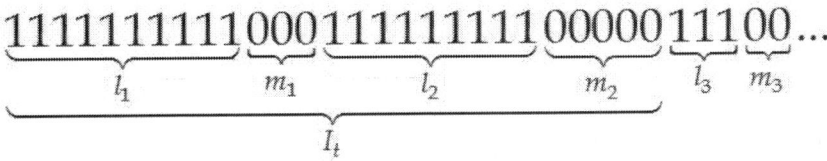

Figure 1. The recognition results in form of binary sequence for kNN classifier

Using kN classifier it is important that among k nearest neighbours we have the related positive objects majority or the absolute one. Let us consider the simpler case meaning the related majority. The kNN classifier correct work consists in fact that for k nearest neighbours it has to be executed the condition

$$\left| \bigcup_i \tilde{l}_i \right| > \left| \bigcup_i \tilde{m}_i \right|, \ i = 1, 2, 3... ,$$

(9)

where $\tilde{l}_i$, $\tilde{m}_i$ are the groups that appear after class size decrease. Under the group one understands the homogeneous sequence of elements. In such sequence (see Fig.1) there exist patterns of all classes. In general case there is no direct conformity between the group number and the class number although.

Let us consider the case of non-pair k value in kNN classifier only. This means that we have the case of synonymous classification. Such univocacy could disappear in case of pair k value and votes equality for different classes.

Let us estimate the effect of class size reduction in case of kNN classifier. Note that reduced class sizes are equal to each other and equal s^*. Let us consider the kNN classifier correct work condition: $ENT\left(\dfrac{k}{2}\right) + 1 \leq s^*$. In contradistinction to 1NN classifier there is no such an importance of the first nearest patterns of the true class. Thus all such sequences one could denote as l_i. Let us determine the probabilities that it will be selected s^* patterns from the true class by the combinatorial approach. These probabilities have fiducial sense. This means that for the given part of positive objects there will be no selections among the patterns of the false classes by the correspondent combinatorial way. The multiplication of pointed two probabilities determines the probability of kNN classifier correct work. Let assign q_j as the recognition error probability for the corresponding m_i groups:

The combinatorial expression for q_j probability could be written in form of:

$$q_1 = P\left(\inf\left(\left|\bigcup_i m_i\right| \right) \ge ENT\left(\frac{k}{2}\right) + 1 \right);$$

$$q_2 = P\left(\inf\left(\left|\bigcup_i m_i\right| \right) + |m_{i+1}| \ge ENT\left(\frac{k}{2}\right) + 1 \right);$$

$$q_3 = P\left(\begin{array}{c} \inf\left(\left|\bigcup_i m_i\right| \right) + |m_{i+1}| + |m_{i+2}| \ge \\ \ge ENT\left(\frac{k}{2}\right) + 1 \end{array} \right);\dots$$

$$q_j = P\left(\begin{array}{c} \inf\left(\left|\bigcup_i m_i\right| \right) + \left|\bigcup_j m_{i+j-1}\right| \ge \\ \ge ENT\left(\frac{k}{2}\right) + 1 \end{array} \right);\dots \tag{10}$$

$$q_j = \frac{\displaystyle\sum_{j=ENT\left(\frac{k}{2}\right)+1}^{s^*} C^j_{\left|\bigcup_{i,j} m_{i+j-1}\right|} \, C^{s^*-j}_{s-\left|\bigcup_{i,j} m_{i+j-1}\right|}}{C^{s^*}_s}, \quad \left|\bigcup_{i,j} m_{i+j-1}\right| \ge ENT\left(\frac{k}{2}\right) + 1. \tag{11}$$

The fiducial probability for arbitrary true pattern sequence is equal:

$$P_{q_j} = \frac{\displaystyle\sum_{j=ENT\left(\frac{k}{2}\right)+1}^{s^*} C^j_{\left|\bigcup_i l_i\right|} \, C^{s^*-j}_{s-\left|\bigcup_i l_i\right|}}{C^{s^*}_s}. \tag{12}$$

Thus the correct recognition probability for kNN classifier has been determined by probability (Eq. 12) and addition to probability (Eq. 11):

$$P_j = P_{q_j}(1-q_j) = \frac{\displaystyle\sum_{j=ENT\left(\frac{k}{2}\right)+1}^{s^*} C^j_{\left|\bigcup_i l_i\right|} \, C^{s^*-j}_{s-\left|\bigcup_i l_i\right|}}{C^{s^*}_s} -$$

$$\frac{\left(\displaystyle\sum_{j=ENT\left(\frac{k}{2}\right)+1}^{s^*} C^j_{\left|\bigcup_i l_i\right|} \, C^{s^*-j}_{s-\left|\bigcup_i l_i\right|} \right)\left(\displaystyle\sum_{j=ENT\left(\frac{k}{2}\right)+1}^{s^*} C^j_{\left|\bigcup_{i,j} m_{i+j-1}\right|} \, C^{s^*-j}_{s-\left|\bigcup_{i,j} m_{i+j-1}\right|} \right)}{\left(C^{s^*}_s\right)^2}. \tag{13}$$

It is modelled the recognition process with different sequences of patterns of true and false classes for the 1NN and kNN classifiers in case of absolute majority. For modelling the face recognition system has been taken. The class size (training set) has been taken as 18 according to the database that it was made. On the Fig.1 the results of modelling of the training set decrease influence on the recognition results for the 1NN classifier have been presented. On the Fig.2 the similar results for kNN classifier under condition $ENT\left(\dfrac{k}{2}\right)+1=s^{*}$ have been presented.

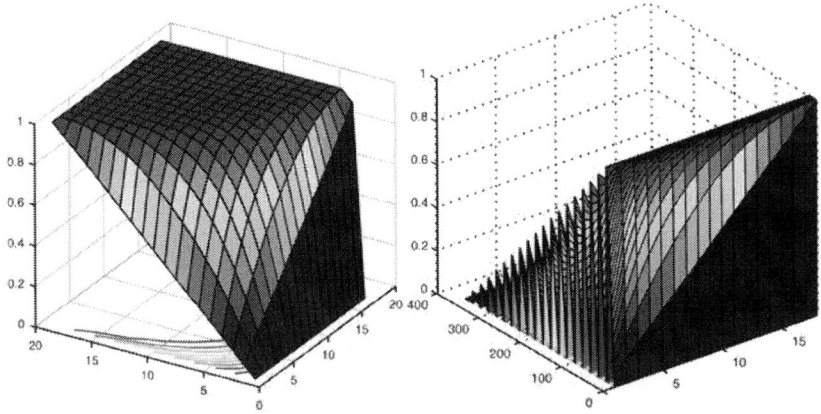

Figure 2. The probability of correct recognition as function of training set (x axis) and number of true/false objects in the target sequence (y axis) for the 1NN classifier

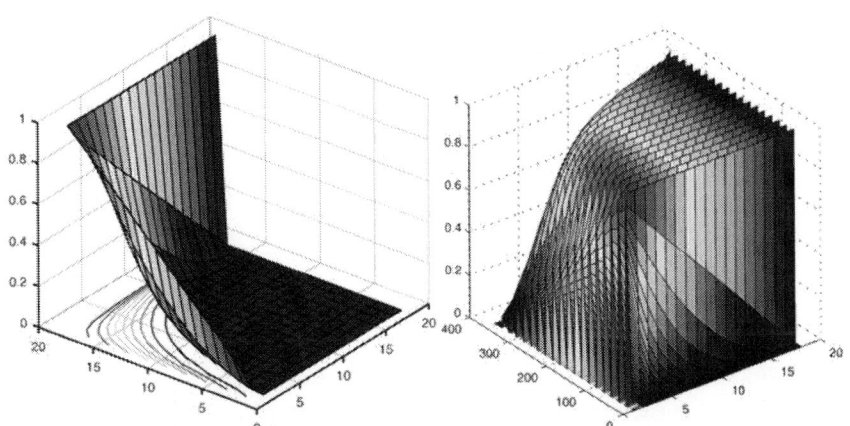

Figure 3. The probability of correct recognition as function of training set (x axis) and number of true/false objects in the target sequence (y axis) for the kNN classifier

On the Fig.1, 2 x axis means the size of the training set and the y axis means the size of the true patterns sequence (left picture) and sequence of both true and false patterns (right picture). The y axis has been formed by the following way. We organized 2 cycles where we changed the number of true and false patterns. For every combination of these patterns and different class sizes we calculate the probability of correct recognition.

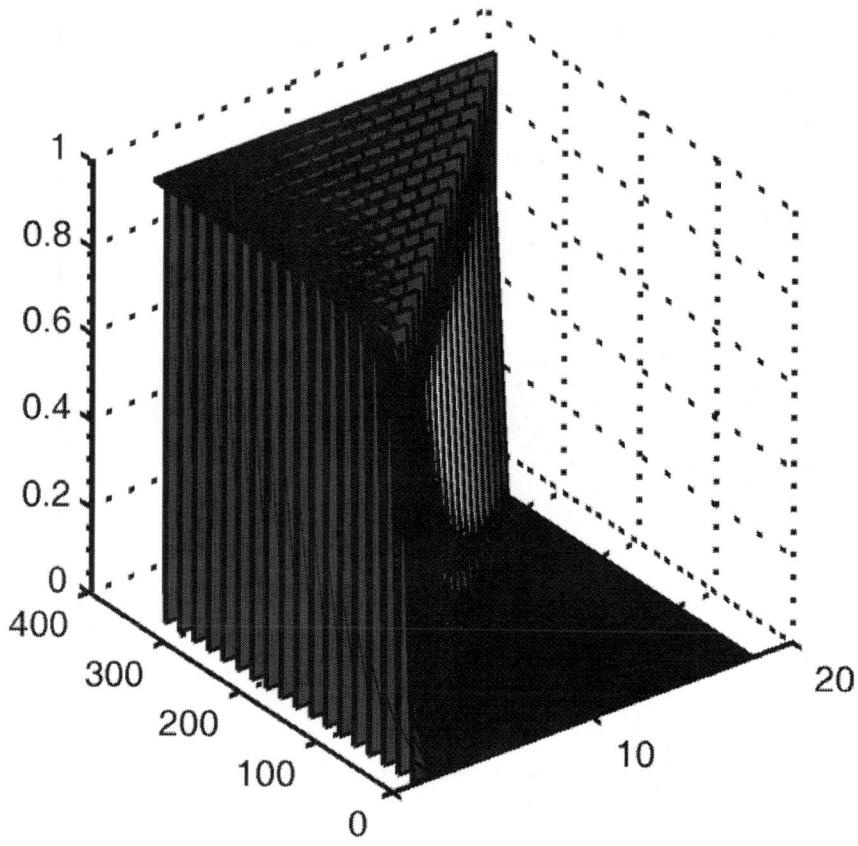

Figure 4. The probability of correct recognition as function of training set (x axis) and $ENT\left(\frac{k}{2}\right)+1$ value (y axis)

On the Fig.4, 5 the results of kNN classifier modelling have been presented. Here it has been satisfied the following condition: $ENT\left(\frac{k}{2}\right)+1 \leq s^a$. On the Fig. 5 the fiducial probability as function of training set size (xx axis) and $ENT\left(\frac{k}{2}\right)+1$ value (yy axis).

The probability part of proposed approach is based in following idea. Despite of combinatorial approach, where the recognition results were determined precisely, we define only the probability of the initial sequence existence. Due to low probability of arbitrary sequence existing (especially for the large sequences) it has been determined the probability of homogeneous sequences existing of the type {0} or {1}. This probability has been determined on the basis of the last object in given sequence as probability of replacing this object (the object from the true class {1} by the others objects of the false classes from the database. This means that the size of homogeneous sequence has been determined by the most "week" object in the homogeneous pattern sequence. The probability of existing of the non-homogeneous sequences is inversely proportional to the $2^{|l+m|}$ value, where $|l+m|$ is the sequence size. This procedure could be realized using distribution function (fatigue function) of the distances between the objects. This approach has been developed for metrical classifiers and classifiers on the basis of distance function in [Kapustii et al., 2008; Tayanov & Lutsyk, 2009]. Thus we need to calculate the probability of sequence with true patterns existing that has definite size or for the given probability rate we need to calculate the maximal size of the sequence that satisfies this probability. For the binary sequence the sum of the weights of the lower order bits is always less than the next most significant bit.

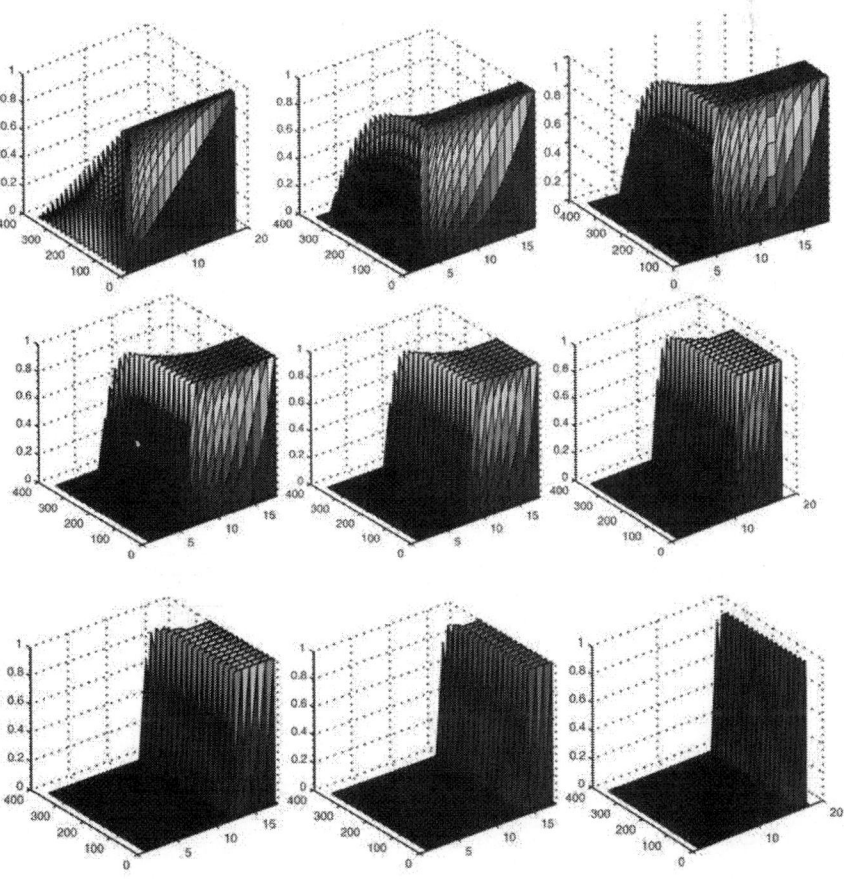

Figure 5. The probability of correct recognition as function of $^{ENT\left(\frac{k}{2}\right)+1}$ (x axis) and number of true/false objects in the target sequence (y axis)

The difference is equal to 1. This means that arbitrary pattern replacement of the true class in the fiducial interval is equivalent to the alternate replacement of the previous ones. The minimal whole order of the scale of notation that has such peculiarity is equal to 2. Thus we need to calculate the weights of the true patterns position and compare them with binary digit. Such representation of the model allows us to simplify the probability calculation of the patterns replacement from the true sequences by the patterns of false classes. On the other side the arbitrary weights can be expressed through the exponent of number 2 that also simplifies the presentation and calculation of these probabilities. So the probability of the homogeneous sequence of the true patterns existence has been calculated on the basis of distance distribution function and is the function of the algorithm parameters. We should select the sequence of the size that has been provided by the corresponding probability. We after apply the combinatorial approach that allows us to calculate the influence effect of the class size decrease on the recognition probability rate. Thus the probabilistic part of the given approach has been determined by the recognition algorithm parameters. So the integration of both probabilistic and combinatorial parts allows us to define more precise the influence of the effect of the training set reduction.

Let us consider step by step the example of fast computing of replacement of true pattern probability from the sequence where relation between weights of the objects is whole exponent of number 2. Thus for example the weights can be presented by the following way: $w = \{2^9, 2^6, 2^4, 2^3, 2^2, 2^1, 2^0\}$. As known the probability of replacement of true object from the sequence by the false one when it is known that replacement is true event is inversely proportional to the weights of these objects. Let define the probability of replacement of the object having the 2^9 weight comparatively to the object with 2^6 weight. As far as we do not know what object has been replaced the total weight of the fact that there will not be replaced the objects with 2^6 weight and lower is equal: $w = \{2^9, 2^6, 2^4, 2^3, 2^2, 2^1, 2^0\}$. This weight can be expressed trough 2^6 weight accurate within 1 by following way: $2^6(1 + 0.5) = 1.5*2^6$. In case of large sequences this one has week influence on the accuracy. The relation between 2^9 and 2^6 is equal to 8. In case of divisible group of events we obtain the $8\lambda + 1.5\lambda=1$ equation, where the proportional coefficient λ approximately equal to 0.11. So the probability of non-replacement of the object with 2^9 weight is equal $8*0.11=0.88$. The object with 2^6 weight has the corresponding probability equal to $1 - 0.88 = 0.12$. Since we know exactly that replacement is the true event and the last object has weight equal to 1 the accuracy correction that equal to 1 makes the appropriate correction of probability calculation.

3. CLASSIFICATION ON THE BASIS OF DIVISION OF OBJECTS INTO FUNCTIONAL GROUPS

Algorithms of decision making are used in such tasks of pattern recognition as supervised pattern recognition and unsupervised pattern recognition. Clustering tasks belong to unsupervised pattern recognition. They are related to the problems of cluster analysis. Tasks where one provides the operator intervention in the recognition process belong to the learning theory or machine learning theory. The wide direction in the theory of machine learning has the name of statistical machine learning. It was founded by V.Vapnik and Ja. Chervonenkis in the sixties-seventies of the last century and continued in nineties of the same century and has the name of Vapnik-Chervonenkis theory (VC theory) [Vapnik, 2000]. It should be noted that classification algorithms built on the basis of training sets are mostly unstable because learning set is not regular (in general). That is why it has been appeared the idea of development of algorithms that partially use statistical machine learning but have essentially less sensitivity to irregularity of the training sets.

This chapter focuses on tasks that partially use learning or machine learning. According to the general concept of machine learning a set is divided into general training and test (control) subsets. For the training subset one assumes that the class labels are known for every object. Using test subsets one verifies the reliability of the classification. The reliability of algorithms has been tested by methods of cross-validation [Kohavi, 1995; Mullin, 2000].

Depending on the complexity of the classification all objects can be divided into three groups: items that are stable and are classified with high reliability ("easy" objects), objects belonging to the borderline area between classes ("ambiguous" objects) and objects belonging to one class, and deeply immersed inside another one ("misclassified" objects). Among those objects that may cause an error the largest part consists of terminal facilities. Therefore it is important to develop an algorithm that allows one to determine the largest number of frontier facilities. The principal idea of this approach consists in preclassification of objects by dividing them into three functional groups. Because of this it is possible to achieve much more reliable results of classification. This could be done by applying the appropriate algorithms for every of obtained groups of objects.

3.1. The Most Stable Objects Determination

The idea of the model building is as follows. The general object set that have to be classified is divided on three functional groups. To the first group of objects the algorithm selects the objects with high level of classification reliability. The high level of reliability means that objects are classified correctly under the strong (maximal) deviations of the parameters from optimal ones. From the point of view of classification complexity these objects belongs to the group of so called "easy" objects. The second group includes objects, on which there is no consensus. If one selects two algorithms in a composition of algorithms, they should be as dissimilar as possible and they should not be a consensus. If one uses larger number of algorithms, the object belongs to the second group if there is no consensus in all algorithms. If consensus building uses intermediate algorithms, parameters of which are within the intervals between the parameters of two the most dissimilar algorithms, this makes it impossible to allocate a larger number of objects, on which there is no consensus. Dissimilarity between algorithms is determined on the basis of the Hamming distance between results of two algorithms defined as binary sequences [Kyrgyzov, 2008; Vorontsov,

2008]. In practice this also means that in general it will not be detected the new objects, if one uses composition of more than two algorithms, on which one builds the consensus. The third group consists of those objects, on which both algorithms have errors, while they are in consensus. The error caused by these objects can not be reduced at all. Thus the error can not be less than the value determined by the relative amount of objects from the third group. The next step will be the reclassification of the second group of objects. This special procedure allows us to determine the true class, to which a particular object belongs to. Reclassifying the second group of objects we can also have some level of error. This error together with the error caused by the third group will give the total error of all proposed algorithms.

The research carried out in this paper concerns the analysis of statistical characteristics of the results of a consensus generating by two algorithms. The objective of the task analysis is a statistical regularity of characteristics of various subsets taken by division of the general set into blocks of different size. Probability distribution by the consensus for three groups of objects has been carried out by nonparametric estimation using Parzen window with Gauss kernels.

3.1.1. Experimental Results
3.1.2. Case of Three Classifiers in the Consensus Composition

Figs 6-11 show the parametrically estimated pdfs for the probability of a correct consensus, the probability of incorrect consensus and the probability that consensus will not be reached.

As can be seen from the figures these distributions can be represented using one-component, two-component or multicomponent Gauss mixture models (GMM). In multicomponent GMM weights determined according to their impact factors. Distribution parameters (mean and variance) and weights of impact in the model are estimated using EM algorithm. Estimation of corresponding probability values was carried out by blocks with a minimal size of $Q=30$ and $Q=200$ elements. The size of these blocks has been driven by a small sample size which according to various criteria ranges from 30 to 200 items. According to the standard definition of a small sample it is assumed that sample is small when it is characterized by irregular statistical characteristics.

As seen from all figures the estimates obtained by blocks with a minimal size of 30 elements and some more are irregular. This means that for these tasks the sub-sample size of 30 items and some more is small. This has been indicated by long tails in the corresponding probability distributions. The maximum in zero point for two-component model is characterized by a large number of zero probabilities. This can be possible if there are no mistakes in the consensus of two algorithms. Estimates of probabilities on the basis of average values and the corresponding maximum probability distributions (for maximum likelihood estimation (MLE)) are not much different, which gives an additional guarantee for the corresponding probability estimates. Significance of obtained consensus

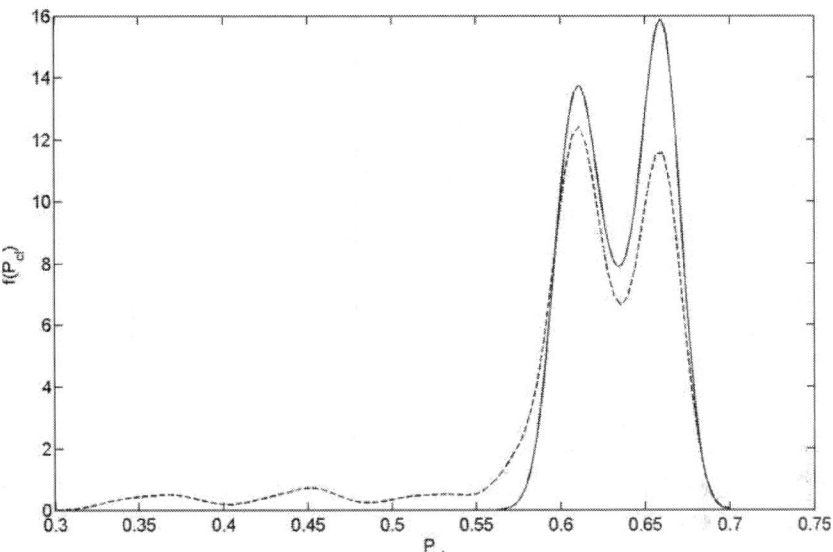

Figure 6. Task "pima" from UCI repository: non-parametrical estimation of the pdf of the correct consensus that consists of two algorithms (solid line is used for the set of 200 objects and dot line is used for set of 30 objects correspondingly).

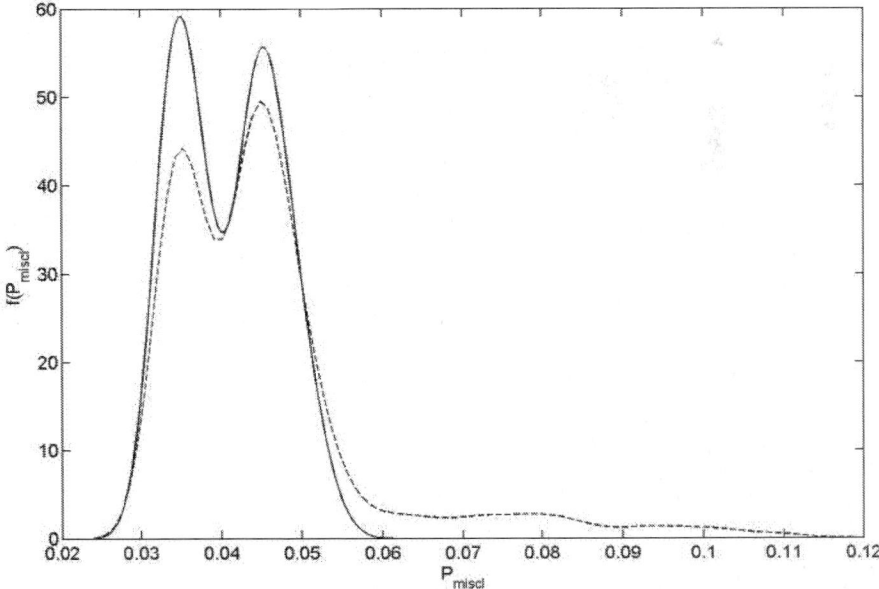

Figure 7. Task "pima" from UCI repository: non-parametrical estimation of the pdf of the incorrect consensus that consists of two algorithms (solid line is used for the set of 200 objects and dot line is used for set of 30 objects correspondingly)

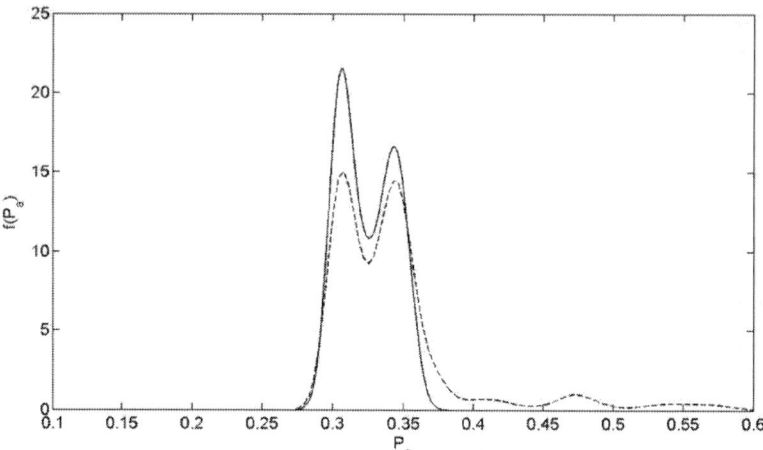

Figure 8. Task "pima" from UCI repository: non-parametrical estimation of the pdf of no consensus between two algorithms (solid line is used for the set of 200 objects and dot line is used for set of 30 objects correspondingly)

estimates of probabilities of correct consensus, incorrect consensus and probability that consensus will not be achieved, provides a classification complexity estimate. Problems and algorithms for the complexity estimation of classification task is discussed in [Basu, 2006]. For example, tasks "pima" and "bupa" are about the same level of complexity because values of three probabilities are approximately equal. Tab. 1 shows that all algorithms excepting proposed new one have large enough sensitivity to the equal by the classification complexity tasks they work with. Mathematical analysis of composition building of algorithms has been considered in details in [Zhuravlev, 1978].

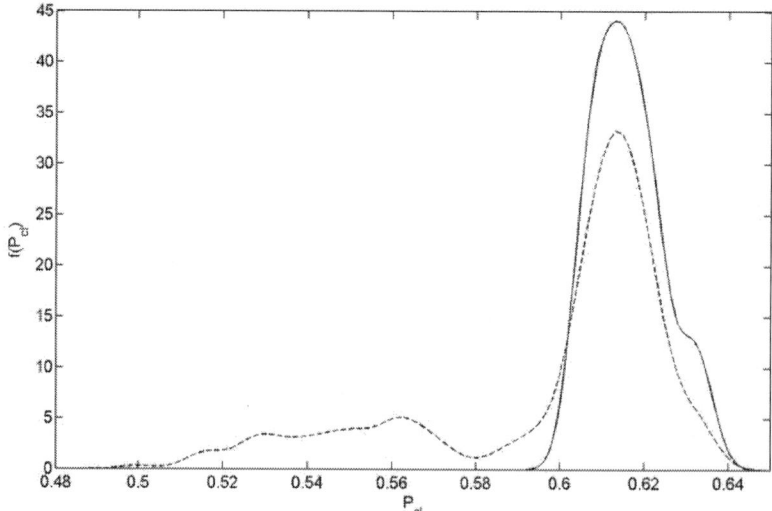

Figure 9. Task "bupa" from UCI repository: non-parametrical estimation of the pdf of the correct consensus that consists of two algorithms (solid line is used for the set of 200 objects and dot line is used for set of 30 objects correspondingly)

Figs 6-11 show graphic dependencies of consensus results for problems taken from repository UCI. This repository is formed at the Irvin's University of California. The data structure of the test tasks from this repository is as follows. Each task is written as a text file where columns are attributes of the object and rows consist of a number of different attributes for every object. Thus the number of rows corresponds to the number of objects and the number of columns corresponds to the number of attributes for each object. A separate column consists of labels of classes, which mark each object. A lot of data within this repository has been related to biology and medicine. Also all these tasks could be divided according to the classification complexity. In the data base of repository there exists a number of tasks with strongly overlapped classes. Some of them will be used for research.

In Table 1. one gives the probabilities of errors obtained on the test data for different classifiers or classifier compositions. All these algorithms were verified on two tasks that are difficult enough from the classification point of view. For the proposed algorithm it has been given the minimal and maximal errors that can be obtained on given tested data.

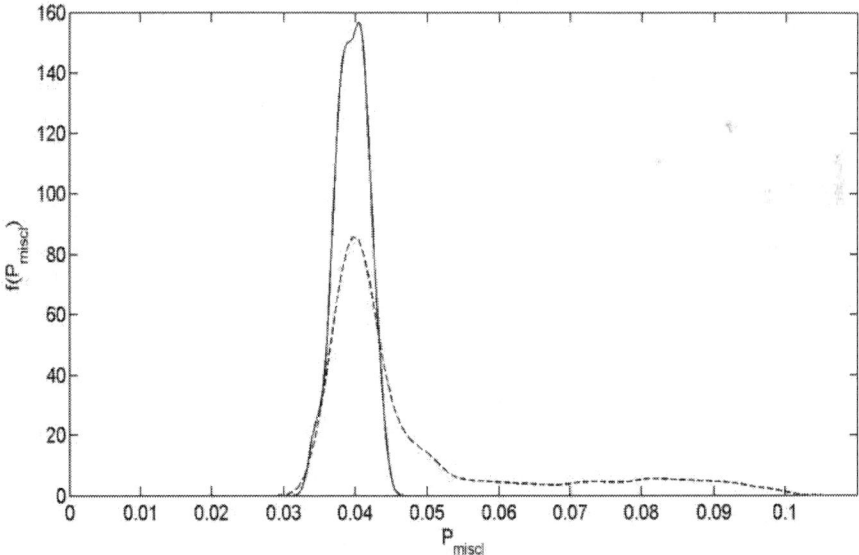

Figure 10. Task "bupa" from UCI repository: non-parametrical estimation of the pdf of the incorrect consensus that consists of two algorithms (solid line is used for the set of 200 objects and dot line is used for set of 30 objects correspondingly)

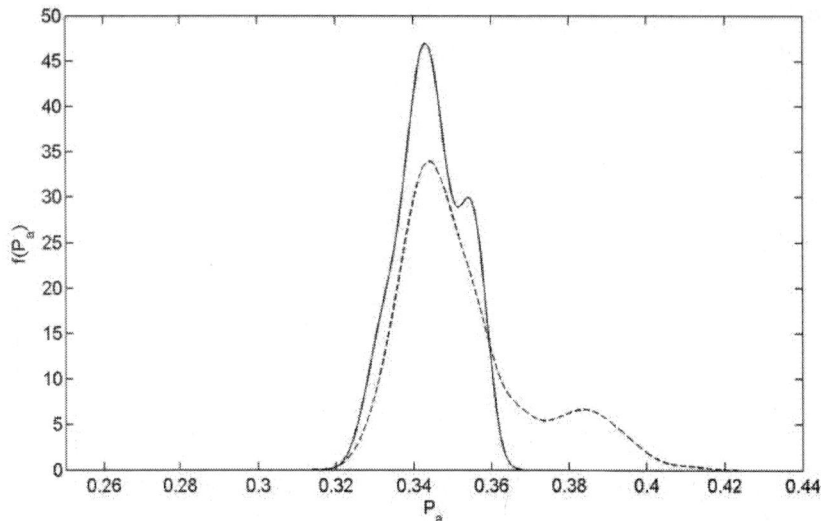

Figure 11. Task "bupa" from UCI repository: non-parametrical estimation of the pdf of no consensus between two algorithms(solid line is used for the set of 200 objects and dot line is used for set of 30 objects correspondingly)

In Table 1. the value of minimal error is equal to consensus error for the proposed algorithm. The value of maximal error has been calculated as sum of minimal error and the half of the related amount of objects, on which there is no consensus (fifty-fifty principle). As seen from the table the value of maximal error is much less than the least value of error of all given algorithms for two tasks from UCI repository. In comparison with some algorithms given in the table the value of minimal error is approximately 10 times less for the proposed algorithm then the error of some other algorithms from the table. The proposed algorithms are characterized by much more stability of the classification error in comparison with other algorithms. It can be seen from corresponding error comparison for two tasks from the UCI repository.

Table 1. Error of classification for different algorithms

Algorithm Task	bupa	pima
Monotone (SVM)	0.313	0.236
Monotone (Parzen)	0.327	0.302
AdaBoost (SVM)	0.307	0.227
AdaBoost (Parzen)	0.33	0.290
SVM	0.422	0.230
Parzen	0.338	0.307
RVM	0.333	-
Proposed algorithm (min/max)	0.040/0.212	0.041/0.203

Table 2. Task "pima" from UCI repository

	Q=200		Q=30	
	μ	σ	μ	σ
P_c	0.635	0.024	0.611	0.064
P_e	0.041	0.006	0.046	0.013
$P_{\bar{c}}$	0.324	0.019	0.344	0.052

Table 3. Task "bupa" from UCI repository

	Q=200		Q=30	
	μ	σ	μ	σ
P_c	0.635	0.024	0.611	0.064
P_e	0.041	0.006	0.046	0.013
$P_{\bar{c}}$	0.324	0.019	0.344	0.052

In tabs 2-3 the estimates of probability of belonging of every object from the task of repository UCI to every of three functional groups of objects have been given. In this case the objects, on which consensus of the most dissimilar algorithms exists (Pc), belong to the class of so called "easy" objects. Then objects, on which both of algorithms that are in consensus make errors (Pe), belong to the class of objects that cause uncorrected error and this error can not be reduced at all. The last class of objects consists of objects, on which there is no consensus of the most dissimilar algorithms (Pc-). This group of objects also belongs to the class of border objects. In the tables one gives variances of corresponding probabilities too. Minimal size of the blocks, on which one builds estimates using algorithms of cross-validation changes from 30 to 200.

In the previous case we analysed the classifier composition that consists of two the most dissimilar algorithms. Now we are going to build the classifier composition that consists of three algorithms. The third algorithms we choose considering the following requirements. These algorithms have to be exactly in the middle of two the most dissimilar algorithms. This means that the Hamming distance between the third algorithm and one of the most dissimilar algorithms is equal to the distance between the "middle" algorithm and the second algorithm in the consensus composition of two the most dissimilar algorithms. In Tabs 4 and 5 the results of comparison of two consensus compositions have been given. The first composition consists of two algorithms and the second one consists of three algorithms correspondingly. As in the previous case we used "pima" and "bupa" testing tasks from UCI repository.

Table 4. Task "pima" from UCI repository

	consensus of two classifiers		consensus of tree classifiers	
	μ	σ	μ	σ
P_c	0.635	0.024	0.607	0.021
P_e	0.041	0.006	0.0347	0.006
$P_{\bar{c}}$	0.324	0.019	0.358	0.017

As seen from the both tabs there is no big difference between two cases. Consensus of two algorithms can detect a bit lager quantity of correctly classified objects that means a bit more reliable detection of correctly classified objects. Consensus of three algorithms can detect a bit larger quantity of objects on which we have no consensus (the third group of objects). But if we will use the "fifty-fifty" principle for detection objects from the third group the general error of classification will be the same. We can also note that the variances of two consensuses compositions have no large differences between each other.

Table 5. Task "bupa" from UCI repository

	consensus of two classifiers		consensus of tree classifiers	
	μ	σ	μ	σ
P_c	0.616	0.008	0.586	0.012
P_e	0.040	0.002	0.037	0.002
$P_{\bar{c}}$	0.344	0.008	0.377	0.013

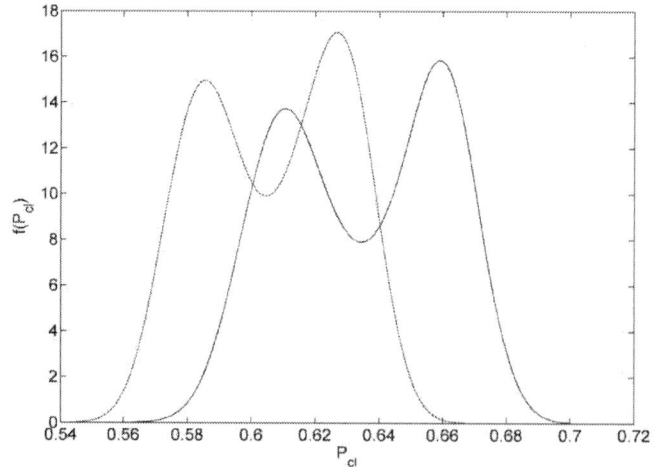

Figure 12. Task "bupa" from UCI repository: non-parametrical estimation of the pdf of correct consensus between two (solid line) and tree (dot-line) algorithms

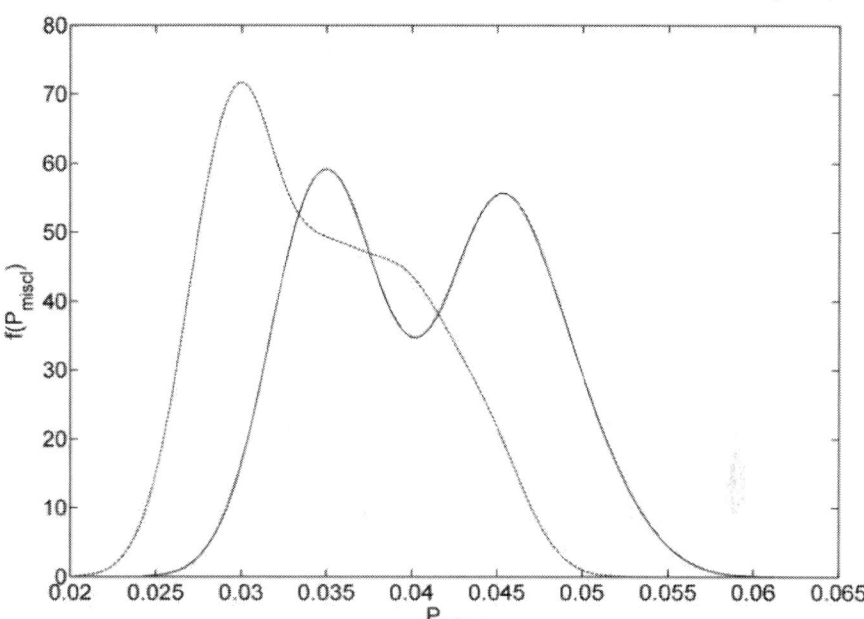

Figure 13. Task "bupa" from UCI repository: non-parametrical estimation of the pdf of incorrect consensus between too (solid line) and tree (dot-line) algorithms

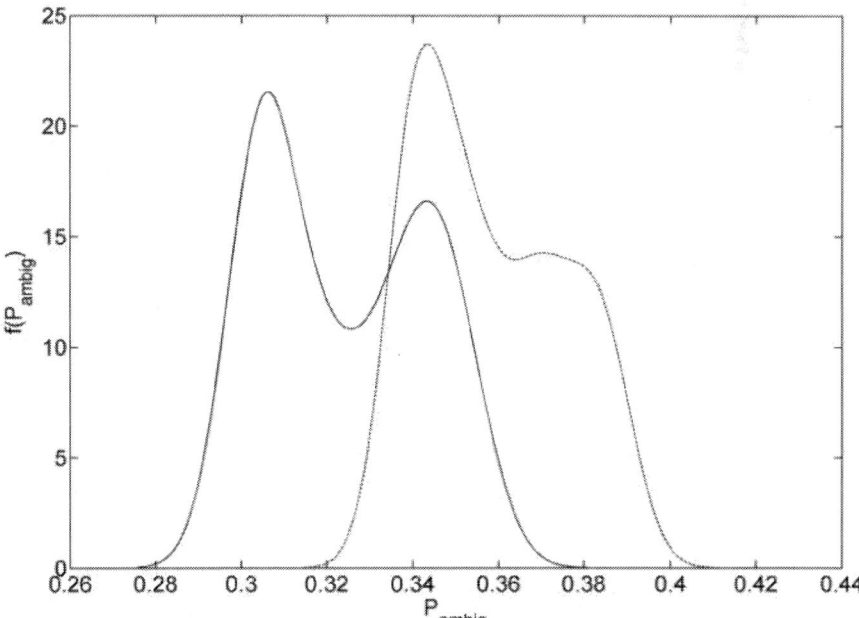

Figure 14. Task "bupa" from UCI repository: non-parametrical estimation of the pdf of no consensus between too (solid line) and tree (dot-line) algorithms

On figs 12-17 the results of consensus building for three algorithms have been given. Here we also use two tasks from the UCI repository as in the case of two algorithms. According to figs corresponding to the task of "bupa" we can make the following conclusions. In comparison to the case of two algorithms we can see that for the number of preclassified groups of objects we have just shifts between the corresponding pdfs and the form of curves is approximately the same. We can also note that relative value of the shift is rather small (about 5% for the pdf of correct probability). This shift is almost conditioned by the statistical error of determining of the most different algorithms.

According to figs corresponding to the task of "pima" we can mark that differences in forms of pdfs are more essential than in previous task. This circumstance could be used for comparison of the task complexity using the value of overtraining as stability to learning. Using such approach it is possible to obtain much more precise and informative estimations of the complexity from the learning process point of view.

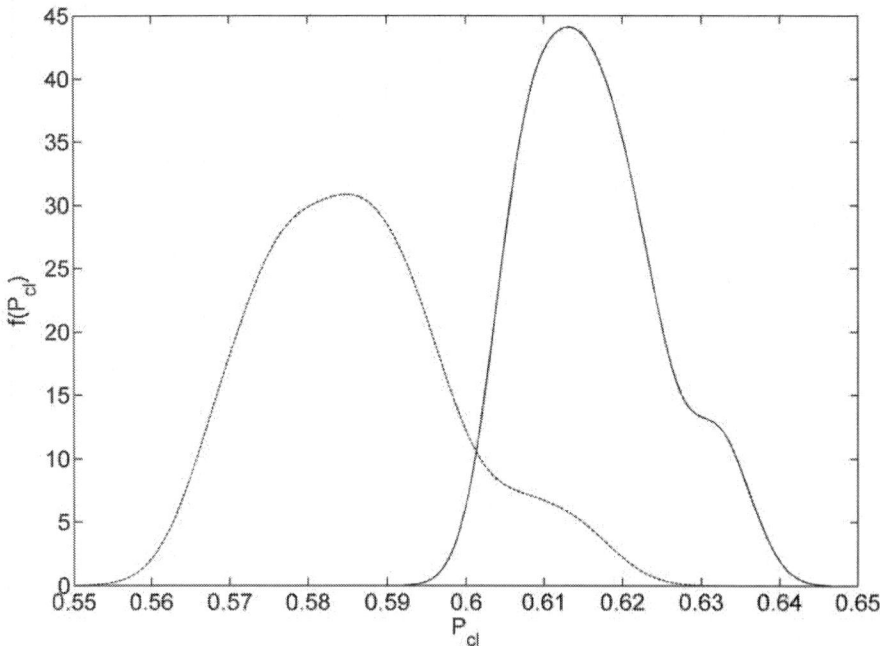

Figure 15. Task "pima" from UCI repository: non-parametrical estimation of the pdf of correct consensus between too (solid line) and tree (dot-line) algorithms

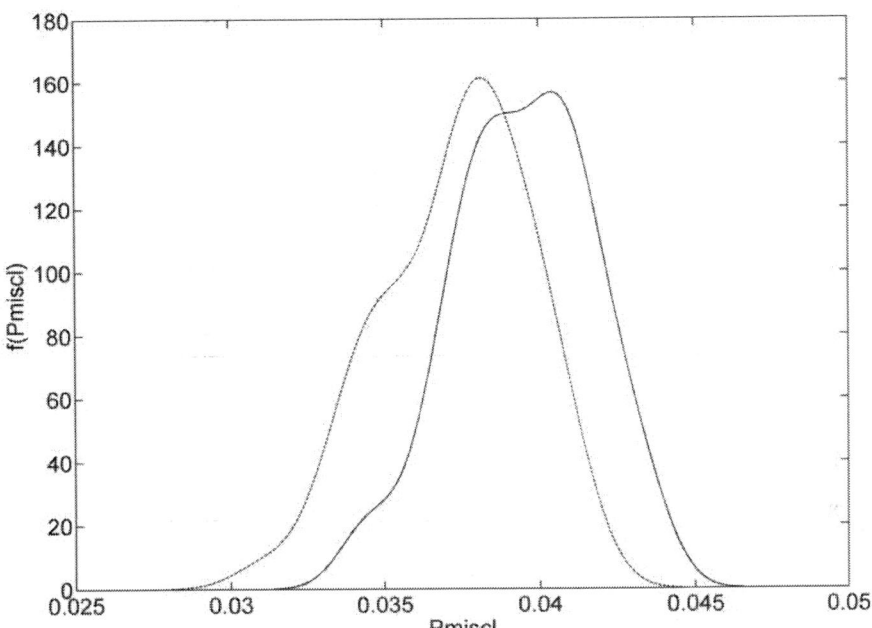

Figure 16. Task "pima" from UCI repository: non-parametrical estimation of the pdf of incorrect consensus between too (solid line) and tree (dot-line) algorithms

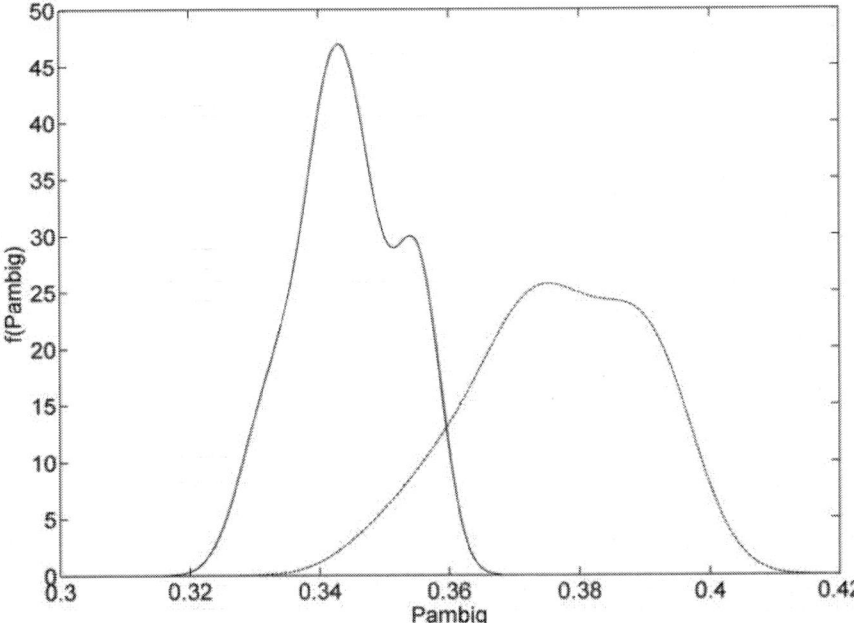

Figure 17. Task "pima" from UCI repository: non-parametrical estimation of the pdf of no consensus between too (solid line) and tree (dot-line) algorithms

4. SPECIFIC OF USAGE OF THE PROPOSED APPROACH FOR PROBLEMS OF FACE RECOGNITION

The problem of face recognitions is one of the principle tasks of the large project connected with determining of human behaviour and psychoanalysis of the human, based on the face expression and body movement. Such type of systems belongs to the class of no contact systems. Unlike to the human recognition systems based on fingerprints or images of iris these systems do not require human to keep finger of eyes near (or on) the scanner. This is also very important from law point of view. It is impossible to force a human to put the finger on the scanner if he does not want to do this and if this is not a criminal case. The same fact is concerned the case of iris recognition systems. To take a picture of somebody is not forbidden and this or that person could not be familiar with the fact that somebody took already picture of the face of such person. This is really important when creating the training and test databases. Face recognitions systems can be joined with hidden video cameras installed in shops, supermarkets, banks and other public places. Here it is important to hide the fact of video surveillance. This could be done with help of no contact recognition systems only. On other hand the facial information and mimicry could be used for the human behaviour determination and psychophysical state of the human. This is important to avoid and predict of acts of terrorism. Here it is very important information about dynamics of face expression and movement of the separate parts of the face.

In spite of the fact that face recognition systems has larger value of error of both of the types than finger print recognition systems, iris recognition systems and others, they find a lot of different applications because of their flexibility of installation, training and testing. In this situation it is very important to make research in the field of recognition probability estimation, overtraining estimation, model parameters estimation, etc. to find the most optimal parameters of the face recognition systems. To build very reliable recognition systems it is important to use proposed approach that allows us to build hierarchical recognition on the basis of objects division into functional groups and due to this to use the effect of preclassification.

For the procedure of decision making one proposes to use the notation of fiducial interval. By the fiducial interval one understands the list of possible candidates for the classification. Usage of the fiducial interval is very useful for the decision making support systems with presence of an operator. The result of the system work is the list of candidates that are the most similar to the object to be recognized. In this case the final decision about the object will be made by operator. The system can work as completely autonomous one with using of the fiducial interval for the decision making. In fiducial interval there exist several group of objects that belong to their own class. Our task is to find the group of objects that corresponds to the object to be recognised or to make decision that there is no corresponding objects in fiducial interval. The idea of fiducial interval consists in following concepts. The size of the fiducial interval (the number of possible candidates) has to be enough to be sure that if the corresponding objects are in the database of the recognition system they will drop into this interval. The size of the fiducial interval corresponds to the fiducial probability. The larger is fiducial interval the larger is fiducial probability. That is why it is convenient to use the notation of fiducial interval for the probability of the fact that corresponding objects will drop in the list of possible candidates. The second paragraph of this chapter has been devoted to the problems of forming of some types of fiducial intervals.

5. DISCUSSION AND FUTURE WORK

In this chapter we shortly considered some approaches for solution of such important problems as recognition reliability estimation and advanced classification on the basis of division of objects into three functional groups. In domain of reliability estimation there exist two principal problems. First problem concerns the tasks of statistical estimation of the probability of correct recognition especially for small training sets. This is very important when we can not achieve additional objects so fast and make our training set more representative. That could be in situations when we work with data slowly changing in time.

Another important problem concerns the effect of overestimation in pattern classification. The value of overestimation could be found as difference between the recognition results on training and test sets. In the beginning of the chapter one mentioned the main problems of the statistical learning theory and overestimation as one of the most principal problems. One did not pay attention to this problem in this chapter but it is planned to do in future research. The attention has been payed to the problems of recognition reliability estimation. In this chapter the results of both combinatorial and probabilistic approach to recognition reliability estimation have been presented. As seen from the figures there was realized the advanced analysis and estimation of the recognition results when the training set is decreased. So we can make the prognosis of the recognition probability for reduced training sets using combinatorial approach. The reliability of such approach can be provided on the basis of probabilistic approach.

It was considered some methods of the reliability estimation for some types of classifiers. Such of the classifiers belongs to the group of so called metrical classifiers or classifiers on the basis of dissimilarity functions or distance functions. It will be interesting to consider the proposed methods in case of other types of classifiers e.g. classifiers using separating hyperplane, classifiers built on logic functions and others. It will be interesting to consider the idea of how to express one classifier through another or to build relations between the different types of classifiers. All this could give us the possibility to use one approach to reliability estimation for any type of the classifiers.

In the second part of the chapter the probability of belonging of every object to each of the three groups of objects: a group of "easy" objects, on which it is reached the correct consensus of two algorithms, a group of objects, on which two the most dissimilar algorithms have an incorrect consensus and a group of objects, on which one does not achieve consensus have been considered. The analysis shows that there are probability distributions of data that can be presented as a multicomponent models including GMM. All this makes it possible to analyze the proposed algorithms by means of mathematical statistics and probability theory. From the figures and tables one can see that the probability estimations using methods of cross-validation with averaged blocks of 30 and 200 elements minimum differ a little among themselves, which makes it possible to conclude that this method of consensus building, where consensus consists in the most dissimilar algorithms, is quite regular and does not have such sensitivity to the samples as other algorithms that use training. As seen from the corresponding tables, the minimum classification error is almost less by order of magnitude than error for the best of existing algorithms. The maximal error is less of 1.5 to 2 times in comparison with other algorithms. Also, the corresponding errors are much more stable both relatively to the task, on which one tests the algorithm and the series of given algorithms where the error value has significantly large variance. Moreover, since the minimal value of error is quite small and stable, it guaranties the stability of receipt of correct classification results on objects, on which consensus is reached by the most dissimilar algorithms. Relatively to other algorithms such a confidence can not be achieved. Indeed, the error value at 30-40% (as compared to 4%) gives no confidence in results of classification. The fact that the number of ambiguous objects selected by two the most different algorithms is less than the number of objects selected by three algorithms conditioned by the overtraining of two the most dissimilar algorithms. So the future research in this domain should be devoted to the problem of overtraining of the ensemble of two the most dissimilar algorithms. This means that it should be reduced the overtraining of the preclassification that allows us to reduce the error of classification gradually and due to this to satisfy much more reliable classification.

REFERENCES

1. M. Basu, T. Ho, 2006 Data complexity in pattern recognition, Springer-Verlag, 1-84628-171-7
2. C. Bishop, 2006 Pattern recognition and machine learning, Springer-Verlag, 0-38731-073-8 York
3. B. Kapustii, B. Rusyn, V. Tayanov, 2007 Mathematical model of recognition systems with smalldatabases. Journal of Automation and Information Sciences, 39 10 7080 , 1064-2315
4. B. Kapustii, B. Rusyn, V. Tayanov, 2008 Features in the Design of Optimal Recognition Systems. Automatic control and computer sciences, 42 2 6470 , 0146-4116
5. B. Kapustii, B. Rusyn, V. Tayanov, 2008 Estimation of the Influence of Information Class Coverage on Generalized Ability of the k-Nearest-Neighbors Classifier. Automatic control and computer sciences, 42 6 283287 , 0146-4116

6. R. Kohavi, 1995 A study of cross-validation and bootstrap for accuracy estimation and model selection, Proceedings of 14th International Joint Conference on Artificial Intelligence, 11371145 , Palais de Congres Montreal, Quebec, Canada, 2010

7. I. Kyrgyzov, 2008 Recherche dans les bases de donnes satellitaires des paysages et application au milieu urban :clustering, consensus et categorisation: Ph.D. thesis, l'ecole Nationale Superiere des Telcommunications, Paris

8. T. Moon, S. Stirling, 2000 Mathematical methods and algorithms for signal processing, Prentice-Hall, 0-20136-186-8 Jersey.

9. M. Mullin, R. Sukthankar, 2000 Complete cross-validation for nearest neighbour classifiers, Proceedings of International Conference on Machine Learning, 639646 , 2000

10. V. Tayanov, O. Lutsyk, 2009 Classifier Quality Definition on the Basis of the Estimation Calculation Approach. Computers & Simulations in Modern Science, Mathematical and Computers in Science and Engineering, A series of Reference Books and Textbooks, 166171 , 1790-2769

11. V. Vapnik, 2000 The Nature of Statistical Learning Theory (2), Springer-Verlag, 0-38798-780-0 York

12. K. Vorontsov, 2010 Exact combibatorial bounds on the probability of overfitting for empirical risk minimization. Pattern Recognition and Image Analysis, 20 3 269285 1054-6618

13. K. Vorontsov, 2008 On the influence of similarity of classifiers on the probability of overfitting pattern recognition and image analysis: new information technologies, Proceedings of International Conference on Pattern Recognition and Image Analysis: new information technologies (PRIA-9), 2 , Nizhni Novgorod, Russian Federation, 303306 , 2000

14. E. . Weinstein, , 2011 Gauss's Inequality, In: A Wolfram Web Resource, 16.03.2011, Available from http://mathworld.wolfram.com

15. J. Zhuravlev, 1978 An Algebraic Approach to Recognition and Classification Problems. Problems of cybernetics, 33 568 (in Russian

CHAPTER 2

3D Multi-Spectrum Sensor System with Face Recognition

Joongrock Kim[1], Sunjin Yu[2], Ig-Jae Kim[3] and Sangyoun Lee[1]

[1]Department of Electrical and Electronic Engineering, Yonsei University, 134 Shinchon-dong, Seodaemun-gu, Seoul 120-749, Korea
[2]Department of Broadcasting and Film, Cheju Halla University, 38, Halladaehak-ro, Jeju-si, Jeju-do 690-708, Korea
[3]Imaging Media Research Center, Korea Institute of Science and Technology, Hwarang-ro 14-gil 5, Seongbuk-gu, Seoul 136-791, Korea

ABSTRACT

This paper presents a novel three-dimensional (3D) multi-spectrum sensor system, which combines a 3D depth sensor and multiple optical sensors for different wavelengths. Various image sensors, such as visible, infrared (IR) and 3D sensors, have been introduced into the commercial market. Since each sensor has its own advantages under various environmental conditions, the performance of an application depends highly on selecting the correct sensor or combination of sensors. In this paper, a sensor system, which we will refer to as a 3D multi-spectrum sensor system, which comprises three types of sensors, visible, thermal-IR and time-of-flight (ToF), is proposed. Since the proposed system integrates information from each sensor into one calibrated framework, the optimal sensor combination for an application can be easily selected, taking into account all combinations of sensors information. To demonstrate the effectiveness of the proposed system, a face recognition system with light and pose variation is designed. With the proposed sensor system, the optimal sensor combination, which provides new effectively fused features for a face recognition system, is obtained.

Keywords: image sensor; depth sensor; sensor fusion; face recognition

1. INTRODUCTION

Progress in computer vision and sensor technology has made it possible to acquire various types of information, such as two-dimensional (2D) data at different wavelengths, as well as three-dimensional (3D) information in real-time [1–4]. Since each type of sensor provides different features under various environmental conditions, the performance of an application depends highly on selecting a sensor or multi-sensor combination. Therefore, the selection of appropriate sensors has become one of the most significant factors for high-performance vision systems [2,5,6].

Most of the early computer vision applications were based on visible (RGB) sensors, since a visible image represents what humans perceive visually [1,6]. The range of wavelengths correspond to a frequency range from approximately 380 nm to 700 nm. Images also contain texture information, which is one of the most important types of information for object detection, tracking and classification [7,8]. However, since visible sensor images are strongly distorted by changes in illumination, the performance of visible-sensor-based systems also depends highly on illumination conditions.

The wavelength range of near-infrared (IR) and thermal-IR sensors are from 0.7 μm to 1.0 μm and from 9 μm to 14 μm, respectively. Although these wavelengths are not perceived by the human visual system, IR images contain distinctive features, such as thermal radiation emitted by objects. Furthermore, compared to visible sensor images, infrared images are more invariant to visible illumination changes [9,10]. However, neither near-IR nor thermal-IR images include color (RGB) information. In addition, thermal-IR images do not contain texture information and can be affected by ambient temperature.

Recently, real-time 3D depth sensors, such as the Kinect and time-of-flight (ToF) sensors, have been introduced and have become some of the most useful sensors for vision applications [3,4]. Real-time 3D sensors make it possible to analyze detailed 3D shape information that cannot be acquired by 2D sensors. Since the pixel value of a depth image represents the distance between a camera and an object, many researchers have attempted to apply these sensors to 3D-based applications, such as 3D-based recognition and 3D object modeling. However the distance information from such sensors is highly noisy and cannot support texture information.

As explained above, each sensor provides different information and has advantages and disadvantages in different environments. Therefore, to design more flexible and robust systems, the selection of sensors for a specific application is a very important problem. However, obtaining the optimal sensor combination based on calibrated and fused information from all sensors is very difficult, because of the heterogeneous characteristics of sensors.

In this paper, we propose a sensor system, which we refer to as a 3D multi-spectrum sensor system, consisting of three types of sensors: visible, thermal-IR and ToF sensors. Through the registration of all sensors, visible, near-IR, thermal-IR and 3D information is integrated into a 3D multi-spectrum data framework in real-time. In the data framework, all information from all sensors is calibrated and can be easily fused. With the data framework, we can easily select optimal sensor combinations considering all combinations of sensor information. To show the effectiveness of the proposed system, we apply the system to face recognition in the presence of light and pose variations. With the proposed system, we can obtain fused optimal features for high-performance face recognition. In addition, the proposed system can also be used for surveillance, 3D object modeling and object and human recognition.

This paper is organized as follows. In the next section, we briefly review the related state-of-the art research areas. The proposed sensor system and its application with recognition methods are discussed in Section 3. Section 4 shows experimental results on the use of 3D multi-spectrum face data in face recognition. Conclusions are given in Section 5.

2. RELATED WORKS

Previous works related to this paper are roughly divided into three research areas: sensor registration, applications of time-of-flight cameras and face recognition. Since the proposed system consists of three different sensors (visible, thermal-IR and ToF sensors) at different locations, the images from the sensors need to be registered. Therefore, sensor registration is introduced in this chapter. In addition, recent studies using ToF camera and face recognition in terms of light and pose variation are presented as applications of the proposed system.

2.1. Sensors Fusion

Various sensors, such as visible, thermal-IR and 3D sensors, have been introduced into the commercial market. Since each sensor provides different features, many approaches for combining features to improve the performance of the system have been proposed. In order to fuse sensor data, registration to transform different sets of data into one coordinate system should be accomplished.

Many approaches have been proposed for registering different types of images, such as visible, IR and 3D. In [11], a simple registration between IR and visible images using SIFTis proposed. In [12,13], registration between visible and thermal-IR image data for face recognition involving illumination variations is described. A real-time fusion method of multiple passive imaging sensors, visible, IR, and 3D LADARimaging, is presented in [14]. In [15], a method to register a pair of images captured from visible and IR sensors by line and point matching is presented. In addition, calibration between depth and color sensors by a maximum likelihood solution by using checker board is performed in [16,17]. In [18], a multiple sensor fusion system, which combines RGB-Dvision, lasers and a thermal sensor, in order to detect people, in a mobile robot.

Even though there have been many attempts to develop registration between different types of 2D sensors, perfect registration cannot be achieved using only 2D information. Additionally, even though many approaches for calibration between color and depth images have been proposed, high-performance registration between IR and 3D is still a challenging problem.

Moreover, there have thus far been no attempts to register three types of sensors: IR, visible and 3D sensors.

2.2. Applications of Time-of-Flight Camera

Recently, a ToF camera, which generates full-range distance data in real-time, has been used to extend the application range of 3D data to real-time systems, such as human-computer interaction (HCI) [19,20], surveillance [21,22] and robotics [23–26]. One of the most useful applications of the ToF camera is 3D object modeling in order to minimize errors in 3D data, since this data contains noise, due to the motion and orientation of the object to be acquired, as well as the reflectivity of surfaces [27–29]. In [27], 3D object reconstruction exploits sequential distance images captured at different positions. In [29], a 3D shape scanning method with a ToF camera is presented to improve the quality of 3D scans based on filtering and scan alignment techniques. In [28], a method to generate spatially consistent 3D object models by registering 3D data from multiple views is described. In addition, the lack of information from ToF cameras can be compensated for by using additional sensors [30–32]. A real-time segmentation and tracking technique that fuses depth and RGB color data proposed in [30] solves some of the problems in RGB image-based segmentation and tracking, such as occlusions and fast motion. In [31], a real-time 3D hand gesture method using calibration ToF and RGB cameras improves the detection rate, as well as the handling of hand overlap with the face to allow for complex 3D gestures. In [32], both a ToF camera and stereo vision are used to make more accurate depth images, and a generated depth map for augmented reality scenarios is applied.

2.3. FACE RECOGNITION

Many approaches for face recognition to handle pose and light variation have attempted to use various source data, such as 2D, 3D, thermal-IR and near-IR face data, for computer vision and pattern recognition. Face recognition can be roughly classified into two categories: 2D and 3D data-based face recognition.

In 2D-based face recognition, features invariant under visible light changes are extracted from 2D visible, thermal-IR and near-IR face images. In [33], the techniques for decomposing 2D visible face images into non-negative factors to address illumination changes are presented. In [34], a local ternary pattern (LTP) is proposed that can compensate for the main weaknesses of LBP, including sensitivity to large variations in illumination and to random and quantization noise in uniform and near-uniform image regions. A novel solution for illumination invariant face recognition using near-infrared images and LBP features is proposed in [9]. In [35], a comprehensive and timely review of the literature on the use of infrared imaging for face recognition is presented. In [36], the active appearance model (AAM) is applied to normalize pose and facial expression changes on thermal-IR images, and anatomical features invariant to the exact pattern of facial temperature emissions are extracted for face recognition. In [10], image fusion between visible, near-IR and thermal-IR images is presented, which can enhance the performance of face recognition under uncontrolled illumination conditions. Although IR-image-based face recognition is an effective approach for eliminating visible light changes, the performance still depends on the pose of the face.

One way of dealing with pose and light variations is to use 3D face data, since any face pose can be generated by simple transformations, such as translation, rotation and scaling of 3D face models. In addition, 3D face shape information, such as curvature [37,38], profile [39,40] and range image [41,42], can be extracted from 3D face models, since those features are invariant to pose and light variations. A face recognition system using a combination of color and depth images is proposed in [43,44]. In [45], a novel thermal 3D modeling system using 3D shape, visible and thermal infrared information is proposed that addresses the head pose variation problem in face recognition systems. However, the system cannot acquire thermal 3D data in real-time.

3. PROPOSED 3D MULTI-SPECTRUM SENSOR SYSTEM

In order to integrate various sensor information into one calibrated datum, we propose a novel 3D multi-spectrum sensor system that can provide 3D, visible, near-IR and thermal-IR information. The system consists of ToF, color and thermal-IR sensors. Through the registration step between sensors, we generate calibrated 3D multi-spectrum data in real-time. As an application using the 3D multi-spectrum data, we apply it to a face-recognition system that can address variations in light and pose, as these are the most significant factors causing performance decline [46].

3.1. Proposed System

Our proposed system consists of ToF, color and thermal-IR cameras, as shown in Figure 1. Although the ToF camera provides depth information and a near-IR (gray-scale) image simultaneously in real-time, it does not supply color (RGB) or thermal-IR information. Therefore, we propose a system that can generate 3D multi-spectrum data that include 3D shape, visible and thermal-IR information by registering three different kinds of cameras in real-time. The generated 3D multi-spectrum data created by the system are used for face recognition and to solve problems associated with variations in pose and light. With the proposed system, we can use four different kinds of information: (1) 3D depth data from the ToF camera; (2) near-infrared data from the ToF camera; (3) visible (RGB) data from the color camera and (4) thermal-IR data from the thermal-IR camera.

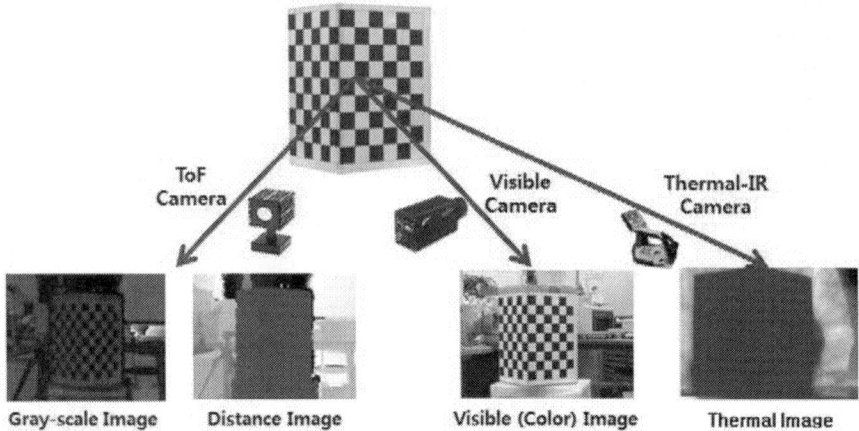

Figure 1. Proposed system comprising ToF, visible and thermal-IR cameras.

Since the thermal-IR and ToF cameras can capture thermal-IR (3–5 µm, 8–12 µm) and NIR(750 nm–1,400 nm) ranges regardless of the visible range (360 nm–820 nm), they can be used in extremely low light conditions. As shown in Figure 2, even though the image captured by the visible range camera is almost black in the dark, the images from the ToF and thermal-IR cameras are not affected by changes in external light. Therefore, there may be a wider range of uses for these cameras beyond the visible range, e.g., in surveillance applications, such as human detection and tracking at night. Even though IR and ToF cameras are almost invariant to light variation, they do not provide color or detailed texture information. Therefore, the proposed system has many advantages in terms of surveillance, robot vision and HCI, which require 3D information, as well as color and thermal-IR information in real-time.

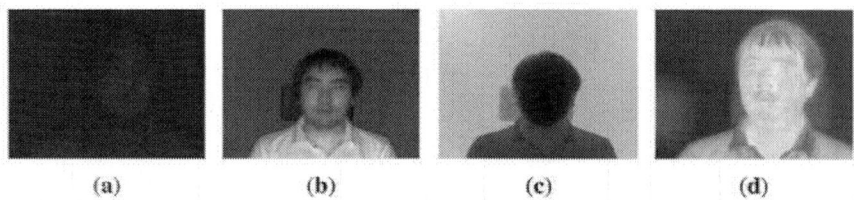

Figure 2. Captured images in a lightless condition; (**a**) visible range image; (**b**) near-infrared (IR) image from the time-of-flight (ToF) camera; (**c**) distance image from the ToF camera; and (**d**) thermal-IR image.

3.2. Registration of T of, Color and Thermal-Ir Cameras

In this step, the registration is used to find visible and thermal-IR information corresponding to 3D data. Since the proposed system consists of three cameras located at different positions, each image from the sensors has different image coordinates. In addition, since each camera provides different features, such as color, thermal-IR, near-IR and 3D information, coordination among cameras is difficult.

Three-dimensional geometry, in which a 3D point (X_W) in a world coordinate system is projected onto a 2D point (x_I) in an image coordinate system, is used for registration between visible or thermal-IR and ToF cameras. As shown in Figure 3, since the origin of the world coordinate (O_W) is the same as the origin of the ToF camera, a point in the world coordinate system can be represented as the camera coordinate system of the ToF camera. In other words, a world coordinate of a point in 3D space is directly acquired as the 3D point of the ToF camera, as described in Equation (1). Additionally, x_I represents a 2D coordinate of the image (pixel) coordinate system of the visible camera. x′$_I$ is a 2D coordinate of the image coordinate system of the ToF camera.

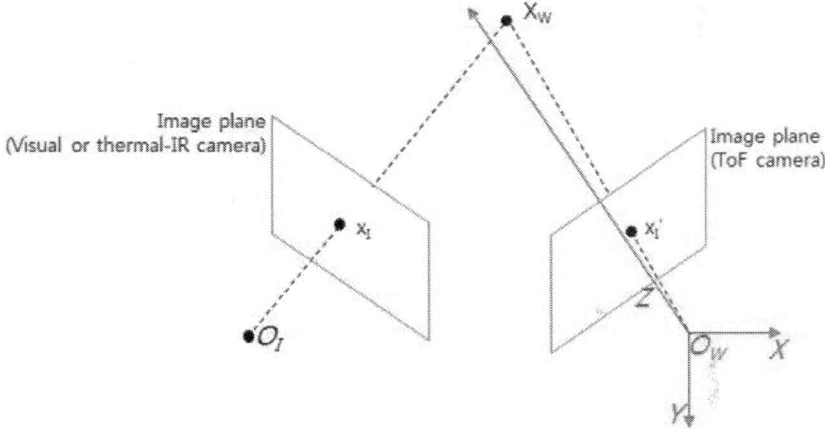

Figure 3. Relationship between a different coordinate system (world (X_W)) and image (x_I, x′I) coordinate system).

$$X_{ToF} = X_W = \begin{bmatrix} x & y & z \end{bmatrix}^T \qquad (1)$$

$$x_I = \begin{bmatrix} u & v \end{bmatrix}^T \qquad (2)$$

$$x_I' = \begin{bmatrix} u' & v' \end{bmatrix}^T \qquad (3)$$

First, we estimate the world coordinate of a 3D point (X_{ToF}) from the image coordinate of a 2D point (x′I) in the image plane of the ToF camera, as shown in Figure 4. In Figure 4, (u_0,v_0) is the principal point of the image plane and (u_{res}, v_{res}) represents the resolution of the distance image.

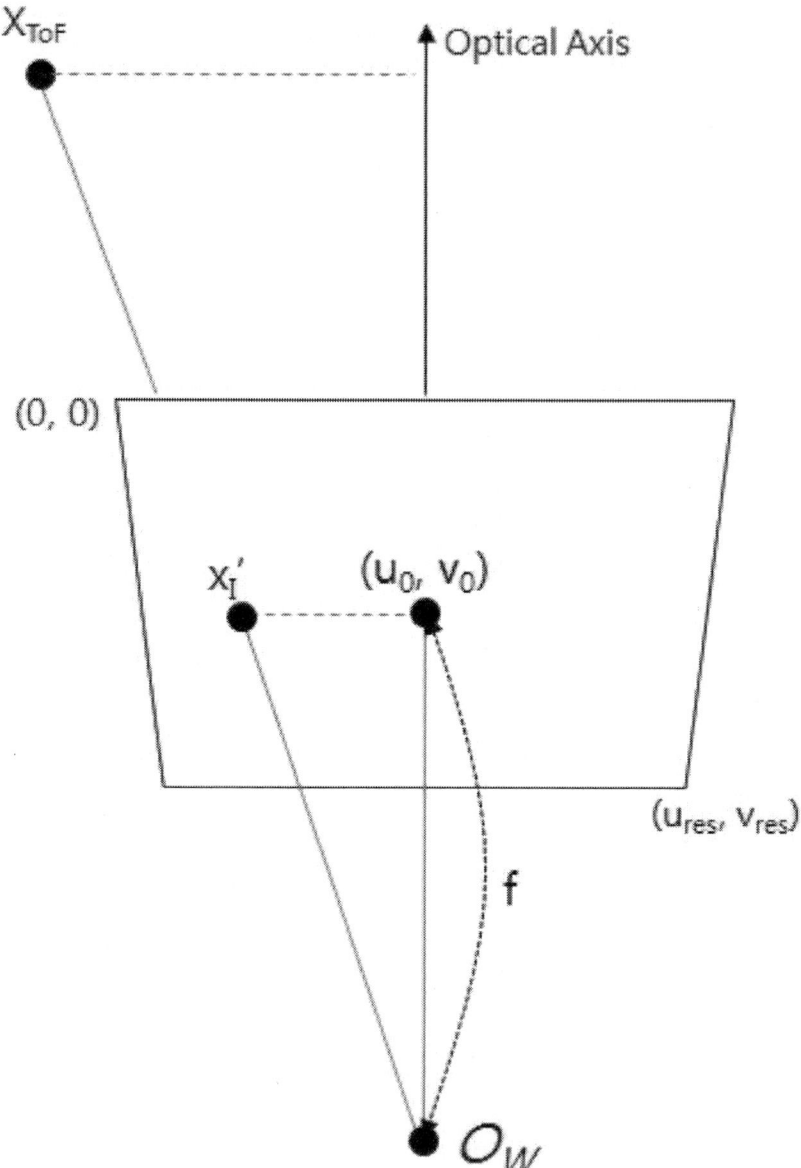

Figure 4. Structure of ToF camera: a 3D point (X_{ToF}) is projected on a 2D point (x'_I) in the image plane of a ToF camera.

A 3D point (X_{ToF}) is projected onto a 2D point ($x'I$) on the image plane. Since the pixel value of a distance image from the ToF camera represents the distance between the camera and an object, we can obtain the distance information (z) from the images. Therefore, we can estimate the x-coordinate of the 3D point (X_{ToF}) according to Equations (4) and (5). f is the focal length of the ToF camera, which can be calculated from the camera calibration [47].

$$\frac{f}{z} = \frac{(u' - u_0)}{x} \tag{4}$$

$$x = \frac{(u' - u_0) \times z}{f} \tag{5}$$

The coordinate of the 3D point (X_{ToF}) can be estimated using the same method as in Equation (6).

$$y = \frac{(v_0 - v') \times z}{f} \tag{6}$$

Therefore, the world coordinate of the 3D point (X_{ToF}) based on the distance image can be calculated using Equation (7).

$$X_{ToF} = \begin{bmatrix} x & y & z \end{bmatrix}^T = \begin{bmatrix} \frac{(u' - u_0) \times z}{f} & \frac{(v_0 - v') \times z}{f} & z \end{bmatrix}^T \tag{7}$$

The correspondence between X_{ToF} and x_I can be represented by Equation (8). C and R indicate the translation and rotation of the camera coordinate with respect to the world coordinate system, respectively. C is a 3 × 1 vector, and R is a 3 × 3 matrix. K is a 3 × 3 camera calibration matrix that includes the internal parameters of the camera [15].

$$x_I = KR^T (X_{ToF} - C) \tag{8}$$

Equation (8) can be represented as in Equation (9), which is a linear equation with a homogeneous coordinate.

$$x_I = (KR^T| - KR^TC) \begin{pmatrix} X_{ToF} \\ 1 \end{pmatrix} \tag{9}$$

In summary, the relationship between X_{ToF} and x_I is given by Equation (10). The 3 × 4 matrix $P = (KT^T| - KR^TC)$ is the projection matrix of the camera.

$$\tilde{x}_I = P\tilde{X}_{ToF} \tag{10}$$

where $\tilde{X}_{ToF} = [x\ y\ z\ 1]^T$ is the homogeneous coordinate of the 3D point ($X_{ToF} = [x\ y\ z]^T$), which is a 4 × 1 column vector with one added as the last element of the vector, and $\tilde{x}_I = [u\ v\ 1]^T$ is the homogeneous coordinate of the 2D point ($x_I = [u\ v]^T$), which is a vector.

We divide the proposed system into on-line and off-line steps, as shown in Figure 5. Before the on-line process, the projection matrices, which represent the relationship between the 3D coordinates of the ToF camera and the 2D projective coordinate of the visible (P_C) and thermal-IR (P_T) cameras, should be estimated off-line. Subsequently, 3D multi-spectrum data are generated and applications, such as face recognition, are completed on-line.

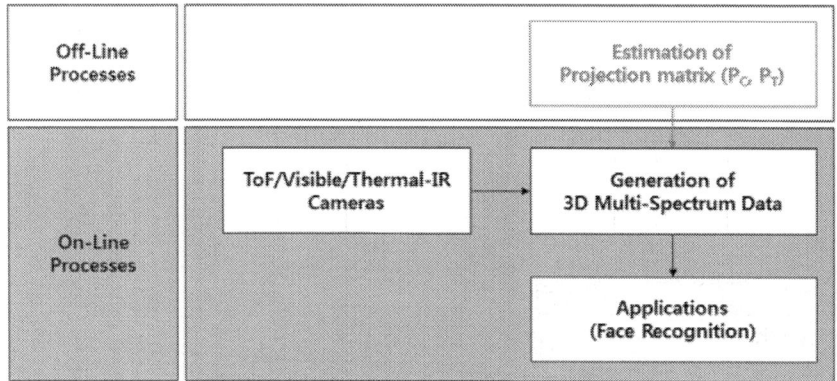

Figure 5. The proposed system can be divided into off-line and on-line processes.

In the proposed system, 3D points can be acquired from the ToF camera, and two kinds of 2D points can be obtained from the color and thermal-IR cameras. Therefore, the two camera projection matrices can be correctly estimated. One is a matrix (P_C) representing the relationship between the 3D points and 2D points of the color images, as in Equation (11), and the other is a matrix (P_T) based on the 2D points of the thermal-IR images, as in Equation (12).

$$\tilde{x}_C = P_C \tilde{X}_{ToF} \tag{11}$$

$$\tilde{x}_T = P_T \tilde{X}_{ToF} \tag{12}$$

where $\tilde{x}_C$ and $\tilde{x}_T$ are homogeneous coordinate representations of 2D points in the color image and the thermal-IR image, respectively, and $\tilde{X}_{ToF}$ is a homogeneous coordinate representation of a 3D point from the ToF camera. Figure 6 illustrates the main concept of the registration between cameras.

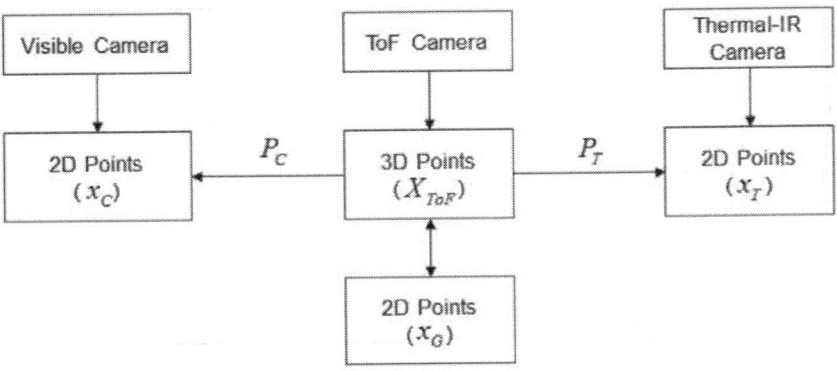

Figure 6. Registration between cameras.

To estimate each projection matrix in an off-line process, the corresponding points having the same feature point in all three cameras should be extracted. To identify the corresponding points between different cameras, we use a calibration rig to extract feature points at the corners of a check pattern in a color image. We also make holes in the corners of the check pattern as feature points for the thermal-IR image. The 2D points (x_C, x_T) for color and thermal-IR can be extracted from each image of the calibration rig shown in Figure 7.

Figure 7. Images of the calibration rig taken by the visible camera (**left**) and the thermal-IR camera (**right**).

Using a ToF camera, 3D points (X_{ToF}) can be acquired directly, since the distance image has already been calibrated using the 2D points (x_G) of the near-IR image from the ToF camera (Figure 8). Therefore, we first extract the feature points that are also the corner points of the check pattern in the near-IR image, as shown in Figure 8a. By using the points extracted from the near-IR images, the 3D coordinates can be extracted from the distance image, as in Figure 8b, as well as from the ToF camera, as shown in Figure 8c.

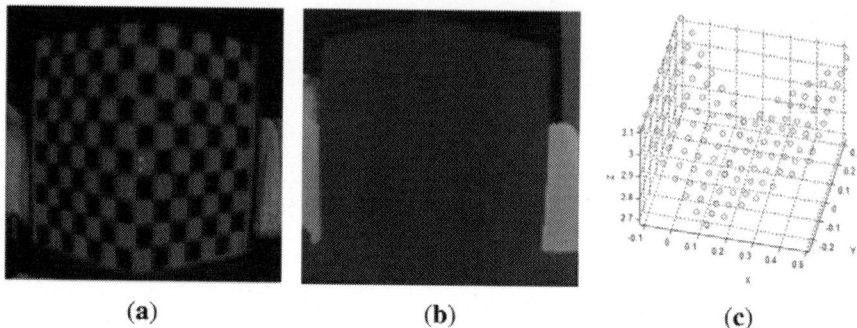

(a)   (b)   (c)

Figure 8. 3D points from the ToF camera: (**a**) near-IR image of the calibration rig from the ToF camera; (**b**) distance image of the calibration rig from the ToF camera; (**c**) extracted 3D coordinates of the calibration rig at the corner point of the check pattern in the near-IR image.

If we know the number of correspondences $x_i \leftrightarrow X_i$, $i = 1, 2, \ldots N$ ($N \geq 6$) between the 3D points (X_i) and the 2D image points (x_i), we can estimate the camera projection matrix using the direct linear transformation (DLT) algorithm, which is a minimization method used to find an approximate linear solution by singular value decomposition (SVD) [47]. Since we can extract a number of correspondences using a calibration rig, we can estimate the camera projection matrices (P_C, P_T). If we know the projection matrix, then we can obtain each corresponding projected 2D image point on the visible and thermal-IR image planes from the 3D coordinates obtained with the ToF camera using Equations (2) and (3), thereby allowing for the acquisition of color and thermal-IR information corresponding to the 3D points.

3.3. Generation Of 3d Multi-Spectrum Face Data

Before 3D multi-spectrum face data can be generated, the noise in 3D distance images from ToF camera must be addressed. We first perform a median filter to remove the salt and pepper type noise in every image. After that, we use an average image of 10 distance images to capture a more precise distance image.

We then apply the previous registration method to find the corresponding color and thermal-IR information to obtain the 3D points of the face to be recognized. In other words, we find the corresponding 2D coordinates of the color and thermal-IR images of the 3D points from the estimated projection matrix (P_C, P_T) in Equations (11) and (12). Finally, we generate 3D multi-spectrum face data, which include visible and thermal-IR textures with 3D shape information.

Figure 9 shows the corresponding distance image (3D shape information) (a) and near-IR image (b) of the thermal-IR image (c) and visible image (d). The red-cross points in Figure 9c,d represent projected correspondence points on the thermal-IR and visible images from the 3D face region in (b).

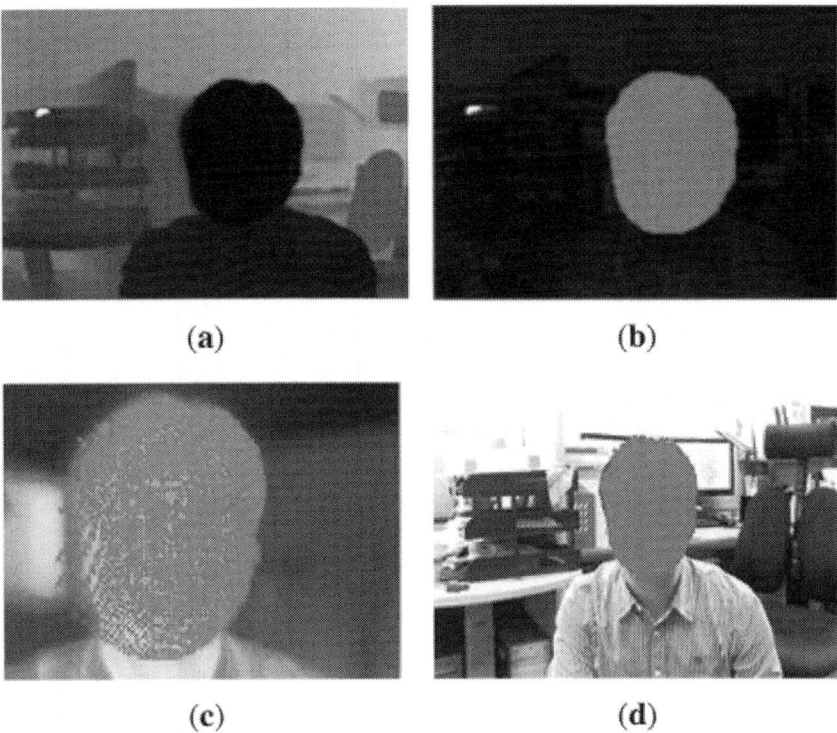

(a)   (b)

(c)   (d)

Figure 9. (a) 3D depth image and corresponding points in the near-IR image (b); thermal-IR image (c); and visible (color) image (d).

As a result of the registration, 3D multi-spectrum face data for face recognition can be generated. Since the resolution of the depth image from the ToF camera is too small, we generate 3D multi-spectrum face data by 3D mesh rendering in OpenGL to create more detailed 3D face data. Figure 10a shows visible and thermal-IR images, as well as two kinds of 3D multi-spectrum data. One image is color textured (Figure 10b) and the other is thermal-IR textured (Figure 10c). Since the coordinate of the distance image corresponds perfectly with the coordinate of the near-IR image, 3D multi-spectrum face data, including near-IR texture information, can also be generated, as shown in Figure 10d.

3.4. Face Recognition

Face recognition is performed using the generated 3D multi-spectrum face data. Then, 3D face data from the ToF camera contains certain distance noise, as shown in Figure 11. The performance of 3D face recognition is highly dependent upon distance noise [48]. Therefore, we perform a 2.5D face recognition step, which includes a 2D frontal face image projected from the transformed 3D multi-spectrum face data by a normalization step.

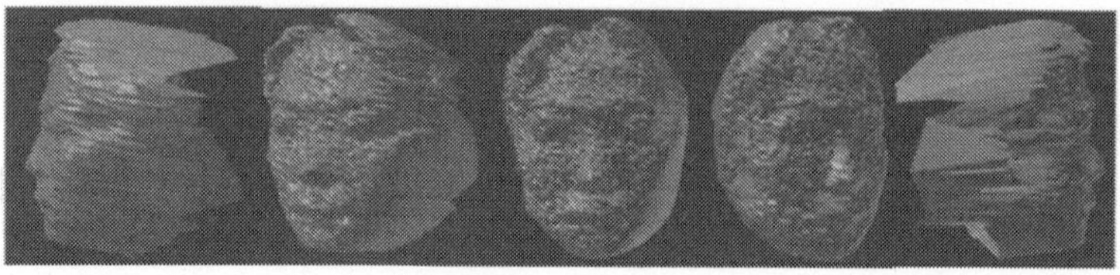

(a)

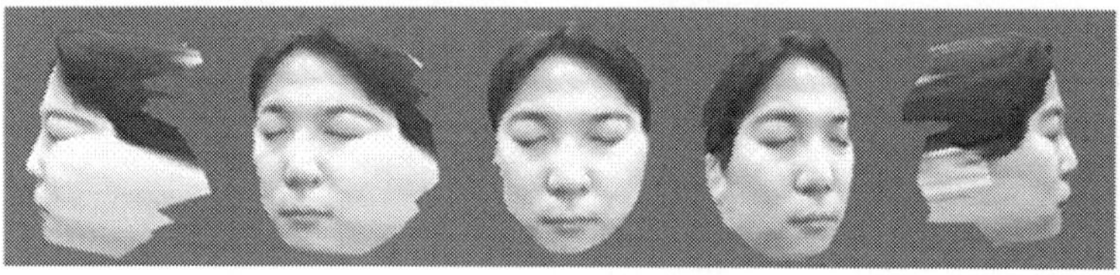

(b)

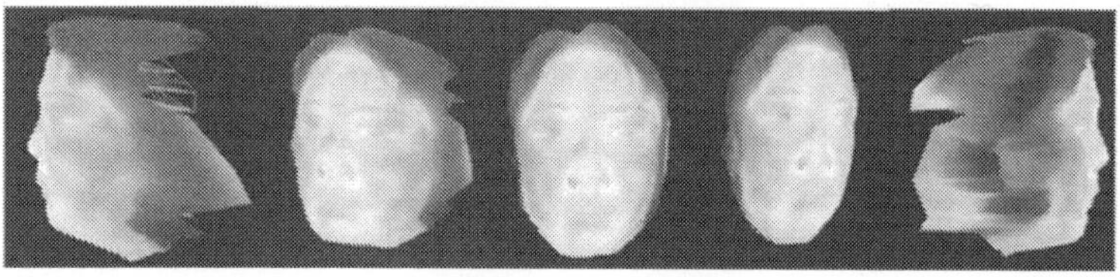

(c)

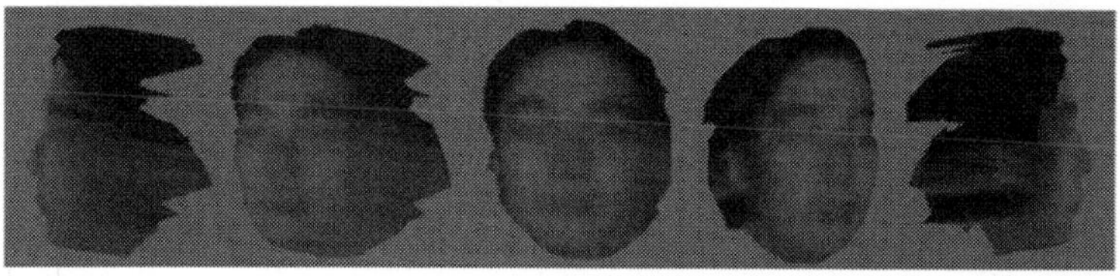

(d)

Figure 10. 3D multi-spectrum face data: (**a**) 3D face shape data from the ToF camera; (**b**) color 3D multi-spectrum data; (**c**) thermal-IR 3D multi-spectrum data; and (**d**) near-IR 3D multi-spectrum data.

Figure 11. Samples of 3D face data acquired by the ToF camera.

Before recognizing a face, a normalization step to transform rotated faces into reference posed faces is necessary to reduce the pose variation. There are many pose estimation algorithms for minimizing the mean square error between points in reference data and the closest points in input data by using translation, rotation and scaling [49]. We use an iterative closest points (ICP) algorithm, which is one of the most commonly used algorithms for registering 3D data [50,51]. In order to evaluate 2.5D face recognition, we apply four kinds of classification methods (listed below) that are commonly used for 2D face recognition.

1. Principal component analysis (PCA)
2. Fisher linear discriminant analysis (FLDA)
3. PCA feature extraction + support vector machine (PCA + SVM)
4. PCA feature extraction + reduced multivariate polynomial pattern classifier (PCA + RM) [52].

4. EXPERIMENTS

In order to evaluate the proposed 3D multi-spectrum face-data-based recognition system, we compare the performance of our proposed system with several existing face recognition methods using different face data, including 2D/3D and color/thermal-IR face data. We separate the experiments into pose and light variations in order to ensure the robustness of our proposed approach.

4.1. Experimental Environments

In order to obtain 3D information and near-IR images in real-time, we use a SR-4000 ToF camera with MESAimaging that can provide a resolution of 176×144 pixels at 30 FPS. Color images of the scene are captured using a FL2 by Point Grey that provides a resolution of $1,024 \times 768$ pixels at 30 FPS. Thermal-IR images are acquired from a ThermaCAM S65 with an FLIRsystem that has a resolution of 320×240 pixels at 50/60 Hz. Since each sensor operates with a different frame rate, we need to set temporal synchronization among the sensors. We adjust the timing for the image capturing of visual and thermal-IR cameras to the timing of the image capturing of the ToF camera. The SR-4000 ToF camera supports a software trigger mode that uses the callback mechanism to capture depth images. When the callback function is called upon, all images are captured by multi-thread. Although we had tried to set the temporal correspondence by multi-thread, 2D image capturing and depth image capturing are, respectively, started with a small time difference, which is about $1\,msec$. Figure 12 shows the installation of the three cameras. The ToF camera in Figure 12c is positioned to the right of the thermal-IR and visible cameras. The visible camera in Figure 12b is set next to the thermal-IR camera, as shown in Figure 12d. Each camera is located as close as possible to reduce occlusion caused by the different camera positions.

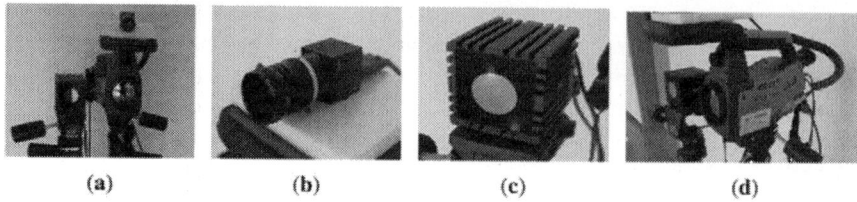

(a)   (b)   (c)   (d)

Figure 12. The proposed system: (**a**) proposed full system; (**b**) visible camera; (**c**) ToF camera; and (**d**) thermal-IR camera.

Figure 13 shows our experimental environments. The distance between the face image to be acquired and the cameras is set at 1 m. Three light sources are used to create light variation. In addition, a dark screen is installed in the background to make it easy to separate face images from the background.

Figure 13. Experimental environments: three light sources and three different kinds of cameras.

Even though the data acquisition system is implemented in C, OpenGLand OpenCV, which is a real-time processing library for computer vision by Intel, other steps, including normalization and recognition, were simulated using Matlab 2009b on a machine with a 2.93 GHz Intel Core i7 870 and 4 GB of physical memory. Since there is no public facing database that includes all registered 3D, visible and thermal-IR information with variable illumination and pose conditions, we created two databases with five different poses and five levels of light variation. All images for the database are captured indoor with daylight conditions. Each database contains 500 3D multi-spectrum face datasets obtained from 100 subjects (5 (variations) × 100 (subjects) = 500). We detect face region using the Viola-Jones face detector, which was implemented using OpenCV [53]. All face images are normalized to 50 × 50 pixels for recognition. Figure 14 shows sample images from our database, which consists of distance, visible and thermal-IR images. The first row in Figure 14shows color (RGB) images from the FL2 camera. The second row shows the thermal-IR images captured by the ThermaCAM S65. The third and fourth rows show the distance and near-IR images, respectively, from the SR-4000 ToF camera.

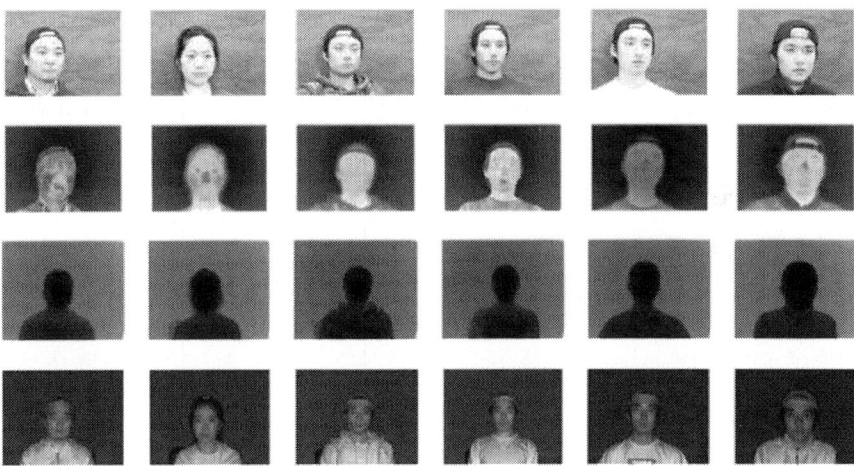

Figure 14. Example images in our face database including 3D, visible and thermal-IR information. The first row shows some color images from the visible camera. The second row shows images from the thermal-IR camera. Distance images from the ToF camera are shown in the last row.

4.2. Estimation Of Projection Matrices of the Visible and Thermal-Ir Cameras

An average image derived from the 100 ToF camera distance images is used to estimate precise projection matrices after applying a median filter to reduce the effects of distance noise. Once the projection matrices have been estimated in the off-line process, it does not need to be operated again. To estimate the projection matrix, we extract 77 corresponding points from the visible, thermal-IR and distance images. After that, we estimate the projection matrices of the visible and thermal-IR images with respect to the 3D points from the ToF camera. The accuracy of the projection matrices is evaluated as re-projection error, which is the Euclidean distance between the projected 2D coordinates from Equations (11) and (12) and the 2D coordinates of extracted points in the visible and thermal-IR images, as shown in Figure 15.

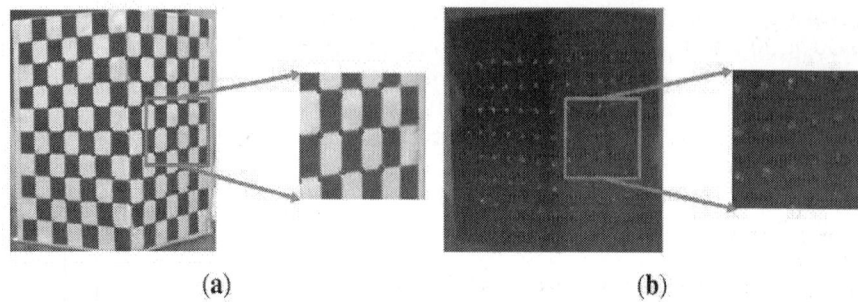

(a)   (b)

Figure 15. Re-projection error between projected 2D coordinates from 3D points and 2D points extracted from the **(a)** visual image and **(b)** thermal-IR image. The red and blue points represent extracted 2D points and projected 2D points, respectively.

We measured re-projection errors at different visual and thermal-IR camera positions, but with the ToF camera fixed. The mean values of 30 repetitions of re-projection errors with the visual and thermal-IR camera are 2.7308 pixels and 1.5629 pixels, respectively. Since the resolution of the visual image is larger than that of the thermal-IR image, the re-projection error of the visual image is more sensitive to noise than the thermal-IR image. There are many reasons why re-projection error is generated. First, the low resolution (176×144) of the ToF camera causes more error in high-resolution images during the feature extraction step. Second, even though an average of 30 distance images is used to extract 3D points, the values still might contain distance noise. Therefore, distance noise estimation and precise feature extraction algorithms are needed to estimate precise projection matrices.

4.3. Face Recognition with Pose Variation

In this experiment, we only consider pose variation in a face regardless of illumination change. Therefore, only pose differences in the face database are used. Lighting conditions remain constant. To verify the robustness against pose variation, we generated a database consisting of 100 subjects, each exhibiting five different poses (left, right, up, down and front) with respect to the ToF camera for a total of 500 images. Figure 16 shows example images of a few subjects in various poses.

Using those images, we can generate 3D multi-spectrum face data and perform normalization by ICP. Normalized face data are shown in Figures 17, 18 and 19, which were constructed using visible, thermal-IR and near-IR multi-spectrum data, respectively.

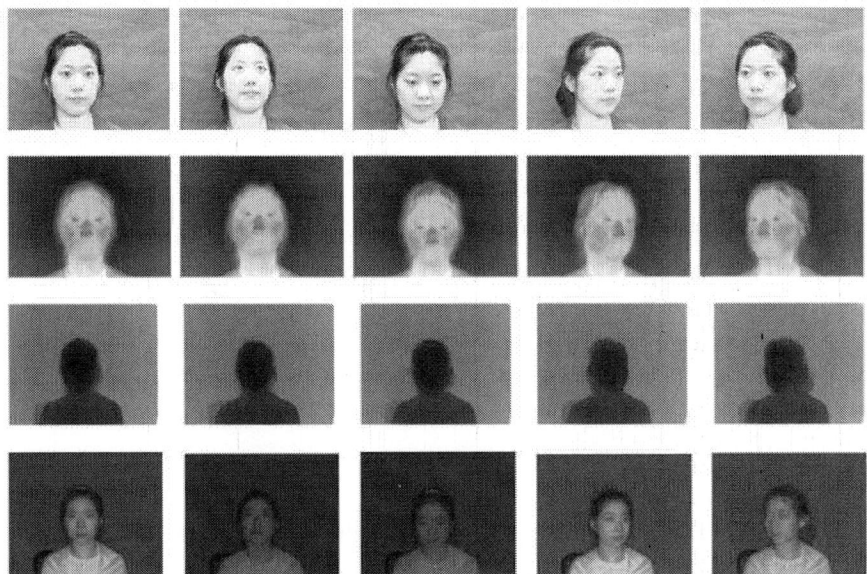

Figure 16. Example images from the database of five different facial poses: front, up, down, right and left.

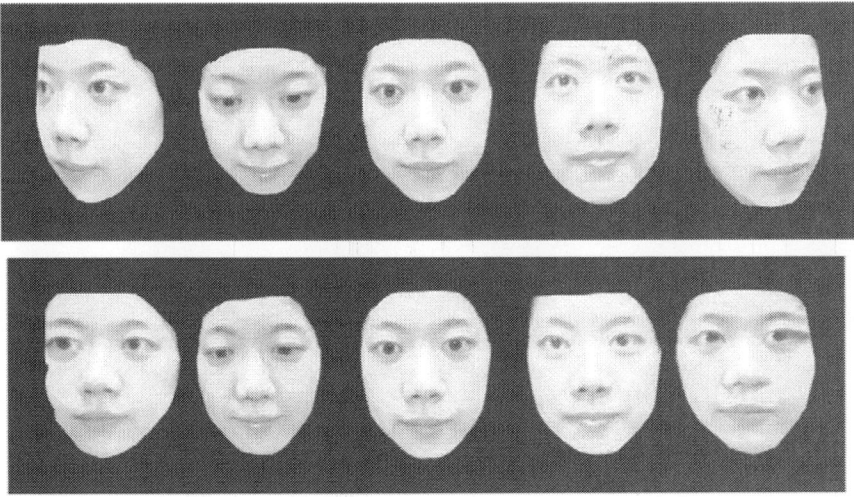

Figure 17. The first row shows 3D visible multi-spectrum face data before normalization and the second row shows normalized 3D visible multi-spectrum face data.

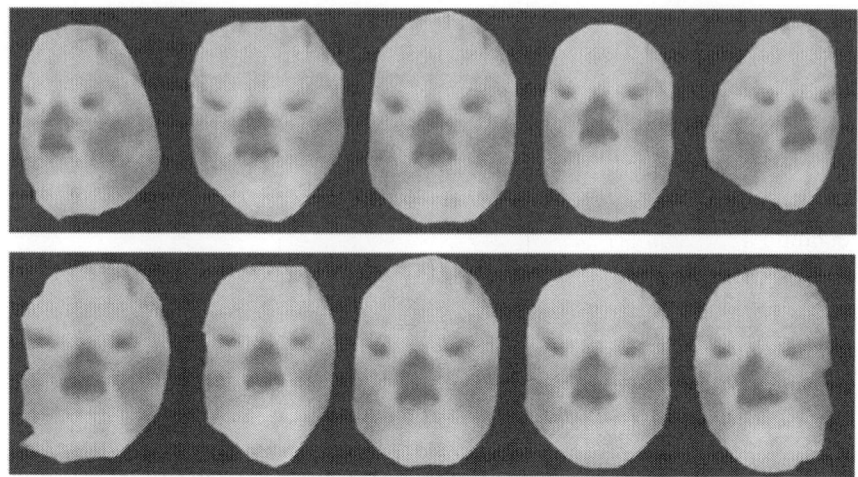

Figure 18. The first row shows 3D thermal-IR multi-spectrum face data before normalization, and the second row shows normalized 3D thermal-IR multi-spectrum face data.

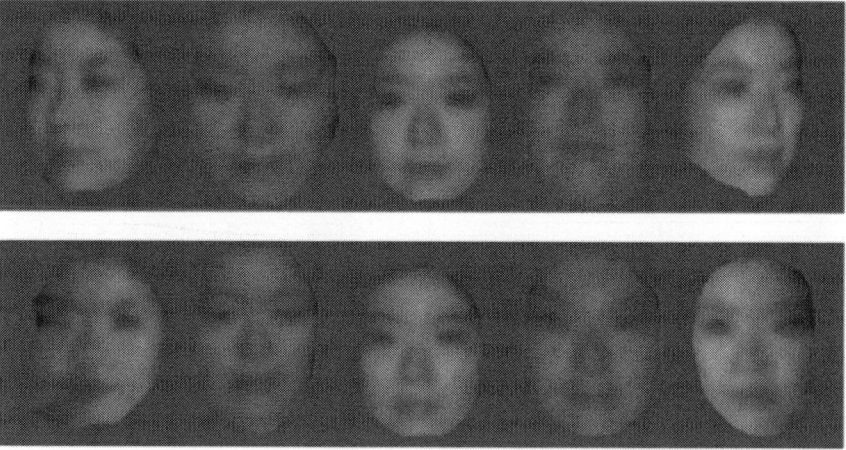

Figure 19. The first row shows 3D near-IR multi-spectrum face data before normalization, and the second row shows normalized 3D near-IR multi-spectrum face data.

Experiments using five-fold cross validation are performed to verify the face images. That is, four face images of a person are used for training, and then, one face image is used for testing. We first train each classifier using 400 face data images, with four posed data images per subject, and perform the test using 100 other face data images in different poses. Recognition is performed using a nearest neighbor classifier with PCA, FDA, PCA + SVM and PCA + RM.

To compare the proposed 2.5D face recognition using 3D multi-spectrum face data with other recognition methods, we calculate the recognition rate for different face data by PCA, FDA, PCA + SVM and PCA + RM. In this experiment, we have used different polynomial orders 1 (RM_1), 2 (RM_2) and 3 (RM_3). Since order 3 shows saturated performance, the experiment stops at order 3. In the SVM and RM experiments, we extracted features using PCA and adopt SVM and RM as classifiers. In the SVM experiments, we adopt a linear model, a polynomial model and a radial basis function as kernels. The parameters, such as the number of principal components in the eigenface and the number of support vectors with SVM, are experimentally selected to achieve the lowest error rate with each method. Five types of experiments are performed to observe the performance and the robustness against pose variation using 3D multi-spectrum face data.

1. 2D visible face images with pose variation (2D-vis)
2. 3D range images with pose variation (3D)
3. 3D range images after ICP normalization (3D + ICP)
4. Projected 2D visible face images from 3D multi-spectrum data after ICP normalization (2D-vis + ICP)
5. Projected 2D thermal-IR face images from 3D multi-spectrum data after ICP normalization (2D-the + 3D + ICP)
6. Projected 2D near-IR face images from 3D multi-spectrum data after ICP normalization (2D-NIR + 3D + ICP)

The approach in (5) is our proposed method. All experiment results are shown in Table 1.

Table 1. Recognition rate with respect to pose variation. ICP, iterative closest points; PCA, principal component analysis; FLDA, Fisher linear discriminant analysis; SVM, support vector machine.

Case	2D-vis	3D	3D + ICP	2D-vis + 3D + ICP	2D-the + 3D + ICP	2D-NIR + 3D + ICP
PCA	48.6%	44.2%	74.6%	87%	**88.4%**	83.8%
	(243/500)	(221/500)	(373/500)	(435/500)	(442/500)	(419/500)
FLDA	50.8%	48.6%	83.8%	**93.2%**	91.2%	88.6%
	(254/500)	(243/500)	(419/500)	(466/500)	(456/500)	(443/500)
PCA + SVM (linear)	53.2%	50.6%	87.8%	**94.4%**	93.2%	91.2%
	(266/500)	(253/500)	(439/500)	(472/500)	(466/500)	(456/500)
PCA + SVM (poly)	57.2%	52.6%	91.2%	**96.4%**	94.4%	93.2%
	(286/500)	(263/500)	(456/500)	(482/500)	(472/500)	(466/500)
PCA + SVM (Rbf)	55.8%	54.6%	87%	90.6%	**98.4%**	92.4%
	(279/500)	(273/500)	(435/500)	(453/500)	(492/500)	(462/500)
PCA + RM_1	53.8%	52.6%	84.4%	92.4%	**93.2%**	89.8%
	(269/500)	(263/500)	(422/500)	(462/500)	(466/500)	(449/500)
PCA + RM_2	55.2%	49.8%	87.2%	94.4%	**96.4%**	91.8%
	(276/500)	(249/500)	(436/500)	(472/500)	(482/500)	(459/500)
PCA + RM_3	57.2%	54%	87.8%	**94.4%**	93.2%	91.2%
	(276/500)	(270/500)	(439/500)	(472/500)	(466/500)	(456/500)

Based on these experiments, we can solve the pose variation problem by normalizing the 3D face data. Since there is no light variation to evaluate in this experiment, the result of using 3D visible face data (4), and 3D thermal-IR face data (5) yield similar recognition rates as the highest recognition rate, as shown in Figure 20. Even though the result of (3) uses 3D information, the recognition rate is not as high, because the range images from the ToF camera contain more distance noise. Among the classifiers, the SVM and RM classifiers show the best recognition rate.

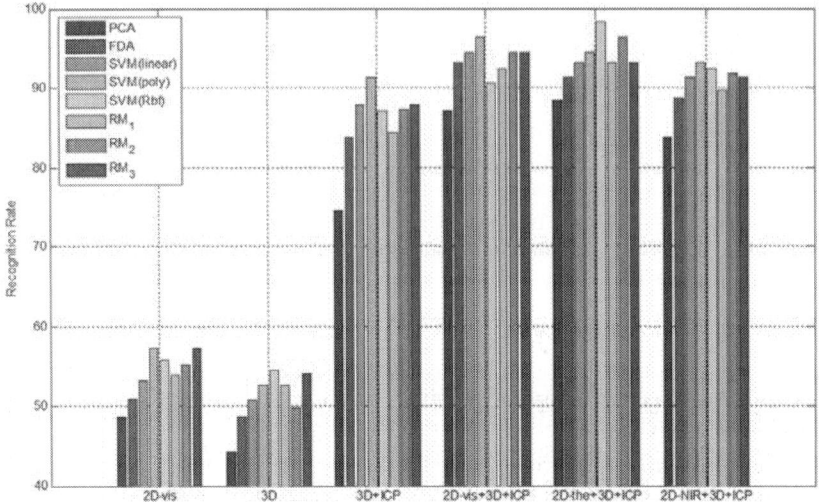

Figure 20. Recognition rate in terms of recognition approaches.

4.4. Face Recognition with Light Variation

We verify performance with light variation using a database consisting of images of 100 subjects having five different levels of light variation (30° left, 15° left, front, 30° right and 15° right) without pose variations, as shown in Figure 14. This experiment does not require a normalization step for 3D multi-spectrum face data. Figure 21 shows the invariance of the thermal-IR images with illumination change, which strongly influences the visible image.

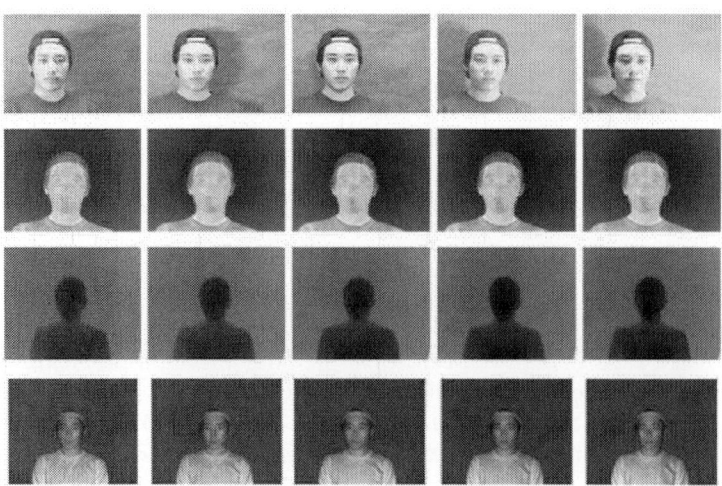

Figure 21. An example from the face database having five different light angles: 90° left, 45° left, front, 90° right and 45° right.

The experiments are performed using five-fold cross-validation, meaning that one face image is used for testing, and the other four face images are used for training to verify the face image. Therefore, 400 images were used for training, and 100 were used for testing. The experimental methods are the same as for face recognition with pose variation.

To compare the proposed 2.5D face recognition using 3D multi-spectrum face data with other methods, we calculate the recognition rate of each method by PCA, FDA, PCA + SVM and PCA + RM. Five types of experiments are performed to show the robustness of the proposed method against light variation using 3D multi-spectrum face data as follows:

1. 2D visible face images with light variation (2D-vis)
2. 2D thermal-IR face images with light variation (2D-the)
3. 3D range images with light variation (3D)
4. Projected 2D visible face images from 3D multi-spectrum data with light variation (2D-vis + 3D)
5. Projected 2D thermal-IR face images from 3D multi-spectrum data with light variation (2D-the + 3D)
6. Projected 2D near-IR face images from 3D multi-spectrum data with light variation (2D-NIR + 3D)

All recognition rates are shown in Table 2. The results of the experiments indicate that thermal-IR texture is invariant to light variation, even though visible texture is strongly affected by surrounding light conditions. In addition, the range image, which is not changed by illumination, shows a reliable recognition rate, but not as high as the recognition rate of thermal-IR images, as the range data includes distance noise. As shown in Figure 22, the result using the proposed method (5) and the 2D thermal face image (2) shows the robustness against light variation.

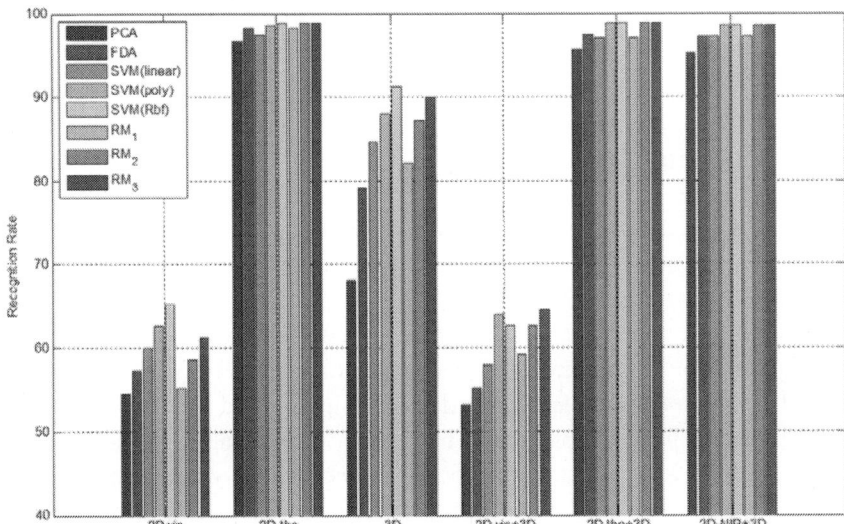

Figure 22. Recognition rate in terms of recognition approaches.

Table 2. Recognition rate with respect to light variation.

Case	2D-vis	2D-ther	3D	2D-vis + 3D	2D-ther + 3D	2D-NIR + 3D
PCA	54.6%	**96.6%**	68.0%	53.2%	95.6%	95.2%
	(273/500)	(483/500)	(340/500)	(266/500)	(478/500)	(476/500)
FLDA	57.2%	**98.2%**	79.2%	55.2%	97.4%	97.2%
	(286/500)	(491/500)	(396/500)	(276/500)	(487/500)	(486/500)
PCA + SVM (linear)	60.0%	**97.4%**	84.6%	58.0%	97.0%	97.2%
	(300/500)	(491/500)	(423/500)	(290/500)	(485/500)	(486/500)
PCA + SVM (poly)	62.6%	98.6%	88.0%	64.0%	**98.8%**	98.6%
	(313/500)	(493/500)	(440/500)	(320/500)	(494/500)	(493/500)
PCA + SVM (Rbf)	65.2%	**98.8%**	91.2%	62.6%	**98.8%**	98.6%
	(326/500)	(494/500)	(456/500)	(313/500)	(494/500)	(493/500)
PCA + RM$_1$	55.2%	**98.2%**	82.0%	59.2%	97.0%	97.2%
	(276/500)	(491/500)	(410/500)	(296/500)	(485/500)	(486/500)
PCA + RM$_2$	58.6%	**98.8%**	87.2%	62.6%	**98.8%**	98.6%
	(293/500)	(494/500)	(436/500)	(313/500)	(494/500)	(493/500)
PCA + RM$_3$	61.2%	**98.8%**	90.0%	64.6%	**98.8%**	98.6%
	(306/500)	(494/500)	(450/500)	(323/500)	(494/500)	(493/500)

Based on these experiments, the proposed 2.5D face recognition technique with 3D multi-spectrum face data shows better performance than 2D and 3D face recognition with variation in pose and light. The proposed face recognition approach improves the recognition rate with light variation, since the thermal-IR pattern of the face is not changed by visible illumination changes. Even though normalized frontal face data can be generated from 3D information, the occluded region between the cameras and the face cannot be reconstructed. Therefore, some distortions in normalized face images can be generated by normalization of the occluded region. This may cause a slight performance reduction with pose variation. This problem can be solved by using additional ToF cameras at different locations to cover the occluded face regions.

4.5. Processing Time

The processing time is an important factor for a real-time sensor system. The processing time for each step is shown in Table 3. Each processing time is calculated by averaging of 100 attempts.

The whole system can be divided into two subsystems: a 3D multi-spectrum sensor system and a face recognition system. The image acquisition and 3D multi-spectrum data generation are implemented by C language, and the recognition part is implemented by Matlab. The 3D multi-spectrum sensor system takes 0.07 s (about 15 data points per second) to generate a 3D multi-spectrum data. In recognition, the processing time is less than 1 msec. Therefore, the proposed 3D multi-spectrum sensor system can be used with various real-time applications, such as robots and surveillance.

Table 3. Processing time for each process.

Content	Processing Time ($msec$)
Image acquisition	29
3D multi-spectrum data generation	41
PCA	0.31
FLDA	0.38
PCA + SVM (linear)	0.62
PCA + SVM (poly)	0.75
Recognition — PCA + SVM (rbf)	0.89
PCA + RM$_1$	0.55
PCA + RM$_2$	0.63
PCA + RM$_3$	0.74

5. CONCLUSIONS

In this paper, we propose a novel 3D multi-spectrum sensor system that provides registered visible, near-IR, thermal-IR and 3D information in real time. By using this information, we can design more flexible and robust systems in terms of selecting sensor combinations and more effective fused features. We showed the usefulness of the proposed system for a face recognition system design with variations in pose and illumination. This system may also be very useful for designing vision systems for surveillance, 3D object modeling and object recognition.

ACKNOWLEDGMENTS

This work was supported by a National Research Foundation of Korea (NRF) grant funded by the Korean government (MEST) (NRF-2011-0016302). Additionally, this work was supported by the Industrial Strategic technology development program (10040018, Development of 3D Montage Creation and Age-specific Facial Prediction System) funded by the Ministry of Knowledge Economy (MKE, Korea).

REFERENCES

1. Ukimura, O. *Image Fusion*; InTech: Rijeka, Croatia, 2011.
2. Thomas, C. *Sensor Fusion and its Applications*; Sciyo: Vienna, Austria, 2011.
3. Kolb, A.; Barth, E.; Koch, R.; Larsen, R. *Time-of-Flight Sensors in Computer Graphics*; EUROGRAPHICS STAR Report; Munich: Germany, 2009.

4. Lange, R. 3D Time-of-Flight Distance Measurement with Custom Solid-State Image Sensors in CMOS/CCD-Technology. Ph.D. Dissertation, University Siegen, Siegen, Germany, 2000.

5. Wilfried, E. *An Introduction to Sensor Fusion*; Research Report 47/2001; Vienna University of Technology: Vienna, Austria, 2001.

6. Andrea, F.; Nappi, A.M.; Riccioa, D.; Sabatinoa, G. 2D and 3D face recognition: A survey.*Patt. Recogn. Lett.* **2007**, *28*, 1885–1906.

7. Zhong, Y.; Jain, A.K. Object localization using color, texture and shape. *Pattern Recognit.***2000**, *33*, 671–684.

8. Mirmehdi, M.; Maria, P. Segmentation of color textures. *IEEE Trans. Pattern Anal. Mach. Intell.* **2000**, *22*, 142–159.

9. Stan, Z.; Li, R.; Chu, S.; Liao, S.; Zhang, L. Illumination invariant face recognition using near-infrared images. *IEEE Trans. PAMI* **2007**, *29*, 627–639.

10. Chang, H.; Koschan, A.; Abidi, M.; Kong, S.G.; Won, C.H. Multispectral Visible and Infrared Imaging for Face Recognition. Proceedings of the IEEE Conference on Computer Vision and Pattern Recognition, Anchorage, AK, USA, 23–28 June, 2008; pp. 1–6.

11. Rui, T.; Zhang, S.A.; Zhou, Y.; Jianchun, X.; Jian, D. Registration of Infrared and Visible Images Based on Improved SIFT. Proceedings of the 4th International Conference on Internet Multimedia Computing and Service, Wuhan, China, 9–11 September 2012; pp. 144–147.

12. Kong, S.G.; Heo, J.; Boughorbel, F.; Zheng, Y.; Abidi, B.R.; Koschan, A.; Yi, M.; Abidi, M.A. Multiscale fusion of visible and thermal IR images for illumination-invariant face recognition. *Int. J. Comput. Vision* **2007**, *71*, 215–233.

13. Arandjelovic, O.; Hammoud, R.; Cipolla, R. Thermal and reflectance based personal identification methodology under variable illumination. *Pattern Recognit.* **2010**, *43*, 1801–1813.

14. Fay, D.A.; Waxman, A.M.; Verly, J.G.; Braun, M.I.; Racamato, J.P.; Frost, C. Fusion of Visible, Infrared and LADAR Imagery. Proceedings of the 4th International Conference on Information Fusion, Montreal, Canada, 7–10 August 2001.

15. Han, J.; Pauwels, E.J.; de Zeeuw, P. Visible and infrared image registration in man-made environments employing hybrid visual features. *Pattern Recognit. Lett.* **2012**, *34*, 42–51.

16. Zhang, C.; Zhang, Z. Calibration between depth and color sensors for commodity depth cameras. In *Multimed. Expo(ICME)*; Barcelona, Spain; 11–15; July; 2011; pp. 1–6.

17. Herrera, C.; Kannala, J. Joint depth and color camera calibration with distortion correction.*IEEE Trans. Pattern Anal. Mach. Intell.* **2012**, *34*, 2058–2064.

18. Martinez-Otzeta, J.M.; Ansuategui, A.; Ibarguren, A.; Sierra, B. RGB-D, laser and thermal sensor fusion for people following in a mobile robot. *Int. J. Adv. Robot. Syst.* **2013**, *10*.

19. Kollorz, E.; Penne, J.; Hornegger, J.; Barke, A. Gesture recognition with a time-of-flight camera. *Int. J. Intell. Syst. Technol. Appl.* **2008**, *5*, 334–343.

20. Holte, M.B.; Moeslund, T.B.; Fihl, P. View-invariant gesture recognition using 3D optical flow and harmonic motion context. *Comput. Vision Image Underst.* **2010**, *114*, 1353–1361.

21. Falie, D.; Buzuloiu, V. Wide Range Time of Flight Camera for Outdoor Surveillance. Proceedings of the Microwaves, Radar and Remote Sensing Symposium, Kiev, Ukraine, 22–24 September 2008; pp. 79–82.

22. Silvestre, D. Video Surveillance Using a Time-of-Light Camera. Ph.D. Thesis, Technical University of Denmark, Lyngby, Denmark, 2007.

23. Fransen, B.R.; Herbst, E.V.; Harrison, A.M.; Adams, W.; Trafton, J.G. Real-time Face and Object Tracking. Proceedings of the Conference on Intelligent Robots and Systems 2009, St Louis, MO, USA, 11–15 October 2009; pp. 2483–2488.

24. Dorrington, A.; Kelly, C.; McClure, S.; Payne, A.; Cree, M. Advantages of 3d Time-of-Flight Range Imaging Cameras in Machine Vision Applications. Proceedings of the 16th Electronics New Zealand Conference (ENZCon), North Dunedin, New Zealand, 23 November 2009; pp. 95–99.

25. Chen, H. Oliver Wulf and Bernardo Wagner, Object detection for a mobile robot using mixed reality. *Interact. Technol. Sociotech. Syst.* **2006**, *4270*, 466–475.

26. Prusak, A.; Melnychuk, O.; Schiller, I.; Roth, H.; Koch, R. Pose estimation and map building with a PMD-camera for robot navigation. *Int. J. Intell. Syst. Technol. Appl.* **2008**, *5*, 355–364.

27. Foix, S.; Aleny, G.; Andrade-Cetto, J.; Torras, C. Object Modeling Using a ToF Camera under an Uncertainty Reduction Approach. Proceedings of the 2010 IEEE International Conference on Robotics and Automation, Anchorage, AK, USA, 3–8 May 2010; pp. 1306–1312.

28. Kim, Y.M.; Theobalt, C.; Diebel, J.; Kosecka, J.; Miscusik, B.; Thrun, S. Multi-View Image and ToF Sensor Fusion for Dense 3D Reconstruction. Proceedings of the 2009 IEEE 12th International Conference on Computer Vision Workshops, Jena, Germany, 9 September 2009; pp. 1542–1549.

29. Cui, Y.; Schuon, S.; Derek, C.; Thrun, S.; Theobalt, C. 3D Shape Scanning with a Time-of-Flight Camera. Proceedings of the IEEE Conference on Computer Vision and Pattern Recognition, San Francisco, CA, USA, 13–18 June 2010; pp. 1173–1180.

30. Bleiweiss, A.; Werman, M. Fusing Time-of-Flight Depth and Color for Real-Time Segmentation and Tracking. Proceedings of the DAGM 2009 Workshop on Dynamic 3D Imaging, Jena, Germany, 9 September 2009.

31. Van den Bergh, M.; van Gool, L. Combining RGB and ToF Cameras for Real-time 3D Hand Gesture Interaction. Proceedings of the 2011 IEEE Workshop on Applications of Computer Vision, Kona, HI, USA, 5–6 January 2011; pp. 66–72.

32. Hahne, U.; Alexa, M. Depth Imaging by Combining Time-of-Flight and On-Demand Stereo. Proceedings of the 2009 Workshop on Dynamic 3D Imaging, Jena, Germany, 9 September 2009; pp. 70–83.

33. Buciu, I.; Nafornita, I. Non-negative matrix factorization methods for face recognition under extreme lighting variations. Proceedings of International Symposium on Signals, Circuits and Systems, (ISSCS 2009), Iasi, Rumania, 9–10 July 2009; pp. 1–4.

34. Tan, X.; Triggs, B. Enhanced local texture feature sets for face recognition under difficult lighting conditions. *Trans. Image Process.* **2010**, *19*, 1635–1650.

35. Ghiass, R.S.; Arandjelovic, O.; Bendada, H.; Maldague, X. Infrared Face Recognition: A Literature Review. Proceedings of the International Joint Conference on Neural Networks, Dallas, TX, USA, 4–9 August 2013.

36. Ghiass, R.S.; Arandjelovic, O.; Bendada, H.; Maldague, X. Vesselness Features and the Inverse Compositional AAM for Robust Face Recognition Using Thermal IR. Proceedings of the Twenty-Seventh AAAI Conference on Artificial Intelligence, Bellevue, WA, USA, 14–18 July 2013; pp. 357–364.

37. Colombo, A.; Cusano, C.; Schettini, R. 3D face detection using curvature analysis. *Pattern Recognit.* **2006**, *39*, 444–455.

38. Sun, D.; Sung, W.-P.; Chen, R. 3D face recognition based on local curvature feature matching. *Appl. Mech. Mater.* **2011**, *121–126*, 609–616.

39. Li, C.; Barreto, A. Profile-Based 3D Face Registration and Recognition. Proceedings of the Information Security and Cryptology—ICISC 2004, Seoul, Korea, 2–3 December 2004; Volume 3506, pp. 478–488.

40. Beumier, C.; Acheroy, M. Automatic 3D face authentication. *Image Vision Comput.* **2001**, *18*, 315–321.

41. Achermann, B.; Jiang, X.; Bunke, H. Face Recognition Using Range Images. Proceedings of the International Conference on Virtual Systems and Multimedia, Geneva, Switzerland, 10–12 September 1997; pp. 129–136.

42. Srivastava, A.; Liu, X.; Hesher, C. Face recognition using optimal linear components of range images. *Image Vision Comput.* **2006**, *24*, 291–299.

43. Malassiotis, S.; Strintzis, M.G. Robust face recognition using 2D and 3D data: Pose and illumination compensation. *Patt. Recogn.* **2005**, *38*, 2537–2548.

44. Godil, A.; Ressler, S.; Grother, P. Face recognition using 3D face shape and color map information: Comparison and combination. *Biom. Technol. Hum. Identif.* **2005**.

45. Yu, S.; Kim, J.; Lee, S. Thermal 3D modeling system based on 3-view geometry. *Opt. Commun.* **2012**, *285*, 5019–5028.

46. Adini, Y.; Moses, Y.; Ullman, S. Face recognition: The problem of compensating for changes in illumination direction. *Pattern Anal. Mach. Intell.* **1997**, *19*, 721–732.

47. Hartley, R.; Zisserman, A. *Multiple View Geometry*; Cambridge University Press: Cambridge, UK, 2000.

48. Ebers, O.; Ebers, T.; Spiridonidou, T.; Plaue, M.; Beckmann, P.; Barwolff, G. Towards Robust 3D Face Recognition from Noisy Range Images with Low Resolution. In *Preprint Series of the Institute of Mathematics*; Technische Universitat Berlin: Berlin, Germany, 2008.

49. Murphy-Chutorian, E.; Trivedi, M.M. Head pose estimation in computer vision: A survey. *Pattern Anal. Mach. Intell.* **2009**, *31*, 607–626.

50. Zhang, Z. Iterative point matching for registration of free-form curves. *Int. J. Comput. Vision* **1994**, *13*, 119–152.

51. Wollner, P.; Arandjelovic, O. Freehand 3D Scanning in a Mobile Environment using Video. Proceedings of the ICCV 2011 Workshops, Barcelona, Spain, 6–13 November 2011; pp. 445–452.

52. Toh, K.-A.; Tran, Q.-L.; Srinivasan, D. Benchmarking a reduced multivariate polynomial pattern classifier. *Pattern Anal. Mach. Intell.* **2004**, *26*, 740–755.

53. Viola, P.; Jones, M.J. Robust real-time face detection. *Int. J. Comput. Vision* **2004**, *57*, 137–154.

CHAPTER 3

Face Liveness Detection Using Defocus

Sooyeon Kim, Yuseok Ban and Sangyoun Lee

Department of Electrical and Electronic Engineering, Yonsei University, 134 Shinchon-dong, Seodaemun-gu, Seoul 120-749, Korea

ABSTRACT

In order to develop security systems for identity authentication, face recognition (FR) technology has been applied. One of the main problems of applying FR technology is that the systems are especially vulnerable to attacks with spoofing faces (e.g., 2D pictures). To defend from these attacks and to enhance the reliability of FR systems, many anti-spoofing approaches have been recently developed. In this paper, we propose a method for face liveness detection using the effect of defocus. From two images sequentially taken at different focuses, three features, focus, power histogram and gradient location and orientation histogram (GLOH), are extracted. Afterwards, we detect forged faces through the feature-level fusion approach. For reliable performance verification, we develop two databases with a handheld digital camera and a webcam. The proposed method achieves a 3.29% half total error rate (HTER) at a given depth of field (DoF) and can be extended to camera-equipped devices, like smartphones.

Keywords: face liveness detection; anti-spoofing; defocus; 2D fake face; webcam

1. INTRODUCTION

At present, many people deal with personal business using portable devices. From unlocking cellular phones to financial business transactions, people can easily conduct their individual business tasks through such a device. Due to this trend, personal authentication has become a significant issue [1]. Instead of using a simple PIN code, industries have developed stronger security systems with biometric authorization technology [2]. Biometric traits, such as face, iris and fingerprint, are very powerful factors to protect one's private information.

However, attempts to invade security systems and steal personal information have been increasing. One type of these attacks involves using fake identities. Spoofing faces and fingerprints are threatening security systems and privacy. This would not matter if current face recognition (FR) systems were secure, but current systems cannot distinguish fake faces from real faces. In some cases, the FR system embedded in cellular phones gives approvals to forged faces. This phenomenon is an example of weakness in the biometric system. If this problem remains unsolved, anyone will be able to easily obtain others' personal information in order to commit identity-related crimes. For this reason, technological defense against spoofing attacks is necessary, so as to protect personal systems and users' private data. Over the last decade, researchers have shown steady progress in developing anti-spoofing technologies [3]. Most of these methods concentrate on exploiting features obtained from the analysis of textures, spectrums and motion in order to detect face liveness.

In this paper, we propose a new method to secure face identification systems from forged 2D photos. The key factor of our methods is that we utilize the camera function, variable focusing. In shape-from-focus, it is possible to construct 3D images using focus measures [4,5]. Even though we need not recover the 3D depth images, we use the characteristics of the defocusing technique in order to predict the existence of the depth information. By adjusting the focusing parameters, parts of the image that are not in focus become blurry. With this function, we can evaluate differences in the degree of focus between real faces and fake faces and use this information to detect face liveness. To evaluate our method, we organized two databases using a handheld digital camera and a webcam.

The remainder of this paper is organized as follows. In Section 2, we discuss previous studies on face liveness detection and the theoretical background of camera focusing. Our proposed methodologies are stated in Section 3. In Section 4, experimental results are shown and the details are discussed. Finally, concluding remarks are provided in Section 5.

2. RELATED WORK

2.1. Countermeasures Against Spoofing Faces

Numerous approaches to minimize vulnerability to attacks using spoofing faces have been proposed. In early research, intrusive methods that request user cooperation, such as speaking phrases and shaking one's head [6], were developed. However, these approaches cause users inconvenience and rely on users' cooperation. For this reason, many researchers have attempted to develop non-intrusive methods.

Depending on the type of attack, methods can be categorized into three groups: 2D static attacks (facial photographs), 2D dynamic attacks (videos) and 3D attacks (masks). Skills and devices for disguising one's identity have evolved gradually. Masks and videos are examples of advanced spoof attacks.

Some studies have focused on protecting FR systems from these advanced attacks [7,8]. However, due to the difficulty and cost of obtaining such advanced tools, 2D static attacks, such as photographs, have been widely used by attackers. In this chapter, we review studies for detecting 2D facial photo-based spoof attacks.

There are three main spoof detection approaches, depending on the characteristics of input faces. The first approach is based on textures. Real and fake faces have different texture characteristics. Some studies have used texture to detect forged faces. Kim et al. [9] applied local binary patterns (LBP) for texture analysis and power spectrum for frequency analysis. Määttä et al. [10] and Bai et al. [11] also detected face liveness by examining micro texture with multiscale LBP. Peixoto et al.[12] proposed a method to detect and maintain edges (high-middle frequencies) with different Gaussian characteristics under poor illumination conditions. In [13], the authors extracted essential information for discrimination using a Lambertian model. Singh et al. [14] proposed a method to classify real faces based on a second-order gradient. This approach focuses on differences between skin surfaces of real and fake faces. Kant et al. [15] presented a real-time solution using the skin elasticity of the human face. Approaches with a single image have advantages in terms of low capacity and simplicity.

The second approach uses motion information. Signs of liveness, such as eye blink and head movements, are clues to distinguish motionless spoofing faces. Image sequences can be used to perceive movements. These factors are exploited intuitively [16–20]. In addition, optical flow and various illumination approaches are helpful to analyze the differences between real and fake faces [21–25]. Applying the entropies of RGB color spaces is one factor in face liveness detection [26]. To make a robust system, several methods use a combination of static and dynamic images [18,27].

The last approach is based on 3D facial information. The obvious difference between a real face and a fake face is the presence or absence of depth information. Human faces have curves, while photos are flat. By considering this feature, researchers have classified spoofing attacks. Wang et al. [28] suggested an approach to detect face liveness by recovering sparse 3D facial models, and Lagorio et al. [29] presented a solution based on 3D facial shape analysis.

2.2. Background Related to Focusing

Unlike previous research, our method utilizes the effect of defocus. Defocusing is exploited to estimate the depth in an image [4,5,30]. The degree of focus is determined by the depth of field (DoF), the range between the nearest and farthest objects in a given focal plane. Entities in the DoF are perceived to be sharp. In order to emphasize the effect of defocus, the DoF should be narrow. There are three parameters that modulate DoF, and Figure 1 shows those conditions for shallow DoF [31]. The first factor is the distance between the camera and the subject; a short distance produces a shallow DoF. The second factor is the focal length, which is adjusted to be longer for a shallow DoF. The last factor is the lens aperture of the camera, which is made wider to produce a shallow DoF. Using these options, we can achieve images with a narrow DoF and a large variation in focus [31].

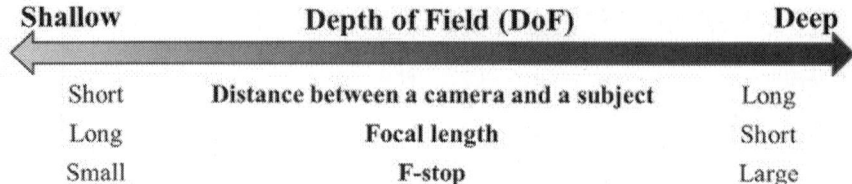

Shallow	Depth of Field (DoF)	Deep
Short	Distance between a camera and a subject	Long
Long	Focal length	Short
Small	F-stop	Large

Figure 1. Factors for the adjustment of the depth of field (DoF).

2.3. Previous Work with Variable Focusing [32]

In the previous work [32], a method for face liveness detection using variable focusing was suggested. Two images sequentially taken at different focuses are used as input images, and focus features are extracted. The focus feature is based on the variation of the sum modified Laplacian (SML) [33] that represents the degrees of focusing. With the focus feature and a simple classifier, fake faces are detected. 2D printed photos are used as spoofing attacks, and a database composed of images with various focuses is produced for evaluation. When DoF is shallow enough to make the only partial area blurred, this method shows good results. However, at a deep DoF, the performance is deteriorated. In order to make up for the weakness of the previous work, we propose an improved method in this paper. Extracting local feature descriptors and frequency characteristics, as well as the focus feature from the defocused images, we detect spoofing faces. Moreover, the quantity of the database is increased, and various experiments are performed to achieve the best result. A detailed explanation will be described in the following sections.

3. PROPOSED METHODOLOGY

In this section, we introduce new FR anti-spoofing methods using defocusing techniques. From partially defocused images, we extract features and classify fake faces. The most significant difference between real and fake faces is the existence of depth information. Real faces have three dimensions, with the nose and ears being relatively far from each other. This distance can be used to adequately represent the depth information. Depending on the object or place of focus, the ear area might or might not be clear, as shown in Figure 2a. Unlike real faces, 2D spoofing faces are flat. There is little difference in clarity, regardless of the focus (Figure 2b). We emphasize this characteristic in order to discriminate real faces from 2D faces.

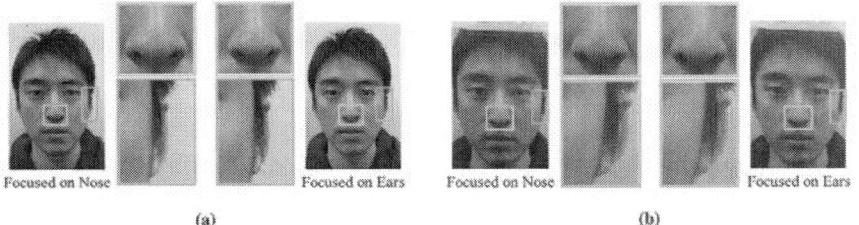

Figure 2. Partially focused images of (**a**) real faces and (**b**) fake faces.

In order to maximize the effect of defocus, we must adjust the DoF to be shallow, as mentioned in Section 2. However, according to the type of camera, the adjustment of DoF may not be possible. Therefore, we obtain input images using two cameras, a handheld digital camera and a webcam. We will explain image acquisition in the following section.

Our system is composed of three steps: image acquisition and preprocessing, feature extraction and classification (Figure 3).

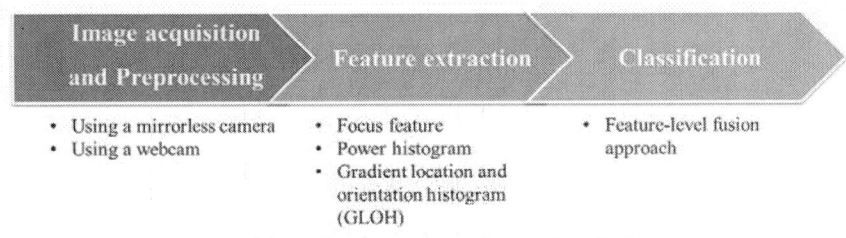

Figure 3. Flowchart of face liveness detection using defocus.

3.1. Image Acquisition and Preprocessing

In our method, image acquisition is an important factor in performance. As mentioned in the previous section, a narrow DoF increases the effect of defocus and assists with detecting fake faces. However, not every camera can easily change its DoF and focal plane. If people use handheld digital cameras, such as DSLR (digital single lens reflex) and mirrorless cameras, the DoF can be made shallow by directly controlling camera settings and the areas of desired focus can be manually selected. However, when users utilize webcams and cameras embedded in cellular phones, they cannot accurately manipulate the DoF. Moreover, the position of the focal plane is inexact with such cameras. Therefore, the process of image acquisition needs to vary with the type of camera. We will introduce two methods appropriate for a handheld digital camera and a webcam, respectively.

3.1.1. Using a Handheld Digital Camera
With handheld digital cameras (DSLR camera, mirrorless camera, compact digital camera, etc.), it is possible to manually control the focal plane and DoF. Hence, two sequential focused facial images are obtained for use in these experiments: one is focused on a nose and the other on ears (Figure 2).

When we set the focus on the ears and nose, we can tap on the LCD panel or turn a focus ring in accordance with the type of handheld digital camera. In this paper, a mirrorless camera (SONY-NEX5) is used, and it has a focus ring. Therefore, we acquire the focused images, turning the focus ring and checking the sharpness in the regions of the ears and nose with our eyes.

In the preprocessing step, we geometrically normalize images based on the location of the eyes [34]. In every image, the positions of faces are slightly different. For accurate comparison, faces must be aligned. Based on the coordinates of the eyes, we translate, rotate and crop facial images. The eyes can be automatically detected by using feature templates. In this paper, however, we select the correct positions of the eyes manually in every image and save the coordinates. Figure 4shows the normalized images produced in the present study. Figure 4a,c is focused on the nose (I_N) and Figure 4b,d on the ears (I_E).

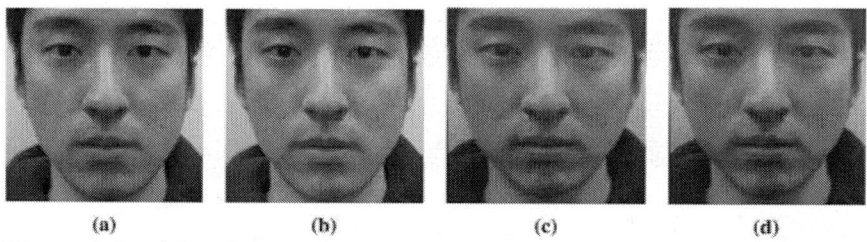

(a)   (b)   (c)   (d)

Figure 4. Real face images focused on (**a**) the nose and (**b**) the ear; and fake face images focused on (**c**) the nose and (**d**) the ear.

3.1.2. Using a Webcam

The focus in a webcam is controlled by adjusting the plastic lens in and out. However, the DoF is unknown, and it is difficult to select the focus area without the use of a supplemental program. Therefore, unless the program is used, it is not easy to obtain images focused on either the nose or ears. In order to acquire input images with a webcam, we approach the problem in a different way.

Although it is not possible to accurately take images focused on either the nose or ears when using a webcam, it is possible to obtain image sequences by changing the lens motor step. Depending on the adjustment of the lens, the focal plane varies, producing images with different focal planes. From the image sequence collected here, we select two images, I_N and I_E. I_N and $I_E$denote the normalized images for which the nose and ear area are in focus, respectively. In order to determine these images, we detect the nose and ears and calculate the degrees of focus in those areas [4]. As mentioned before, the centers of the eyes and the regions of the ears and nose are selected manually in this paper. When the value of a specific area is at a maximum at the k-step, that region is in focus. Figure 5 depicts the changes in focus values in accordance with the lens step. In Figure 5a, the nose area is in focus at the 20th step and the ears area at the 16th step. With fake faces, the steps of the maximum focus values for the nose and ears are same, as shown inFigure 5b.

This allows one to distinguish between real and fake faces. Through this procedure and normalization, we can choose two images as I_N and I_E (Figure 6).

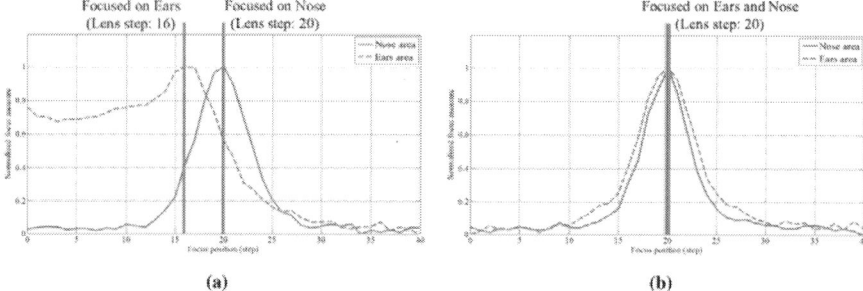

(a)   (b)

Figure 5. Variations of focus measures in accordance with lens steps ((**a**) real face and (**b**) fake face).

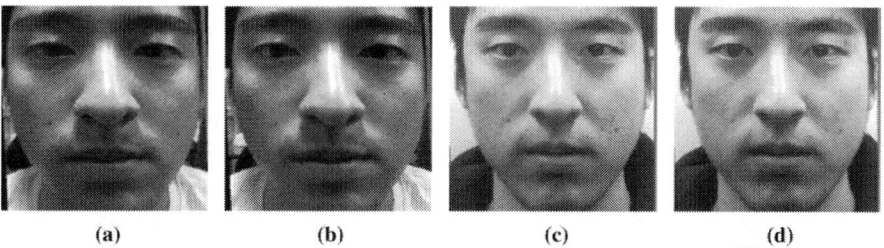

(a)   (b)   (c)   (d)

Figure 6. Normalized webcam images (real face images focused on (**a**) the nose and (**b**) the ear; and fake face images focused on (**c**) the nose and (**d**) the ear).

3.2. Feature Extraction

To detect forged faces, features are extracted from normalized images. In this paper, we use three feature descriptors: focus, power histogram and gradient location and orientation histogram (GLOH) [35].

3.2.1. Focus Feature
The focus feature is related to the degree of focusing. In the previous study [32], this feature was suggested and used for classifying fake faces. Figure 7 shows the flowchart for extracting focus features.

Using several focus measures [4], we can numerically calculate the focus levels in each pixel. There are various focus measures, such as Laplacian-based measures and gradient-based measures. We will show the performance in accordance with the focus measures.

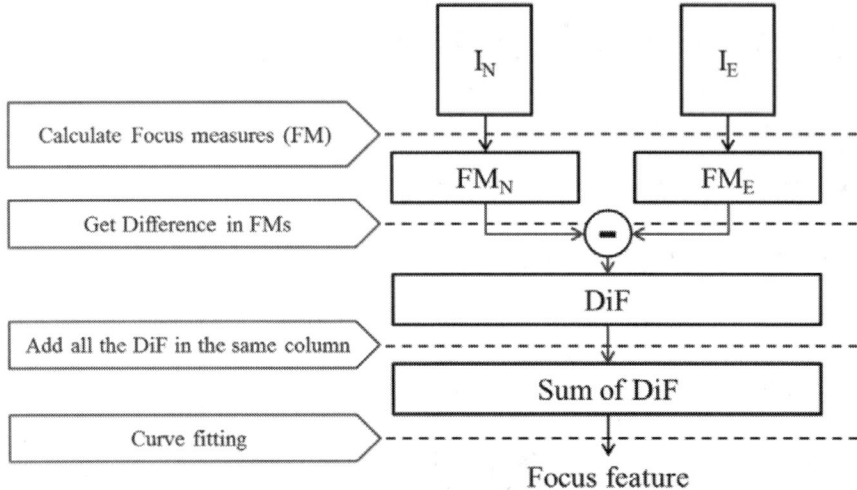

Figure 7. Flowchart of focus feature extraction.

The images in Figure 8 are the results of modified Laplacian (LAPM) focus measure calculations. LAPM is one of the focus measures introduced in [4, 33]. This is presented as the sum of transformed Laplacian filters. Figure 8a, b shows the LAPMs of a real facial image focused on the nose and ears, and Figure 8c, d shows the LAPMs of a fake facial image focused on the nose and ears. We denote the LAPM of nose-focused images by $LAPM_N$ and the LAPM of ear-focused images by $LAPM_E$. In $LAPM_N$ and $LAPM_E$, bright pixels represent high values of LAPM, and those regions are in focus with sharp edges. On the contrary, out-of-focus regions have severe blurring, lose edge information and have low values of LAPM. In the case of real faces, the nose area in $LAPM_N$ (Figure 8a) is brighter than that in $LAPM_E$ (Figure 8b). However, there is little difference between the $LAPM_N$ and $LAPM_E$ of fake faces (Figure 8c, d). Consequently, by computing the variations in focus measures, we can determine the degree of focusing.

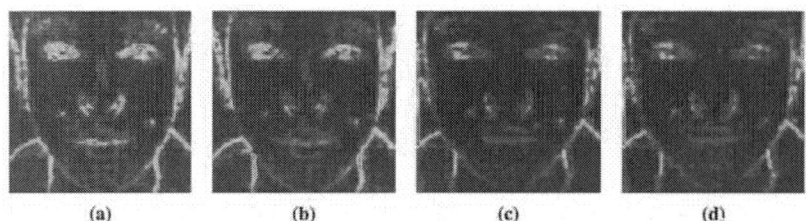

Figure 8. Modified Laplacians (LAPMs) of real face images focused on (**a**) the nose and (**b**) the ear, and LAPMs of fake face images focused on (**c**) the nose and (**d**) the ear.

In order to maximize the LAPM difference between regions of the nose and ears, we subtract $LAPM_E$ from $LAPM_N$ ($= LAPM_N - LAPM_E$). To analyze the differences in LAPMs (DiF, difference in focus measures) in a single dimension, we add all of the DiF in the same column. In Figure 9, blue lines describe the cumulative sums of the DiF of real and fake faces.

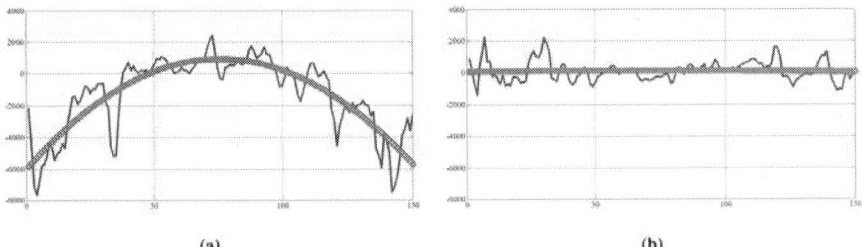

(a)   (b)

Figure 9. Cumulative sums of the differences (DiF) of (a) a real face and (b) a fake face.

However, these distributions are not appropriate to be used for liveness detection without any refinement. The existence of noise affects the results. Therefore, curve fitting is performed to extract meaningful features. The sum of the DiF of real faces has a similar shape to the curvature of a quadratic equation, $y = ax^2 + bx + c$. In the quadratic equation, there are three coefficients, $\mathbf{A} = [a \; b \; c]^T$, and these are exploited as a feature for classification. To calculate the values of these coefficients, we perform error minimization [32].

Figure 9 presents the results of curve fitting (red circles). The curve for the cumulative sum of DiF of the real face is convex, as shown in Figure 9a, while that of the fake face is flat. In Figure 10, coefficients of quadratic equations are plotted. Blue circles are features of real faces, and red crosses are those of spoofing faces. Depending on the range of DoF, the degree of feature overlap will change.

3.2.2. Power Histogram Feature

Out-of-focus images have few edge components because the blurring filter eradicates the boundary. This affects the frequency characteristics of such images. We analyze this feature to identify forged faces. In this section, we introduce another feature, the power histogram feature, which contains spatial frequency information. The process of extracting this feature is presented in Figure 11.

In the first step, we divide a normalized image into three subregions, as shown in Figure 12. When a picture is taken focusing on the ears, we adjust the focal plane to include the ear area. Not only ears, but other components in the DoF are in focus. To analyze those components, we divide the images radially. The first subregion (subR1, Figure 12b) is the nose area, the second subregion (subR2, Figure 12c) includes the eyes and mouth, and the third subregion (subR3, Figure 12d) contains the ears and the contour of the chin.

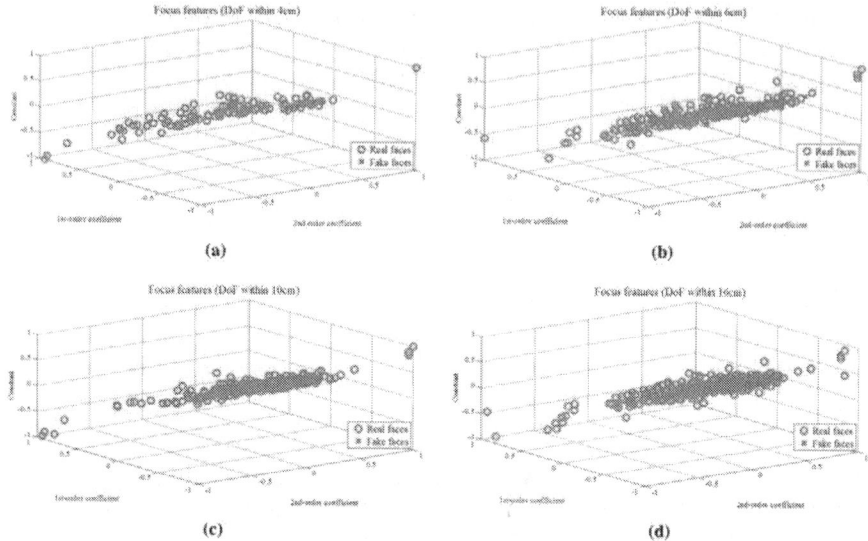

Figure 10. Distributions of focus features (DoF (**a**) within 4 cm, (**b**) within 6 cm, (**c**) within 10 cm and (**d**) within 16 cm).

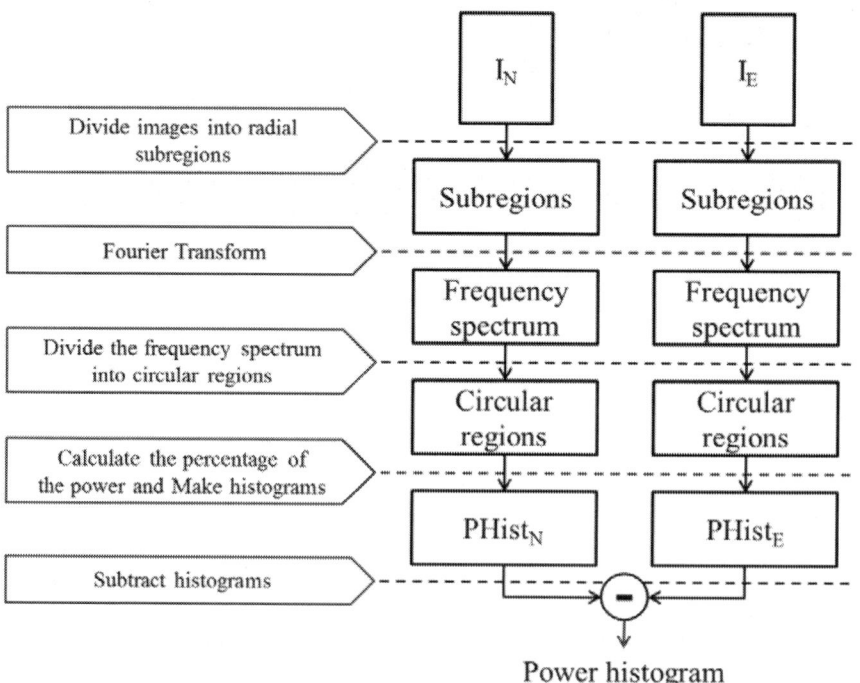

Figure 11. Flowchart of power histogram feature extraction.

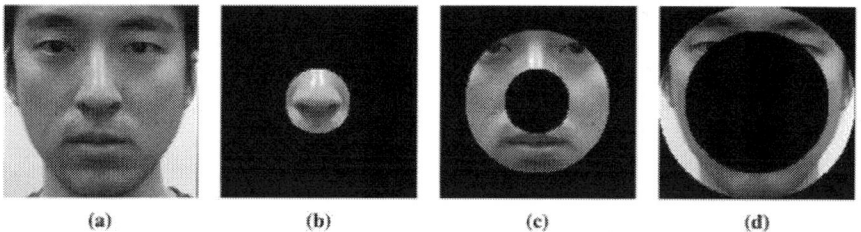

Figure 12. Subregions before extracting the power histogram ((**a**) original image, (**b**) Subregion 1 (subR1), (**c**) subR2 and (**d**) subR3).

Using a Fourier transform, we convert subregions from the spatial domain to the frequency domain. Figure 13 illustrates center-shifted Fourier spectrums of the three described subregions with power being concentrated at the center of each spectrum. According to the subregion, the distributions of power are different. In order to analyze those distributions, we calculate the percentage of power in circular regions. We divide the frequency spectrum into several circles by allowing it to be superimposed. The percentage of power within a circular region is computed by Equation (1) [36],

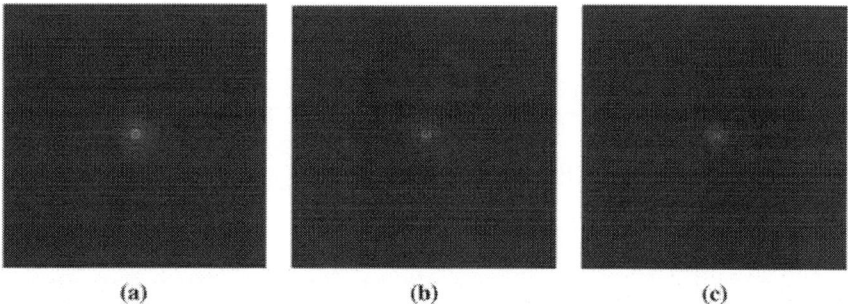

Figure 13. Fourier spectrums of (**a**) subR1, (**b**) subR2 and (**c**) subR3.

$$\alpha(\%) = 100 \times \left[\frac{1}{P_T} \sum_{u,v \in C} P(u, v)\right]$$

$$P_T = \sum_{u=1}^{U} \sum_{v=1}^{V} P(u, v) \tag{1}$$

$$P(u, v) = real(u, v)^2 + imag(u, v)^2$$

where C is a circular region and $real(u, v)$ and $imag(u, v)$ are the real and imaginary parts of the frequency component, respectively. Each spectrum has a histogram, and the value of each bin is the percentage of power in each circular area. By concatenating three histograms, we can obtain a combined histogram from one image. The dimensionality of the histogram is determined by the radii of the circular regions in the frequency spectrum. With real faces, power histograms vary depending on the focus area. However, those of fake faces do not vary. We use the differences in the power histograms as a feature for liveness detection.

3.2.3. GLOH Feature

We extract another feature descriptor, the gradient location and orientation histogram (GLOH) [35], which is an extended version of scale-invariant feature transform (SIFT) [37] and makes it possible to consider more spatial regions, as well as making feature descriptors robust and distinctive. In this paper, we modify and apply this feature locally. Figure 14 shows the flowchart of extracting the GLOH feature.

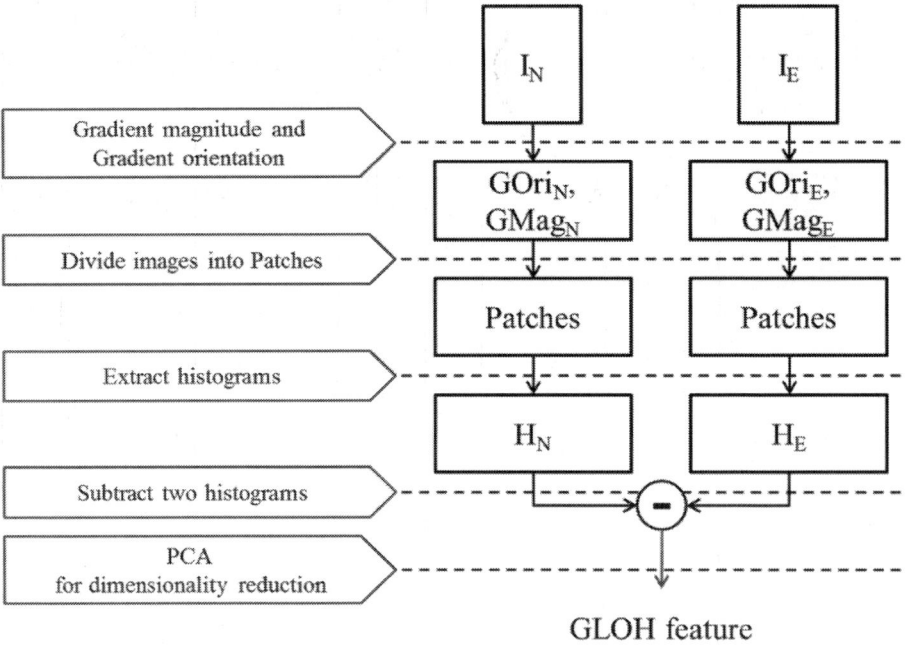

Figure 14. Flowchart of gradient location and orientation histogram (GLOH) feature extraction.

For each Gaussian smoothed image, the gradient magnitude, *GMag*, and orientation, *GOri*, are computed by Equation (2).

$$GOri(x, y) = tan^{-1} \frac{I(x,y+1)-I(x,y-1)}{I(x+1,y)-I(x-1,y)}$$

$$GMag(x, y) = \sqrt{(I(x+1,y) - I(x-1,y))^2 + (I(x,y+1) - I(x,y-1))^2}$$

(2)

Next, we divide the image into P × Q patches in order to draw features locally. Figure 15 shows how to separate the image into patches. GLOH descriptors are derived from polar location grids in patches. As shown in Figure 15, each patch is divided into 17 subregions (three bins in each radial direction and eight bins in each angular direction). Note that the central subregion is not split. In a subregion, the gradient orientations are quantized into 16 bins (Figure 16). From one patch, 17 histograms are created. We reshape these histograms into one column vector, whose dimensionality is 272 (=17 × 16), as illustrated in Figure 17. Finally, a 272 × P × Q-dimensional column vector is extracted from P × Q patches. From I_N and I_E, two vectors, H_N and H_E, are acquired, and the difference between them is determined ($H_N - H_E$). Principal component analysis (PCA) is applied to reduce the final dimensionality.

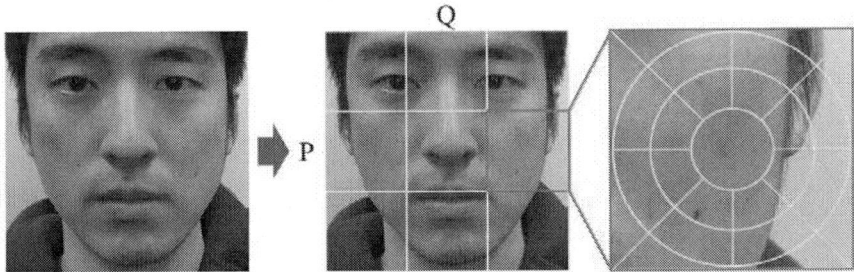

Figure 15. Patches in an image and polar location grids in a patch (patch size: 50 × 50).

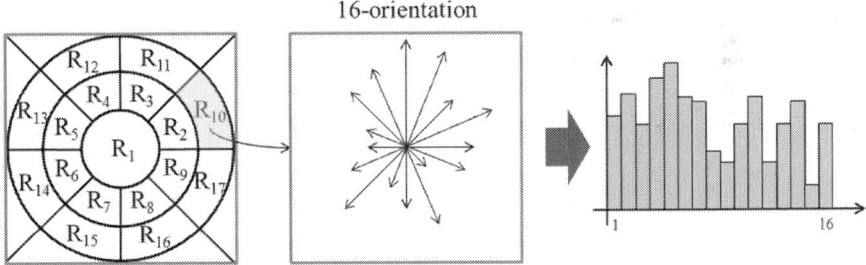

Figure 16. A histogram from one radial subregion.

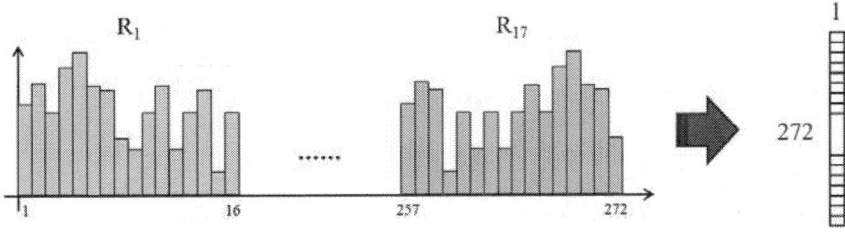

Figure 17. A histogram from one image patch.

3.3. Classification

For classification, the support vector machine-radial basis function (SVM-RBF) is used [38]. The SVM classifier learns normalized focus, power histogram and GLOH features. Furthermore, we carry out fusion-based experiments by concatenating normalized features. Figure 18 shows the flowchart of the feature-level fusion approach. Depending on the training data and the development data, the parameter of the SVM classifier is determined.

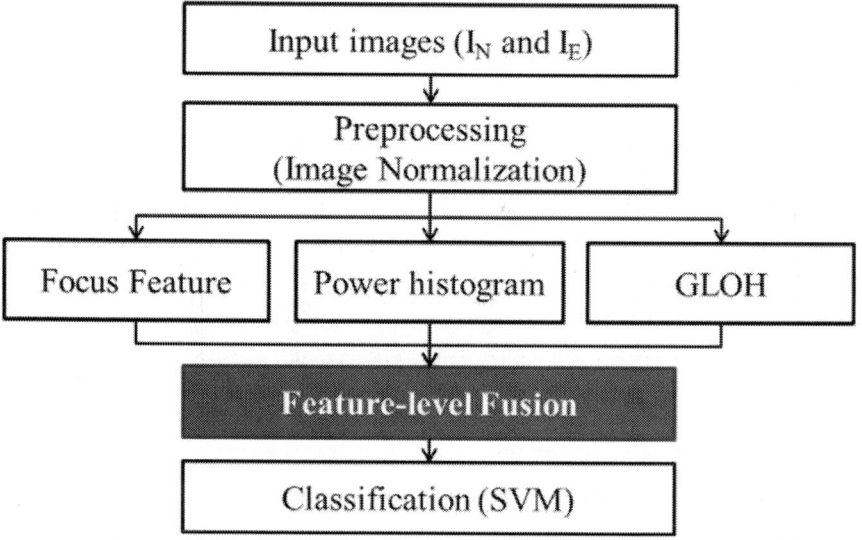

Figure 18. Flowchart of the feature-level fusion approach.

4. EXPERIMENTATION

Before evaluating the performances of our approaches, we collected frontal facial images from 24 subjects, because there is no open facial database that has various focusing areas. Although there are some databases for liveness detection, they do not satisfy our requirements. Therefore, we created two databases, one composed of images taken by a mirrorless camera (SONY-NEX5) and the other containing images taken by a webcam (Microsoft LifeCam Studio). The difference between the two cameras is the possibility of the accurate and delicate control of focus. With the mirrorless camera, it is possible to focus precisely on the nose or ear area. However, the webcam makes it difficult to adjust focus in detail, and users are not able to determine what is in focus. We will explain the processes of acquiring databases in the next section. We printed photos for fake faces with a Fuji Zerox ApeosPort-II C5400 printer.

For evaluations, the following measures are used.

- False acceptance rate (FAR): the proportion of fake images misclassified as real.
- False rejection rate (FRR): the proportion of real images misclassified as fake.
- Total error rate (TER): the sum of FAR and FRR. $TER = FAR + FRR$
- Half total error rate (HTER): half of the TER. $HTER = TER/2$

The performance of the proposed method is evaluated with our own databases. Databases are randomly categorized into 3 groups: training, development and testing sets.

- Training set (30%): to be used for training the classifier.
- Development set (30%): to be used for estimating the threshold of the classifier.
- Testing set (40%): to be used for evaluating the performance.

Thirty percent of the subjects are used for training and development, and forty percent of the subjects are used for testing. Three groups are disjoint. That is, if images of subject 'A' are used for training, they cannot be utilized for development or testing.

4.1. Experiment 1: Using The Mirrorless Camera Database

4.1.1. Data Acquisition

With the mirrorless camera, the nose and ear areas are able to be in focus, and the DoF is manually controlled. In order to obtain images with various DoFs, we adjusted the distance between the camera and the subject, focal length and F-stop. Figure 19 shows the ranges of the parameters. Focal lengths are 16, 28 and 35 mm, and values of F-stop are changed according to the focal lengths from f/3.2 to f/22. Distances between the camera and the subject vary from 20 cm to 55 cm.

Focal length	F-stop	Distance between a camera and a subject			Focal length	F-stop	Distance between a camera and a subject				Focal length	F-stop	Distance between a camera and a subject		
		20cm	25cm	30cm			40cm	45cm	50cm	55cm			45cm	50cm	55cm
16	3.2	1.83	2.91	4.25	28	4.5	3.41	4.36	5.42	6.6	35	4.5	2.74	3.41	4.16
	3.5	2	3.18	4.69		5	3.83	4.9	6.09	7.42		5	3.08	3.83	4.67
	4.5	2.59	4.13	6.04		6.3	4.84	6.18	7.69	9.37		6.3	3.88	4.83	5.89
	5	2.91	4.65	6.79		8	6.11	7.81	9.72	11.9		8	4.99	6.1	7.43
	6.3	3.68	5.88	8.62		9	6.87	8.78	10.9	13.3		9	5.5	6.85	8.35
	8	4.66	7.47	11		10	7.72	9.88	12.3	15		10	6.18	7.7	9.39
	9	5.25	8.44	12.4		14	10.7	13.7				14	8.52	10.6	13
	10	5.92	9.54	14.1		16	12.4	16				16	9.87	12.3	15.1
	14	8.3	13.5			18	14					18	11.1	13.9	
	16	9.71	16			20	15.9					20	12.9	18.7	
	18	11.1													
	20	12.7													
	22	14.6													

0cm 2cm 4cm 6cm 8cm 10cm 12cm 14cm 16cm

DoF

Figure 19. The ranges of the distance between the camera and the subject, focal length, F-stop and DoF.

The total number of images in the mirrorless camera database is 5968 (1492 pairs of real images and 1492 pairs of fake images). The images are categorized into four groups according to the range of DoF and are listed in Table 1. The number of males is 17 and that of females is 7.

Table 1. The number of pairs of images in the database.

Depth of Field	within 4 cm	within 6 cm	within 10 cm	within 16 cm
Real	336	767	1149	1492
Fake	336	767	1149	1492

The size of each normalized image is 150 by 150 pixels, and the distance between the eyes is 70 pixels. Figure 20 shows real (a) and fake (b) samples from the database.

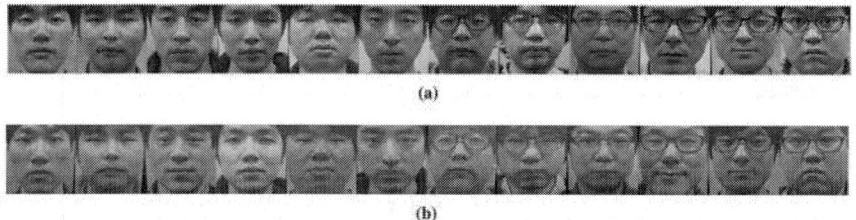

(a)

(b)

Figure 20. Normalized images of (**a**) real and (**b**) fake faces.

4.1.2. Experimental Results

We carry out experiments in accordance with the types of features, and the detailed results are described in Appendix A. The following shows the performance of the concatenated features. The process of combining features is carried out in the feature level. For high performance, we choose features based on the above results. Modified Laplacian (LAPM) and wavelet sum (WAVS) are used as focus features. In the case of the power histogram feature, the radii of the circular regions are 5, 15, 30, 50 and 75. GLOH features are extracted using 75×75 patches without allowing overlap. In order to reduce the dimensionality of the GLOH features, we apply PCA and use several eigenvectors whose variances are 90%. Table 2 shows the denotations of the features.

Table 2. Denotations.

Denotation	Specification
Focus (LAPM)	Modified Laplacian
Focus (WAVS)	Wavelet sum
Power hist	Rad.ver6 (radii = 5, 15, 30, 50, 75)
GLOH	Patch75, No overlapping, PCA 90%
Fusion.ver1	Focus (LAPM) + Power hist + GLOH
Fusion.ver2	Focus (WAVS) + Power hist + GLOH

Table 3 and Figures 21 and 22 illustrate the results of the fusion-based methods. When the DoF is shallow (within 4 cm and 6 cm), the performances of focus features (LAPM and WAVS) are better than those of other features. However, as the DoF becomes deeper, the performances of focus features deteriorate. In the case of the GLOH and fusion-based features, the performances are maintained compared to other features. Especially, the HTERs of the fusion-based features under 16-cm DoF are lower than those of other features (6.27% and 6.08%). These numerical results demonstrate that the fusion-based methods are prominent when the effect of defocusing is low.

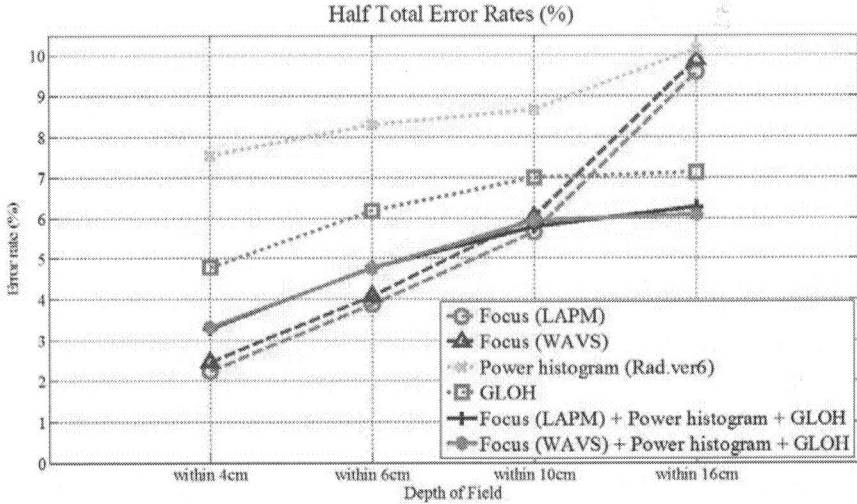

Figure 21. HTERs (%) of features.

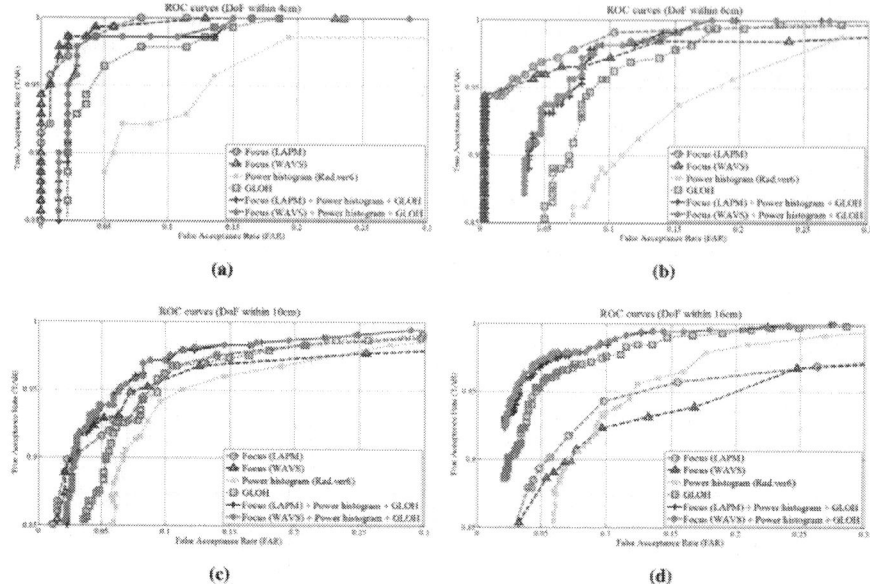

Figure 22. ROC curves of the feature-level fusion (DoF (**a**) within 4 cm, (**b**) within 6 cm, (**c**) within 10 cm and (**d**) within 16 cm).

Table 3. Half total error rates (HTERs) (%) of the experiments with the mirrorless camera database.

	within 4 cm			within 6 cm			within 10 cm			within 16 cm		
	Dev (mean ± std)	Test (mean ± std)	sigma	Dev (mean ± std)	Test (mean ± std)	sigma	Dev (mean ± std)	Test (mean ± std)	sigma	Dev (mean ± std)	Test (mean ± std)	sigma
Focus (LAPM)	0.92 ± 0.40	**2.25 ± 1.54**	0.09	2.72 ± 0.89	**3.89 ± 0.70**	0.09	5.49 ± 0.95	**5.66 ± 0.53**	0.08	7.59 ± 1.43	9.57 ± 1.84	0.08
Focus (WAVS)	1.28 ± 0.97	**2.46 ± 0.91**	0.15	3.37 ± 1.32	**4.09 ± 0.69**	0.095	5.87 ± 0.57	6.04 ± 0.66	0.10	8.33 ± 1.40	9.90 ± 1.20	0.09
Power hist	5.15 ± 1.76	7.54 ± 1.18	0.90	6.50 ± 1.62	8.29 ± 1.42	1.00	6.98 ± 1.13	8.67 ± 1.34	0.70	9.29 ± 1.58	10.1 ± 1.96	0.75
GLOH	3.52 ± 1.24	4.79 ± 1.37	1.75	5.09 ± 0.88	6.19 ± 0.87	3.00	5.08 ± 0.88	7.00 ± 0.99	2.90	6.98 ± 1.26	**7.13 ± 1.31**	3.55
Fusion.ver1	2.45 ± 1.27	**3.29 ± 1.44**	2.30	4.06 ± 0.90	4.78 ± 1.08	3.40	3.92 ± 0.89	**5.79 ± 1.05**	3.65	5.58 ± 1.66	**6.27 ± 1.80**	4.55
Fusion.ver2	2.45 ± 1.27	3.32 ± 1.42	2.40	4.06 ± 0.89	**4.76 ± 0.86**	4.50	3.91 ± 0.96	**5.93 ± 1.23**	4.15	5.57 ± 1.62	**6.08 ± 1.72**	5.25

4.2. Experiment 2: Using the Webcam Database

4.2.1. Data Acquisition

For evaluations, we gathered facial data using Microsoft LifeCam Studio. Using the provided program, we could control the lens motor step from 0 to 40. Therefore, one input sequence is composed of 41 images. Among those, we choose I_N and I_E, as mentioned in Section 3.1.2. The distance between the webcam and the subject is about 20 cm, so that the image can contain a whole face. The number of real face sequences is 94. Normal prints and an HD tablet (iPad 2) are used as spoofing attacks, and the number of sequences is 240 and 120 respectively. Five-fold cross-validation is applied for the evaluation.

4.2.2. Experimental Results

Numerical results are listed in Table 4. The good performance is maintained, even though the webcam database cannot express depth information well compared to the mirrorless camera database. The results of the combined features are the best, and the HTERs of them are 3.02% under normal print attack and 3.15% under HD tablet attack. These experiments show the possibility that our proposed method can be used in security systems at a low cost and with low specification devices. Furthermore, if detailed adjustment of the focus is possible in the device, our method can improve the performance more.

Table 4. HTERs (%) of experiments with the webcam database.

	Normal Print (mean ± std)	**HD Tablet (mean ± std)**
Focus (LAPM)	8.29 ± 0.45	10.0 ± 0.45
Focus (WAVS)	6.48 ± 0.36	7.54 ± 0.49
Power hist	7.93 ± 0.45	7.25 ± 0.43
GLOH	6.09 ± 0.55	5.28 ± 1.06
Fusion.ver1	$\mathbf{3.39 \pm 0.46}$	$\mathbf{3.30 \pm 0.66}$
Fusion.ver2	$\mathbf{3.02 \pm 0.47}$	$\mathbf{3.15 \pm 0.45}$

4.3. Discussion

Due to the characteristic of our proposed method, it is impossible to apply our method to open databases, such as the CASIA database [39] and the Replay-Attack database [40]. Therefore, we conducted comparative experiments by applying other methods to our own database. Table 5 demonstrates the performance comparison between our proposed method and other methods.

Table 5. Performance comparison (HTER (%)).

	within 4 cm			within 6 cm			within 10 cm			within 16 cm		
	Dev (mean ± std)	Test (mean ± std)	sigma	Dev (mean ± std)	Test (mean ± std)	sigma	Dev (mean ± std)	Test (mean ± std)	sigma	Dev (mean ± std)	Test (mean ± std)	sigma
Zhang [41]	29.2 ± 4.12	39.2 ± 3.55	2.00	26.9 ± 4.45	33.9 ± 4.00	2.40	29.3 ± 2.99	35.4 ± 5.70	2.95	25.7 ± 4.70	36.9 ± 4.41	3.85
Kim [9]	12.6 ± 2.86	17.2 ± 3.70	15.0	18.8 ± 5.74	18.8 ± 5.55	18.5	17.65 ± 3.35	20.4 ± 4.05	17.3	23.9 ± 5.28	18.9 ± 6.03	17.2
Määttä [42]	19.7 ± 7.11	22.2 ± 4.35	-	20.4 ± 5.64	21.6 ± 2.52	-	24.4 ± 4.54	23.1 ± 5.40	-	22.6 ± 6.99	21.0 ± 3.14	-
Kim [32]	9.34 ± 4.07	8.39 ± 2.63	-	11.9 ± 1.90	12.1 ± 1.59	-	15.9 ± 1.97	16.3 ± 2.27	-	19.1 ± 2.05	20.8 ± 2.10	-
Fusion.ver1	2.45 ± 1.31	3.39 ± 1.46	2.30	4.51 ± 1.04	4.87 ± 1.13	3.45	4.07 ± 1.03	5.60 ± 0.83	3.45	5.19 ± 1.60	6.71 ± 1.65	4.55
Fusion.ver2	2.50 ± 1.33	3.39 ± 1.45	2.35	4.55 ± 0.92	4.80 ± 0.87	3.65	4.09 ± 1.07	5.82 ± 1.13	4.05	5.20 ± 1.57	6.51 ± 1.57	5.25

Other methods [9,41,42] detect the liveness based on textural analysis (local binary patterns) or frequency components (difference of Gaussian, power spectrum). Even though they have an advantage in terms of using a single image, the performances for our own database are not remarkable, regardless of the DoF; whereas the previous work [32] shows a good result relatively at the within 4-cm DoF. However, when the DoF is deep, the performance of [32] deteriorates. This represents that the performance of the previous system is determined depending on the method of input picture collection with great effects on defocus.

In order to overcome this limitation, we propose our system by considering two factors. The first is by supplementing features. By adding other feature descriptors, we try to maintain good performance, even though the DoF becomes deeper. In the case of the GLOH feature [35], it has high matching scores for images with severe blur, whereas local features used in other methods [9,41,42] are not proper for the defocused images to be compared to the GLOH feature descriptor. The influence of the GLOH feature can be confirmed in the previous Section 4.1.2. In Figure 21, the performances of the focus and power histogram features are deteriorated in accordance with the increase of the DoF. However, the performance of the GLOH feature is maintained. As a result, we can achieve 6.51% HTER (feature-fusion) at a DoF within 16 cm by using additional features specialized for the defocused images. These results are better than the HTERs of other methods and the previous method [32], which uses only a focus feature of 20.8%.

The second way to mitigate the weakness of the previous study [32] is the use of the webcam database. Digital cameras, such as DSLR and mirrorless cameras, have high specifications and make it possible to manually adjust the DoF and focusing areas. However, due to their high cost, people might be unwilling to use digital cameras for image acquisition in anti-spoofing algorithms. Webcams are cheaper than digital cameras and are utilized broadly. With the webcam, we created a database and conducted experiments. As a result, we accomplish 3.02% HTER with the combined feature. The performance with the webcam database is similar to that with the mirrorless camera database.

Even though we show the good performance for liveness detection, our method has a disadvantage in the process of acquiring and normalizing images. In this paper, we set the focus on the ears and nose and find the centers of the eyes manually. In order to apply our proposed method to the security systems at low cost and with low specification devices, like smartphones, facial components must be detected automatically. Recently, many studies for feature point extraction have been in progress, and most cameras and smartphones have a face priority auto focusing function [43–45], which helps to obtain face-focused images by automatically controlling the lens actuator. If these technologies are utilized, the limitation of our method will be settled and applicable to the devices. Moreover, it will strengthen the security of smartphones.

5. CONCLUSION AND FUTURE WORK

We proposed a face liveness detection method based on the characteristics of defocus. Our method pays attention to the difference between the properties of real and 2D fake faces. We use focus, power histogram and GLOH as feature descriptors and classify spoofing faces in terms of the feature-level fusion processes. Our experimental results show 3.29% HTER when the DoF of images is within 4 cm. Moreover, by applying various features, we overcome the limitation of DoF without adding any other sensors. Furthermore, through experiments with a webcam, we confirm that the good performance of our method is maintained.

Even though our proposed method yields good results, it has a limitation for being applied to camera-embedded security systems, such as smartphones, because of the manual processes to acquire the focused images and to detect facial components. Therefore, in future work, we will improve our method in order for it to operate automatically in the image acquisition and preprocessing and to make it possible to embed our method on a smart devices. Furthermore, we will consider more robust countermeasures against videos and 3D attacks by analyzing textural and temporal characteristics. Furthermore, we will advance our method using a light-field camera, which can acquire various focusing information in the spatial domain using a microlens array.

ACKNOWLEDGMENTS

This research was supported by the MSIP (The Ministry of Science, ICT and Future Planning), Korea and Microsoft Research, under ICT/SW Creative research program supervised by the NIPA(National IT Industry Promotion Agency) (NIPA-2014-11-1460). Also, this research was supported by the MSIP (Ministry of Science, ICT & Future Planning), Korea, under the "ITRC (Information Technology Research Center)" support program (NIPA-2014-H0301-14-1012) supervised by the NIPA (National IT Industry Promotion Agency).

AUTHOR CONTRIBUTIONS

Sangyoun Lee and Sooyeon Kim developed the methodology and drafted the manuscript. Moreover, Sooyeon Kim implemented software simulations. Yuseok Ban collected databases and evaluated the performances. The authors approved the final manuscript.

Experiments According to the Type of Features
We carry out experiments in accordance with the types of features. The following shows the performance of our proposed methods.

A.1. Focus Feature

We conduct experiments with eight types of focus features. Eight focus features are categorized into four groups: statistic-based, Laplacian-based, gradient-based and wavelet-based operators [4]. The focus features are listed in Table A1. Related equations are organized in [4].

Table A2 and Figure A1 show HTERs and receiver operating characteristic (ROC) curves of focus features according to the range of the DoF. In general, the performance of the Laplacian group is better than those of other groups. As depicted in Figure A1, focus features in the Laplacian group swarm in the upper side. Especially, modified Laplacian (LAPM) has stable and prominent results all over the DoF (1.64% HTER under the within 4-cm DoF and 8.93% HTER under the within 16-cm DoF). The sum of wavelet coefficients (WAVS) also shows good performance. When the DoF is shallow, the effect of defocusing is great. This makes the focus features of real and fake faces more discriminative. As a result, focus features, except gray-level variance (GLVA), yield the best performances under the within 4-cm DoF. The GLVA focus feature, unusually, has the best performance when the DoF is within 16 cm (12.4% HTER). GLVA is the simple variance of the gray-scale image. Compared to other focus features, it is inadequate to represent the difference between the focused and defocused regions regardless of the DoF.

Table A1. Focus features.

Focus Feature (Abbreviation)	Focus Feature (Full Name)	Category
GLVA	Gray-level variance	Statistic
LAPD	Diagonal Laplacian	Laplacian
LAPM	Modified Laplacian	Laplacian
LAPV	Variance of Laplacian	Laplacian
TENG	Tenengrad	Gradient
TENV	Tenengrad variance	Gradient
WAVS	Sum of wavelet coefficients	Wavelet
WAVV	Variance of wavelet coefficients	Wavelet

Table A2. HTERs (%) of the focus features.

	within 4 cm			within 6 cm			within 10 cm			within 16 cm		
	Dev (mean ± std)	Test (mean ± std)	sigma	Dev (mean ± std)	Test (mean ± std)	sigma	Dev (mean ± std)	Test (mean ± std)	sigma	Dev (mean ± std)	Test (mean ± std)	sigma
GLVA	14.4 ± 2.77	16.6 ± 2.41	0.3	14.1 ± 2.30	13.6 ± 2.20	0.09	11.4 ± 2.34	12.5 ± 1.15	0.08	10.7 ± 2.01	12.4 ± 1.59	0.1
LAPD	1.02 ± 1.15	1.82 ± 0.59	0.15	3.46 ± 1.18	3.97 ± 1.09	0.20	4.91 ± 0.81	5.75 ± 0.56	0.135	8.61 ± 1.78	9.41 ± 1.64	0.09
LAPM	1.07 ± 1.16	1.64 ± 0.59	0.2	3.08 ± 0.99	3.57 ± 1.05	0.09	4.97 ± 0.88	6.10 ± 0.39	0.085	8.56 ± 1.40	8.93 ± 1.90	0.06
LAPV	1.22 ± 0.80	2.50 ± 2.61	0.35	3.06 ± 0.92	3.16 ± 0.86	0.075	5.67 ± 1.05	6.36 ± 0.63	0.25	8.53 ± 1.51	8.48 ± 1.31	0.07
TENG	3.72 ± 1.91	4.68 ± 1.86	0.09	4.27 ± 0.98	4.88 ± 0.55	0.065	5.59 ± 1.07	6.17 ± 1.05	0.065	8.05 ± 1.69	9.14 ± 1.30	0.055
TENV	4.94 ± 1.89	7.11 ± 3.34	0.135	5.41 ± 1.40	6.79 ± 2.03	0.03	7.53 ± 1.25	8.44 ± 1.37	0.06	9.22 ± 2.91	10.8 ± 1.78	0.01
WAVS	1.22 ± 1.11	2.00 ± 0.89	0.3	3.53 ± 1.34	3.50 ± 0.91	0.2	5.53 ± 0.69	6.15 ± 0.72	0.145	8.83 ± 1.73	9.79 ± 1.05	0.08
WAVV	2.65 ± 1.44	4.57 ± 2.44	0.55	5.70 ± 0.84	5.66 ± 1.32	0.09	8.21 ± 0.85	8.96 ± 1.44	0.08	10.7 ± 1.86	11.4 ± 1.73	0.45

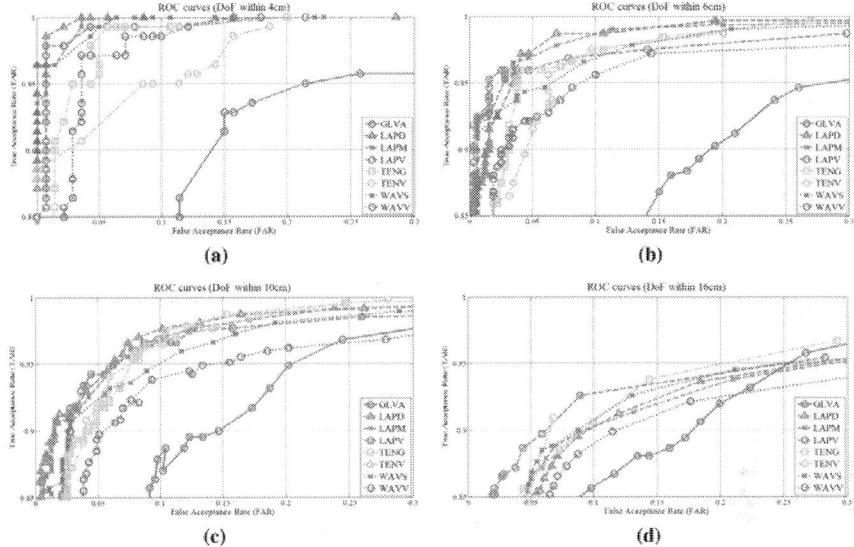

Figure A1. ROC curves of the focus features (DoF (**a**) within 4 cm, (**b**) within 6 cm, (**c**) within 10 cm and (**d**) within 16 cm).

A.2. Power Histogram Feature
In order to find that how to divide the frequency spectrum that yields good performance, we carry out experiments taking the radii of circular regions in Table A3. The dimensionality is the length of the concatenated histograms of the three subregions.

Table A3. Radii of circular regions.

Version	Radii	Dimensionality
Rad.ver1	[1:1:75]	225
Rad.ver2	[3:3:75]	75
Rad.ver3	[5:5:75]	45
Rad.ver4	[10:10:75]	21
Rad.ver5	[15:15:75]	15
Rad.ver6	[5 15 30 50 75]	15

Table A4 describes the numerical results, and Figure A2 illustrates the distributions of the HTERs and ROC curves. When the averages of HTERs are calculated respectively, Rad.ver6 shows a good performance: 7.69% HTER. Even though the dimensionality of the power histogram feature is low, it yields the best performance compared to the others.

Table A4. HTERs (%) of the power histogram features.

	within 4 cm			within 6 cm			within 10 cm			within 16 cm			
	Dev (mean ± std)	Test (mean ± std)	sigma	Dev (mean ± std)	Test (mean ± std)	sigma	Dev (mean ± std)	Test (mean ± std)	sigma	Dev (mean ± std)	Test (mean ± std)	sigma	Avg.
Rad.ver1	5.61 ± 1.73	**7.04 ± 1.86**	2.65	7.76 ± 2.24	**8.19 ± 1.01**	3.05	7.40 ± 1.64	**8.34 ± 1.19**	2.95	8.30 ± 2.13	**10.1 ± 2.04**	3.50	7.84
Rad.ver2	4.64 ± 1.55	**6.93 ± 2.20**	2.60	6.98 ± 1.46	**7.43 ± 1.70**	2.20	7.85 ± 1.51	**8.42 ± 1.29**	1.45	8.54 ± 1.85	**11.5 ± 1.93**	1.55	7.79
Rad.ver3	6.33 ± 2.57	**7.04 ± 1.77**	1.85	7.42 ± 1.51	**8.99 ± 3.32**	1.15	7.30 ± 0.94	**9.32 ± 0.96**	1.25	8.74 ± 1.84	**10.2 ± 1.22**	1.00	8.17
Rad.ver4	5.10 ± 0.96	7.36 ± 2.17	1.05	7.04 ± 1.80	9.71 ± 1.61	0.90	7.50 ± 1.55	9.63 ± 1.71	1.50	9.21 ± 1.75	11.3 ± 0.83	0.65	8.36
Rad.ver5	6.12 ± 2.44	7.36 ± 1.54	1.40	8.84 ± 1.51	9.88 ± 2.81	0.75	8.05 ± 0.96	9.37 ± 0.97	1.00	9.67 ± 1.61	10.9 ± 1.46	0.50	8.77
Rad.ver6	5.15 ± 1.90	**7.07 ± 1.26**	1.50	5.88 ± 1.22	**9.15 ± 2.41**	1.55	6.99 ± 0.76	**8.85 ± 1.92**	1.30	8.50 ± 1.12	**9.93 ± 1.02**	0.90	**7.69**

A.3. GLOH Feature

We perform experiments by altering the size of the patch, the energy percentage in PCA and whether the patches are overlapped or not. In Tables A5, A6, A7, A8, A9 and A10, numerical results are listed, and Figure A3 describes the ROC curves.

As shown in Figure A3, the performances with and without allowing the overlap of patches are similar. In terms of the computational cost, the overlap of patches is not effective. Therefore, it is better not to overlap the patches to extract the GLOH features. In the case of the energy percentage in PCA, when the percentage is 98%, the performance is worse than those under 90% and 95%.

Additionally, experiments are carried out depending on the size of the patch. When the GLOH features are extracted from the whole image, the performance is the worst, and those features cannot represent the spatial properties sufficiently. As the size of the patch is 75 × 75 and the energy percentage in PCA is 90% without allowing the overlap, the performance is the best (7.75% HTER under the within 16-cm DoF).

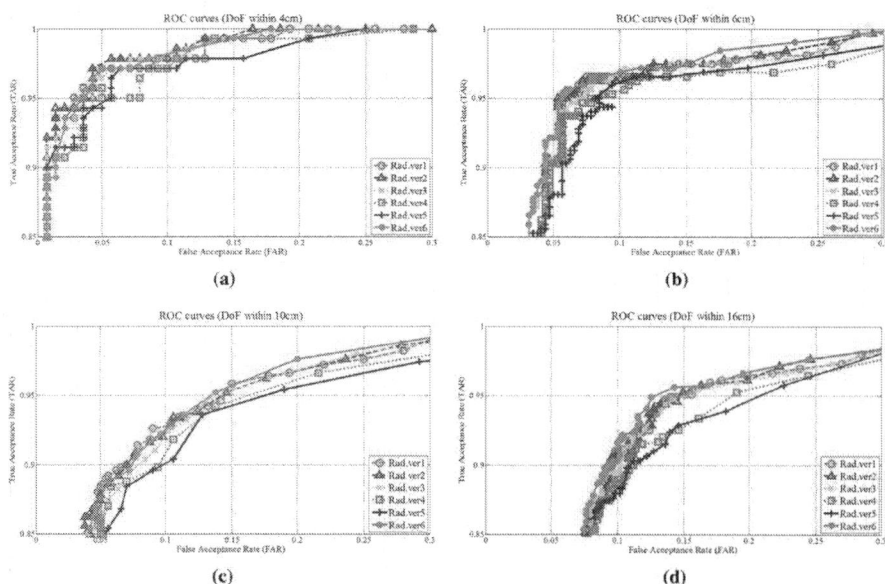

Figure A2. ROC curves of the power histogram features (DoF (**a**) within 4 cm, (**b**) within 6 cm, (**c**) within 10 cm and (**d**) within 16 cm).

Table A5. HTERs (%) of GLOH features (overlap, PCA 90%).

	within 4 cm			within 6 cm			within 10 cm			within 16 cm		
	Dev (mean ± std)	Test (mean ± std)	sigma	Dev (mean ± std)	Test (mean ± std)	sigma	Dev (mean ± std)	Test (mean ± std)	sigma	Dev (mean ± std)	Test (mean ± std)	sigma
Patch25	6.17 ± 1.33	7.61 ± 1.67	0.65	7.42 ± 1.19	8.86 ± 1.11	1.40	7.52 ± 1.64	8.14 ± 1.77	1.70	8.52 ± 1.36	8.92 ± 1.49	2.25
Patch50	4.85 ± 1.62	**7.36 ± 2.21**	1.55	6.74 ± 1.21	**6.47 ± 0.67**	2.30	6.32 ± 1.20	**6.70 ± 0.82**	2.80	6.83 ± 1.51	**8.20 ± 1.72**	3.30
Patch75	5.20 ± 1.65	7.64 ± 1.38	1.80	6.23 ± 0.80	7.01 ± 0.71	2.70	6.58 ± 1.39	6.87 ± 0.78	3.10	6.93 ± 1.20	8.25 ± 0.96	4.20
Patch150	6.48 ± 2.20	9.57 ± 1.86	2.95	8.42 ± 1.07	8.82 ± 2.05	3.70	7.42 ± 1.17	9.36 ± 0.86	3.20	8.83 ± 1.74	10.1 ± 1.52	3.40

Table A6. HTERs (%) of GLOH features (overlap, PCA 95%).

	within 4 cm			within 6 cm			within 10 cm			within 16 cm		
	Dev (mean ± std)	Test (mean ± std)	sigma	Dev (mean ± std)	Test (mean ± std)	sigma	Dev (mean ± std)	Test (mean ± std)	sigma	Dev (mean ± std)	Test (mean ± std)	sigma
Patch25	8.57 ± 1.68	8.64 ± 2.60	0.70	10.0 ± 1.86	9.05 ± 1.12	1.55	10.8 ± 6.04	10.8 ± 4.19	2.25	8.90 ± 1.47	10.3 ± 1.26	2.60
Patch50	6.07 ± 2.97	**7.89 ± 1.81**	1.40	6.50 ± 1.10	**7.49 ± 0.76**	2.50	6.26 ± 1.31	**7.36 ± 1.34**	3.20	7.46 ± 1.52	**8.44 ± 0.92**	4.20
Patch75	7.24 ± 1.84	8.14 ± 1.84	2.15	7.40 ± 1.58	8.58 ± 1.03	3.35	7.99 ± 0.82	7.73 ± 1.06	4.40	7.52 ± 1.57	9.04 ± 1.91	5.90
Patch150	7.60 ± 2.38	10.0 ± 1.34	2.95	10.6 ± 2.36	9.58 ± 1.77	5.30	9.82 ± 1.74	9.67 ± 1.56	3.75	10.7 ± 2.02	10.3 ± 1.63	4.80

Table A7. HTERs (%) of GLOH features (overlap, PCA 98%).

	within 4 cm			within 6 cm			within 10 cm			within 16 cm		
	Dev (mean ± std)	Test (mean ± std)	sigma	Dev (mean ± std)	Test (mean ± std)	sigma	Dev (mean ± std)	Test (mean ± std)	sigma	Dev (mean ± std)	Test (mean ± std)	sigma
Patch25	19.1 ± 11.5	20.3 ± 10.2	0.90	12.5 ± 4.69	12.2 ± 4.32	1.50	9.08 ± 1.29	9.38 ± 1.45	2.45	10.2 ± 2.27	11.0 ± 2.58	3.75
Patch50	6.83 ± 2.37	**7.17 ± 2.13**	1.50	6.79 ± 1.48	**7.61 ± 0.96**	3.10	7.32 ± 0.98	**7.33 ± 1.30**	4.15	7.80 ± 1.92	**9.23 ± 1.37**	5.05
Patch75	7.39 ± 1.58	9.25 ± 2.46	2.05	7.73 ± 1.12	8.94 ± 1.44	4.20	8.43 ± 2.15	9.04 ± 1.20	6.00	8.02 ± 1.48	10.4 ± 2.23	9.55
Patch150	7.70 ± 1.62	10.3 ± 2.23	4.60	10.2 ± 2.24	11.9 ± 1.35	7.30	10.7 ± 2.85	11.7 ± 1.96	7.25	11.2 ± 1.26	12.6 ± 1.86	8.60

Table A8. HTERs (%) of GLOH features (no overlap, PCA 90%).

	within 4 cm			within 6 cm			within 10 cm			within 16 cm		
	Dev (mean ± std)	Test (mean ± std)	sigma	Dev (mean ± std)	Test (mean ± std)	sigma	Dev (mean ± std)	Test (mean ± std)	sigma	Dev (mean ± std)	Test (mean ± std)	sigma
Patch25	5.05 ± 1.00	**5.14 ± 1.37**	0.90	6.90 ± 1.21	7.10 ± 1.26	1.75	6.59 ± 0.82	7.60 ± 1.43	2.45	7.22 ± 1.47	8.89 ± 0.87	3.00
Patch50	4.69 ± 1.89	6.67 ± 1.54	1.55	6.89 ± 1.12	6.63 ± 1.83	2.50	6.07 ± 0.90	7.26 ± 0.68	3.15	7.53 ± 1.09	8.56 ± 1.25	3.65
Patch75	3.11 ± 1.16	5.67 ± 1.74	1.75	5.37 ± 0.83	**6.42 ± 1.73**	2.85	5.60 ± 0.65	**7.07 ± 1.20**	2.85	5.75 ± 1.62	**7.75 ± 1.13**	4.10
Patch150	6.48 ± 2.20	9.57 ± 1.86	2.95	8.42 ± 1.07	8.82 ± 2.05	3.70	7.42 ± 1.17	9.36 ± 0.86	3.20	8.83 ± 1.74	10.1 ± 1.52	3.40

Table A9. HTERs (%) of GLOH features (no overlap, PCA 95%).

	within 4 cm			within 6 cm			within 10 cm			within 16 cm		
	Dev (mean ± std)	Test (mean ± std)	sigma	Dev (mean ± std)	Test (mean ± std)	sigma	Dev (mean ± std)	Test (mean ± std)	sigma	Dev (mean ± std)	Test (mean ± std)	sigma
Patch25	7.85 ± 3.94	8.21 ± 3.54	0.90	7.51 ± 1.11	8.65 ± 1.74	1.95	7.48 ± 1.21	8.35 ± 1.56	3.00	8.44 ± 1.53	9.45 ± 1.92	3.60
Patch50	6.02 ± 1.62	8.67 ± 3.59	1.50	7.49 ± 0.91	**6.73 ± 0.86**	3.35	7.04 ± 0.81	7.54 ± 1.01	5.15	9.40 ± 2.56	**7.91 ± 1.20**	5.40
Patch75	4.69 ± 1.39	**6.10 ± 1.61**	2.20	7.13 ± 2.41	7.16 ± 1.07	3.65	6.74 ± 1.54	**6.62 ± 0.86**	4.00	7.45 ± 1.96	8.52 ± 1.73	5.15
Patch150	7.60 ± 2.38	10.0 ± 1.34	2.95	10.6 ± 2.36	9.58 ± 1.77	5.30	9.82 ± 1.74	9.67 ± 1.56	3.75	10.7 ± 2.02	10.3 ± 1.63	4.80

Table A10. HTERs (%) of GLOH features (no overlap, PCA 98%).

	within 4 cm			within 6 cm			within 10 cm			within 16 cm		
	Dev (mean ± std)	Test (mean ± std)	sigma	Dev (mean ± std)	Test (mean ± std)	sigma	Dev (mean ± std)	Test (mean ± std)	sigma	Dev (mean ± std)	Test (mean ± std)	sigma
Patch25	11.0 ± 5.95	11.4 ± 6.54	1.20	8.10 ± 1.38	8.94 ± 1.95	2.55	7.16 ± 1.24	9.31 ± 1.69	3.60	8.90 ± 2.23	9.92 ± 1.42	4.10
Patch50	6.78 ± 1.96	7.75 ± 1.67	1.80	6.30 ± 1.22	8.88 ± 2.45	3.85	6.30 ± 1.83	**7.39 ± 1.17**	5.40	8.87 ± 2.08	**7.83 ± 1.57**	8.05
Patch75	6.17 ± 1.62	**6.57 ± 1.47**	2.30	6.97 ± 2.07	**7.57 ± 1.76**	6.25	6.90 ± 1.83	7.75 ± 1.17	6.65	9.75 ± 2.28	8.46 ± 1.20	9.60
Patch150	7.70 ± 1.62	10.3 ± 2.23	4.60	10.2 ± 2.24	11.9 ± 1.35	7.30	10.7 ± 2.15	11.7 ± 1.96	7.25	11.2 ± 1.26	12.6 ± 1.86	8.60

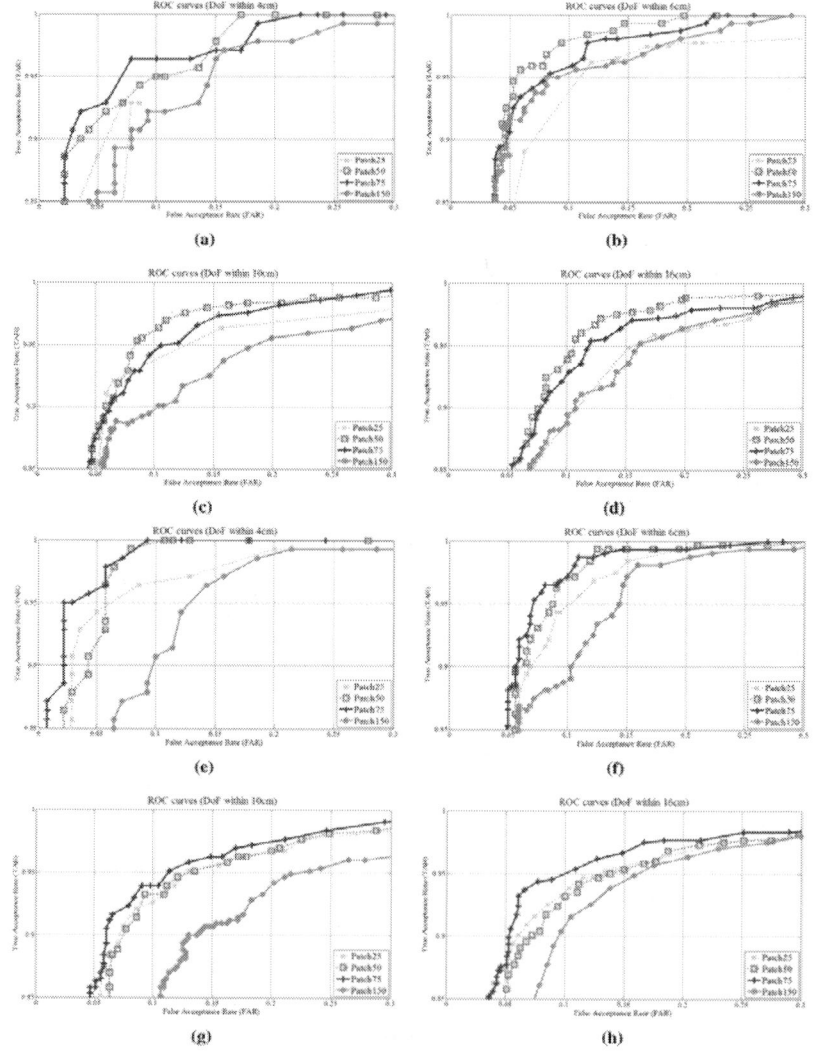

Figure A3. ROC curves of the GLOH features (overlap: PCA 90%, DoF (**a**) within 4 cm, (**b**) within 6 cm, (**c**) within 10 cm and (**d**) within 16 cm; and NO overlap: PCA 90%, DoF (**e**) within 4 cm, (**f**) within 6 cm, (**g**) within 10 cm and (**h**) within 16 cm).

REFERENCES

1. Raty, T.D. Survey on Contemporary Remote Surveillance Systems for Public Safety. *IEEE Trans. Syst. Man Cybern. Part C Appl. Rev* **2010**.

2. Li, S.Z.; Jain, A.K. *Handbook of Face Recognition*; Springer: New York, NY, USA, 2011.

3. Kähm, O.; Damer, N. 2D Face Liveness Detection: An Overview. Proceedings of the International Conference of the Biometrics Special Interest Group (BIOSIG), Darmstadt, Germany, 6–7 September 2012.

4. Pertuz, S.; Puig, D.; Garcia, M.A. Analysis of focus measure operators for shape-from-focus.*Pattern Recognit* **2013**, *46*, 1415–1432.

5. Billiot, B.; Cointault, F.; Journaux, L.; Simon, J.C.; Gouton, P. 3D image acquisition system based on shape from focus technique. *Sensors* **2013**, *13*, 5040–5053.

6. Ali, A.; Deravi, F.; Hoque, S. Liveness Detection Using Gaze Collinearity. Proceedings of the 2012 Third International Conference on Emerging Security Technologies(EST), Lisbon, Portugal, 5–7 September 2012.

7. Zhang, Z.; Yi, D.; Lei, Z.; Li, S.Z. Face liveness detection by learning multispectral reflectance distributions. Proceedings of the 2011 IEEE International Conference on Automatic Face & Gesture Recognition and Workshops (FG 2011), Santa Barbara, CA, USA, 21–25 March 2011.

8. Sun, L.; Huang, W.; Wu, M. TIR/VIS Correlation for Liveness Detection in Face Recognition. In *Computer Analysis of Images and Patterns Lecture Notes in Computer Science*; Springer: Seville, Spain; 29; –31; August; 2011; Volume 6855, pp. 114–121.

9. Kim, G.; Eum, S.; Suhr, J.K.; Kim, D.I.; Park, K.R.; Kim, J. Face Liveness Detection Based on Texture and Frequency Analyses. Proceedings of the 2012 5th IAPR International Conference on Biometrics (ICB), New Delhi, India, 29 March–1 April 2012; pp. 67–72.

10. Määttä, J.; Hadid, A.; Pietikäinen, M. Face Spoofing Detection from Single Images Using Micro-Texture Analysis. Proceedings of the 2011 International Joint Conference on Biometrics (IJCB'11), Washington, DC, USA, 11–13 October 2011.

11. Bai, J.; Ng, T.T.; Gao, X.; Shi, Y.Q. Is Physics-Based Liveness Detection Truly Possible with a Single Image? Proceedings of the 2010 IEEE International Symposium on Circuits and Systems (ISCAS), Paris, France, 30 May–2 June 2010; pp. 3425–3428.

12. Peixoto, B.; Michelassi, C.; Rocha, A. Face liveness detection under bad illumination conditions. Proceedings of the IEEE 2011 International Conference on Image Processing, Brussels, Belgium, 11–14 September 2011; pp. 3557–3560.

13. Tan, X.; Li, Y.; Liu, J.; Jiang, L. Face Liveness Detection from a Single Image with Sparse Low Rank Bilinear Discriminative Model. In *ECCV 2010 Lecture Notes in Computer Science*; Springer: Berlin, Germany, 2010; Volume 6316, pp. 504–517.

14. Singh, A.; Singh, S.K. Effect of Face Tampering on Face Recognition. *Signal Image Process. Int. J* **2013**.

15. Kant, C.; Sharma, N. Fake Face Recognition Using Fusion of Thermal Imaging and Skin Elasticity. *IJCSC* **2013**, *4*, 65–72.

16. Pan, G.; Sun, L.; Zhaohui, W.; Wang, Y. Monocular camera-based face liveness detection by combining eyeblink and scene context. *Telecommun. Syst* **2011**, *47*, 215–225.

17. Jee, H.; Jung, S.; Yoo, J. Liveness Detection for Embedded Face Recognition System. *Int. J. Biol. Life Sci* **2005**, *1*, 235–238.

18. Tronci, R.; Muntoni, D.; Fadda, G.; Pili, M.; Sirena, N.; Murgia, G.; Ristori, M.; Roli, F. Fusion of multiple clues for photo-attack detection in face recognition systems. Proceedings of the 2011 International Joint Conference on Biometrics (IJCB'11), Washington, DC, USA, 11–13 October 2011.

19. Anjos, A.; Marcel, S. Counter-Measures to Photo Attacks in Face Recognition: A Public Database and a Baseline. Proceedings of the 2011 International Joint Conference on Biometrics (IJCB'11), Washington, DC, USA, 11–13 October 2011.

20. Komogortsev, O.V.; Karpov, A. Liveness Detection via Oculomotor Plant Characteristics: Attack of Mechanical Replicas. Proceedings of the 2013 6th International Conference on Biometrics (ICB), Madrid, Spain, 4–7 June 2013.

21. Kollreider, K.; Fronthaler, H.; Bigun, J. Non-intrusive liveness detection by face images.*Image Vis. Comput.* **2009**, *27*, 233–244.

22. Bao, W.; Li, H.; Li, N.; Jiang, W. A liveness detection method for face recognition based on optical flow field. Proceedings of the International Conference on Image Analysis and Signal Processing, Taizhou, China, 11–12 April 2009; pp. 233–236.

23. Huang, C.H.; Wang, J.F. SVM-Based One-Against-Many Algorithm for Liveness Face Authentication. Proceedings of the IEEE International Conference on Systems, Man and Cybernetics (SMC 2008), Singapore, 12–15 October 2008; pp. 744–748.

24. Bharadwaj, S.; Dhamecha, T.I.; Vatsa, M.; Singh, R. Computationally Efficient Face Spoofing Detection with Motion Magnification. Proceedings of the IEEE Computer Society Conference on Computer Vision and Pattern Recognition, Portland, OR, USA, 23–28 June 2013.

25. Kollreider, K.; Fronthaler, H.; Faraj, M.I.; Bigun, J. Real-Time Face Detection and Motion Analysis with Application in "Liveness" Assessment. *IEEE Trans. Inf. Forensics Sec* **2007**, *2*, 548–558.

26. Lee, T.W.; Ju, G.H.; Liu, H.S.; Wu, Y.S. Liveness Detection Using Frequency Entropy of Image Sequences. Proceedings of 2013 IEEE International Conference on Acoustics, Speech and Signal Processing (ICASSP 2013), Vancouver, BC, Canada, 26–31 May 2013.

27. Schwartz, W.R.; Rocha, A.; Edrini, H.P. Face Spoofing Detection through Partial Least Squares and Low-Level Descriptors. Proceedings of 2011 International Joint Conference on Biometrics (IJCB'11), Washington, DC, USA, 11–13 October 2011.

28. Wang, T.; Yang, J.; Lei, Z.; Liao, S.; Li, S.Z. Face Liveness Detection Using 3D Structure Recovered from a Single Camera. Proceedings of the 2013 6th International Conference on Biometrics (ICB), Madrid, Spain, 4–7 June 2013.

29. Lagorio, A.; Tistarelli, M.; Cadoni, M.; Fookes, C.; Sridharan, S. Liveness Detection Based on 3D Face Shape Analysis. Proceedings of the 2013 International Workshop on Biometrics and Forensics (IWBF), Lisbon, Portugal, 4–5 April 2013.

30. Veerender, R.; Acharya, K.; Srinivas, J.; Mohan, D. Depth Estimation Using Blur Estimation in Video. *Int. J. Electron. Comput. Sci. Eng* **2012**, *1*, 2350–2354.

31. Stroebel, L.D. *View Camera Technique*; Focal Press: London, UK, 1999.

32. Kim, S.; Yu, S.; Kim, K.; Ban, Y.; Lee, S. Face Liveness Detection Using Variable Focusing. Proceedings of the 2013 6th International Conference on Biometrics (ICB), Madrid, Spain, 4–7 June 2013.

33. Nayar, S.K.; Nakagawa, Y. Shape from Focus. *IEEE Trans. Pattern Anal. Mach. Intell* **1994**, *16*, 824–831.

34. Brunelli, R.; Poggio, T. Face Recognition: Features versus Templates. *IEEE Trans. Pattern Anal. Mach. Intell.* **1993**, *15*, 1042–1052.

35. Mikolajczyk, K.; Schmid, C. A Performance Evaluation of Local Descriptors. *IEEE Trans. Pattern Anal. Mach. Intell* **2005**, *27*, 1615–1630.

36. Gonzalez, R.C.; Woods, R.E. *Digital Image Processing*, 2nd ed.; Prentice Hall: Upper Saddle River, NJ, USA, 2002.

37. Lowe, D. Distinctive Image Features from Scale-Invariant Keypoints. *Int. J. Comput. Vis* **2004**, *60*, 91–110.

38. Burges, C. A Tutorial on Support Vector Machines for Pattern Recognition. *Data Min. Knowl. Discov* **1998**, *2*, 121–167.

39. Zhang, Z.; Yan, J.; Liu, S.; Lei, Z.; Yi, D.; Li, S. A face antispoofing database with diverse attacks. Proceedings of the 2012 5th IAPR

International Conference on Biometrics (ICB), New Delhi, India, 29 March -1 April 2012; pp. 26–31.

40. Chingovska, I.; Anjos, A.; Marcel, S. On the Effectiveness of Local Binary Patterns in Face Anti-Spoofing. Proceedings of the International Conference of the Biometrics Special Interest Group, 2012 (BIOSIG), Darmstadt, Germany, 6–7 September 2012; pp. 1–7.

41. Zhang, Z.; Yan, J.; Liu, S.; Lei, Z.; Yi, D.; Li, S.Z. A face antispoofing database with diverse attacks. Proceedings of the 2012 5th IAPR International Conference on Biometrics (ICB), New Delhi, India, 29 March – 1 April 2012; pp. 26–31.

42. Määttä, J.; Hadid, A.; Pietikäinen, M. Face spoofing detection from single images using texture and local shape analysis. *IET Biom* **2012**, *1*, 3–10.

43. Cootes, T.; Taylor, C.; Cooper, D.; Graham, J. Active Shape Models— Their Training and Application. *Comput. Vis. Image Underst* **1995**, *61*, 38–59.

44. Nanu, F.; Stan, C.N.; Coreoran, P. Continuous Autofocus Based on Face Detection and Tracking. U.S. Patent 2012/0075492A1, 2 December 2012.

45. Rahman, M.; Kehtarnavaz, N. Real-time face-priority auto focus for digital and cell-phone cameras. *IEEE Trans. Consum. Electron* **2008**, *54*, 1506–1513.

CHAPTER 4

Hidden Markov Models in Automatic Face Recognition - A Review

Claudia Iancu and Peter M. Corcoran

College of Engineering & Informatics National University of Ireland Galway, Ireland

1. INTRODUCTION

Hidden Markov Models (HMMs) are a set of statistical models used to characterize the statistical properties of a signal. An HMM is a doubly stochastic process with an underlying stochastic process that is not observable, but can be observed through another set of stochastic processes that produce a sequence of observed symbols. An HMM has a finite set of states, each of which is associated with a multidimensional probability distribution; transitions between these states are governed by a set of probabilities. Hidden Markov Models are especially known for their application in 1D pattern recognition such as speech recognition, musical score analysis, and sequencing problems in bioinformatics. More recently they have been applied to more complex 2D problems and this review focuses on their use in the field of *automatic face recognition*, tracking the evolution of the use of HMMs from the early-1990's to the present day.

Our goal is to enable the interested reader to quickly review and understand the state-of-art for HMM models applied to face recognition problems and to adopt and apply these techniques in their own work.

2. HISTORICAL OVERVIEW AND INTRODUCTION TO HMM

The underlying mathematical theory of Hidden Markov Models (HMMs) was originally described in a series of papers during the 1960's and early 1970's [Baum & Petrie, 1966; Baum et al., 1970; Baum, 1972]. This technique was subsequently applied in practical pattern recognition applications, more specifically in speech recognition problems [Jelinek et al., 1975]. However, widespread understanding and practical application of HMMs only began a decade later, in the mid-1980s. At this time several tutorials were written [Levinson et al., 1983; Juang, 1984; Rabiner & Juang, 1986; Rabiner, 1989].

The most comprehensive of these was the last, [Rabiner, 1989], and provided sufficient detail for researchers to apply HMMs to solve a broad range of practical problems in speech processing and recognition. The broad adoption of HMMs in automatic speech recognition represented a significant milestone in continuous speech recognition problems [Juang & Rabiner, 2005].

The mathematical sophistication of HMMs combined with their successful application to a wide range of speech processing problems has prompted researchers in pattern recognition to consider their use in other areas, such as character recognition, keyword spotting, lip-reading, gesture and action recognition, bioinformatics and genomics. In this chapter we present a review of the most important variants of HMMs found in the *automatic face recognition literature*. We begin by presenting the initial 1D HMM structures adapted for use in face recognition problems in section 3. Then a number of papers on hybrid approaches used to improve the performance of HMMs for face recognition are discussed in section 4. In section 5 the various 2D variants of HMM are described and evaluated in terms of the recognition rates achieved from each. Finally section 6 includes some recent refinements in the application of HMM techniques to face recognition problems.

3. HMM IN FACE RECOGNITION - INITIAL 1D HMM STRUCTURES

As mentioned in the previous section, HMMs have been used extensively in speech processing, where signal data is naturally one-dimensional. Nevertheless, HMM techniques remain mathematically complex even in the one-dimensional form. The extension of HMM to two-dimensional model structures is exponentially more complex [Park & Lee, 1998]. This consideration has led to a much later adoption of HMM in applications involving two-dimensional pattern processing in general and face recognition in particular.

3.1. Initial Research On Ergodic And Top-To-Bottom 1d Hmm

In 1993, a new approach to the problem of automatic face recognition based on 1D HMMs was proposed by [Samaria & Fallside, 1993]. In this paper faces are treated as two-dimensional objects and the HMM model automatically extracts statistical facial features. For the automatic extraction of features, a 1D observation sequence is obtained from each face image by sampling it using a sliding window. Each element of the observation sequence is a vector of pixel intensities (or greyscale levels).

Two simple 1D HMMs were trained by these authors in order to test the applicability of HMMs in face recognition problems. A test database was used comprising images of 20 individuals with a minimum of 10 images per person. Images were acquired under homogeneous lighting against a constant background, and with very small changes in head pose and facial expressions.

For a first set of tests an ergodic HMM was used. The images were sampled using a rectangular window, size 64 × 64, moving left-to-right horizontally with a 25% overlap (16 pixels), then vertically with 16 pixels overlap and starting again horizontally right-to-left. Using the observation sequence thus extracted, an 8-state ergodic HMM was built to approximately match the 8 distinct regions that seem to appear in the face image (eyes, mouth, forehead, hair, background, shoulders and two extra states for boundary regions).Figure 1 taken from [Samaria & Fallside, 1993] shows the training data used for one subject and the mean vectors for the 8 states found by HMM for that particular subject.

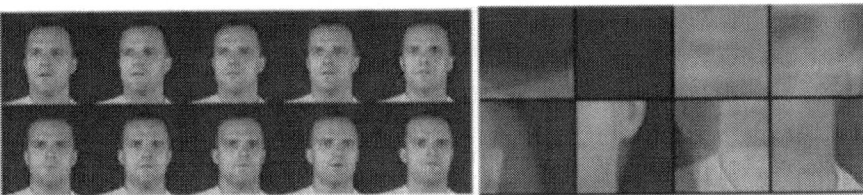

Figure 1. Training data and states for ergodic HMM [Samaria & Fallside, 1993]

In the second set of tests, a left-to-right (top-to-bottom) HMM was used. Each image was sampled using a horizontal stripe 16 pixels high and as wide as the image, moving top-to-bottom with 12 lines overlap. The resulting observation sequence was used to train a 5-state left-to-right HMM where only transitions between adjacent states are allowed. The training images and the mean vectors for the 5 states found by HMM are presented in Figure 2.

Figure 2. Examples of training data and states for top-to-bottom HMM from [Samaria & Fallside, 1993]

In both of these models the statistical determination of model features, yields some states of the HMM which can be directly identified with physical facial features. Training and testing were performed using the HTK toolkit[1] -. According to these authors, successful recognition results were obtained when test images were extracted from the same video sequence as the training images, proving that the proposed approach can cope with variations in facial features due to small orientation changes, provided the lighting and background are constant. Unfortunately these authors did not provide any explicit recognition rates so it is not possible to compare their methods with later research. It is reasonable, however, to surmise that their experimental results were marginal and are improved upon by the later refinements of [Samaria & Harter, 1994].

3.2. Refinement of The Top-To-Bottom 1d Hmm

In a later paper [Samaria & Harter, 1994] refined the work begun in [Samaria & Fallside, 1993] on a top-to-bottom HMM. These new experiments demonstrate how face recognition rates using a top-to-bottom HMM vary with different model parameters. They also indicate the most sensible choice of parameters for this class of HMM. Up until this point, the parameterization of the model had been based on subjective intuition.

For such a 1D top-to-bottom HMM there are three main parameters that affect the performance of the model: the height of the horizontal strip used to extract the observation sequence, L (in pixels), the overlap used, M (in pixels) and the number of states N of the HMM. The height of the strip, L, determines the size of the features and the length of the observation sequence, thus influencing the number of states. The overlap, M, determines how likely feature alignment is and also the length of the observation sequence. A model with no overlap would imply rigid partitioning of the faces with the risk of cutting across potentially discriminating features. The number of states, N, determines the number of features used to characterize the face, and also the computational complexity of the system.

These experiments were performed using the Olivetti Research Lab (ORL) database, containing frontal facial images with limited side movements and head tilt. The database was comprised of 40 subjects with 10 pictures per subject. The experiments used 5 images per person for training and the remaining 5 images for testing. The results were reported as error rates, calculated as the proportion of incorrectly classified images. Three sets of tests were done, varying the values of each of the three parameters as follows: $2 \leq N \leq 10$, $1 \leq L \leq 10$ and $0 \leq M \leq L-1$. For M varied, the number of states was fixed at $N = 5$ and window height L was varied between 2 and 10. According to the tests, the error rates drop as the overlap increases, approximately from 28% to 15%. However a greater overlap implies a bigger computational effort. When L was varied, N was fixed to 5 and the overlaps considered were 0, 1 and L-1. In this case if there is little or no overlap, the smaller the strip height the lower the error rate is, with values between 13% for $L = 1$ up to 28% for $L = 10$. However, for sufficiently large overlap the strip height has marginal effect on the recognition performance, the error rate remaining almost constant around 14%. In the third set of tests N was varied, with $L = 1$ and 0 overlap and $L = 8$ and maximum overlap (M=L-1). The performance is fairly uniform for values of N between 4 and 10, with an increase in error for values smaller than three.

The conclusions of this paper are: (i) a large overlap in the sampling phase (the extraction of observation sequences) yields better recognition rates; the error rate varies from up to 30% for minimum overlap down to 15% for maximum overlap; (ii) for large overlaps the height of the sampling strip has limited effect. The error rate remains almost constant at 15% for maximum overlap, regardless of the value of L, and (iii) best results are obtained with a HMM with 4 or more states. Error rate drops from around 25% for 1-2 states to 15% from 4 states onward. We remark that these early models were relatively unsophisticated and were limited to fully frontal faces with images taken under controlled background and illuminations conditions.

3.3. 1d Hmm with 2d-Dct Features for Face Recognition

In [Nefian & Hayes, May 1998], Samaria's version of 1D HMM, is upgraded using 2D-DCT feature vectors instead of pixel intensities. The face image is divided into 5 significant regions, *viz*: hair, forehead, eyes, nose, and mouth. These regions appear in a natural order, each region being assigned to a state in a top-to-bottom 1D continuous HMM. The state structure of the face model and the non-zero transition probabilities are shown in Figure 3.

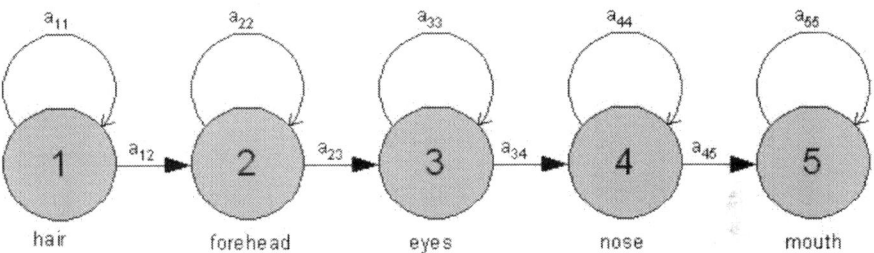

Figure 3. Sequential HMM for face recognition

The feature vectors were extracted using the same technique as in [Samaria & Harter, 1994]. Each face image of height H and width W is divided into overlapping strips of height L and width W, the amount of overlap between consecutive strips being P, see Figure 4. The number of strips extracted from each face image determines the number of observation vectors.

The 2D-DCT transform is applied on each face strip and the observation vectors are determined, comprising the first 39 2D-DCT coefficients. The system is tested on ORL[2] - database containing 400 images of 40 individuals, 10 images per individual, image size 92 × 112, with small variations in facial expressions, pose, hair style and eye wear. Half of the database is used for training and the other half is used for testing. The recognition rate achieved for L=10 and P=9 is 84%. Results are compared with recognition rates obtained using other face recognition methods on the same database: recognition rate for the eigenfaces method is 73%, and for the 1D HMM used by Samaria is also 84%, but the processing time for DCT based HMM is an order of magnitude faster - 2.5 seconds in contrast to 25 seconds required by the pixel intensity method of [Samaria & Harter, 1994].

3.4. 1d Hmm With Klt Features For Face Detection And Recognition

In a second paper [Nefian & Hayes, October 1998] introduce an alternative 1D HMM approach, which performs the face detection function in addition to that of face recognition. This employs the same topology and structure as in the previous work of these authors, described above, but uses different image features. In contrast with the previous paper, the observation vectors used here

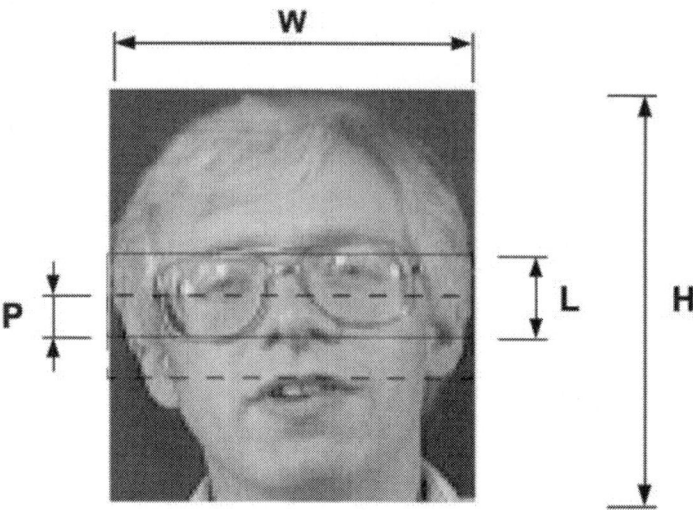

Figure 4. Face image parameterization and blocks extraction [Nefian & Hayes, May 1998].

are the coefficients of Karhunen-Loeve Transform. The KLT compression properties as well as its decorrelation properties make it an attractive technique for the extraction of the observation vectors. Block extraction from the image is achieved in the same way as in the previous paper. The eigenvectors corresponding to the largest eigenvalues of the covariance matrix of the extracted vectors form the KLT basis set. If μ is the mean of the vectors used to compute the covariance matrix, a set of vectors is obtained by subtracting this mean from each of the vectors corresponding to a block in the image. The resulting set of vectors is then projected onto the eigenvectors of the covariance matrix and the resulting coefficients form the observation vectors.

The system is used both for face detection and recognition by the authors. For face detection, the system is first trained with a set of frontal faces of different people taken under different illumination conditions, in order to build a face model. Then, given a test image, face detection begins by scanning the image with horizontally and vertically overlapping rectangular windows, extracting the observation vectors and computing the probability of data inside each window given the face model, using Viterbi algorithm. The windows that have face model likelihood higher than a threshold are selected as possible face locations. The face detection system was tested on MIT database with 48 images of 16 people with background and with different illuminations and head orientations. Manually segmented faces from 9 images were used for training and the remaining images for testing, with a face detection rate of 90%.

For face recognition this system was applied to the ORL database containing 400 images of 40 individuals, 10 images per individual, at a resolution of 92 × 112 pixels, with small variations in facial expressions, pose, hairstyle and eye wear. The system was trained with half of the database and tested with the other half. The accuracy of the system presented in this paper is increased slightly over earlier work to 86% while the recognition time decreases due to use of the KLT features.

3.5. Refinements to 1d Hmm with 2d-Dct Features

Following on the work of [Samaria, 1994] and [Nefian, 1999], Kohir & Desai wrote a series of three papers using the 1D HMM for face recognition problems. In a first paper, [Kohir & Desai, 1998], these authors present a face recognition system based on 1D HMM coupled with 2D-DCT coefficients using a different approach for feature extraction than that employed by [Nefian & Hayes, May 1998 & October 1998]. The extracted features are obtained by sliding square windows in a raster scan fashion over the face image, from left to right and with a predefined overlap. At every position of the window over the image (called sub-image) 2D DCT are computed, and only the first few DCT coefficients are retained by scanning the sub-image in a zigzag fashion. The zigzag scanned DCT coefficients form an observation vector. The sliding procedure and the zigzag scanning are illustrated in Figure 5 [Kohir & Desai, 1998].

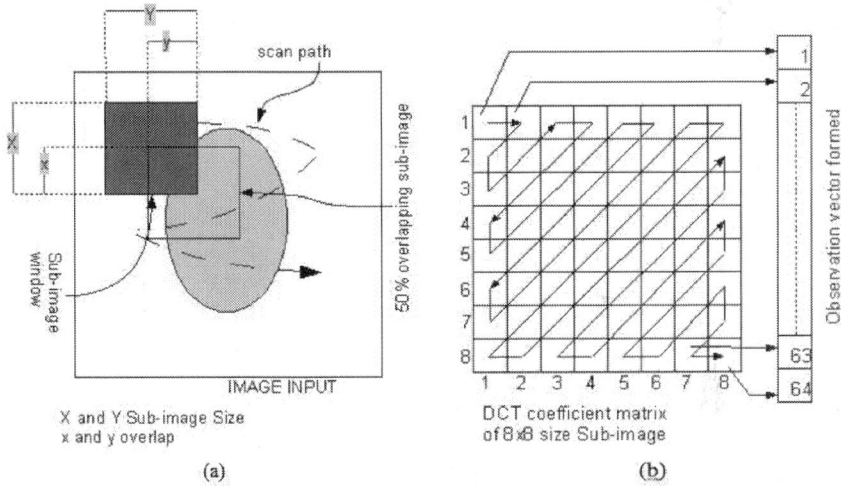

Figure 5. a) Raster scan of face image with sliding window. (b) Construction of 1D observation vector from zigzag scanning of the sliding window [Kohir & Desai, 1998].

The performance of this system is tested using the ORL database. Half of the images were used in the training phase and the other half for testing (5 faces for training and the remaining 5 for testing), sampling windows of 8 × 8 and 16 × 16, were used with 50% and 75% overlaps, and 10, 15 and 21 DCT coefficients were extracted. The number of states in the HMM was fixed at 5 as per the earlier work of [Nefian & Hayes, May 1998]. The recognition rates vary from 74.5% for a 16 × 16 window, with a 50% overlap and 21 DCT coefficients to 99.5% for 16 × 16 window, 75% overlap and 10 DCT coefficients.

In a second paper [Kohir & Desai, 1999] these authors further refined their research contribution. To evaluate the recognition performances of the system, 2 new experiments are performed:

- In a first experiment the proposed method is tested with different numbers of training and testing faces per subject. The tests were performed on the

ORL database, and the number of training faces was increased from 1 to 6, while the remaining faces were used in the testing phase. A sampling window of 16 × 16 with 75% overlap was used with 10 DCT coefficients as these had provided optimal recognition rates in their earlier work. The recognition rates achieved are from 78.33% for a single training image and 9 testing images up to 99.5% which is the rate obtained when 5 or 6 training images and 5 or 4 testing images are used. It is worth noting that the ORL database comprises frontal face images in uniform lighting conditions and that recognition rates close to 100% are often achieved when using such datasets.

- In a second experiment the system was tested while increasing the number of states in the HMM. Again the ORL database is used, with 5 images for training and 5 for testing. The recognition rates vary as follows: 92% for a 2-states HMM, increasing to 99.5% for a 5-states HMM and stabilizing around 97%-98% when using up to 17 states. The system was also tested with the SPANN database[3] - containing 249 persons, each with 7 pictures, with variations in pose, 3 pictures were used for training and the remaining 4 for testing, and the recognition rate achieved was 98.75%.

- A third paper, [Kohir & Desai, 2000] describes the same 1D HMM with DCT features, with a variation in the training phase. In this paper, first a *mean image* is constructed from all the training images, and then each training image is subtracted from the *mean image* to obtain a *mean subtracted image*. The observation vectors are extracted from these *mean subtracted images* using the same window sliding method. The observation vector sequences are then clustered using the K-means technique, and thus an initial state segmentation is obtained. Subsequently, the conventional training steps are followed. In the recognition phase, each test image is first subtracted from the *mean image* obtained during the training phase and recognition is performed on the resulting *mean subtracted image*.

The experiments for face recognition were performed on the same two databases, ORL and SPANN. For ORL database 5 pictures were used for training and the remaining 5 for testing, and the recognition rate obtained is 100%, compared to 88% when the eigenfaces method is used. For SPANN database 3 pictures were used for training and the remaining 4 for testing, the obtained recognition rate was 90%, compared again with the eigenfaces method where a 77% recognition rate was achieved. For the ORL database different resolutions were also tested, the highest recognition rate, 100% being obtained for 96 × 112.

Also, 'new subject rejection' for authentication applications was tested on the ORL database. The database was segmented into 2 sets: 20 subjects corresponding to an 'authorized' subject class - 5 pictures used in training phase and the rest in the testing phase. The remaining 20 subjects are assigned to an 'unauthorized' class - all 10 pictures are used in the testing phase. For each 'authorized' subject a HMM model is built. Also a separate 'common HMM' model is built using all *mean subtracted training images* of all the 'authorized' subjects. For each test face, if the probability of the 'common HMM' is the highest, the input face image is rejected as 'unauthorized', otherwise the input face image is treated as 'authorized'. The results are: 100% rejection of any new subjects and 17% rejection of known subjects (false negatives).

3.6. Refinement of 1d Hmm with Sequential Prunning

As proved by [Samaria & Harter, 1994], the number of states used in a 1D HMM can have a strong influence on recognition rates. The problem of the optimal selection of the structure for an HMM is considered in [Bicego at al., 2003a]. The first part of this paper presents a method of improving the determination of the optimal number of states for an HMM. These authors then proceed to prove the equivalence between (i) a 1D HMM whose observation vectors are modelled with *multiple Gaussians per state* and (ii) a 1D HMM with *one Gaussian per state* but employing a larger number of states. According to the authors, there are several possible methods for solving the first problem, e.g. cross-validation, Bayesian inference criterion (BIC), minimum description length (MDL). These are based on training models with different structures and then choosing the one that optimizes a certain selection criterion. However, these methods involve a considerable computational burden plus they are sensitive to the local-greedy behaviour of the HMM training algorithm, i.e. the successful training of the model is influenced by the initial estimates selected.

The approach proposed by [Bicego et al., 2003a] addresses both the computational burden of model selection, and the initialization phase. The key idea is the use of a decreasing learning strategy, starting each training session from a 'nearly good' situation derived from the previous training session by pruning the 'least probable' state. More specifically, the authors proposed starting the model training with a large number of states. They next run the estimation algorithm and, on convergence, evaluate the model selection criterion. The 'least probable' state is then pruned, and the resulting configuration of the model with one less state is used as a starting point for the next sequence of iterations. In this way, each training session is started from a 'nearly good' estimate. The key observation supporting this approach is that, when the number of states is extremely large, the dependency of the model behaviour on the initial estimates is much weaker. An additional benefit is that using 'nearly good' initializations drastically reduces the number of iterations required by the learning algorithm at each step in this process. Thus the number of model states can be rapidly reduced at low computational cost.

In order to assess the performance of their proposed method, these authors tested the pruning approach and the standard approach (consisting in training one HMM for varying number of states) with BIC criterion and MMDL (mixture minimum description length) [Figueiredo et al., 1999] criterion. These two strategies are compared in terms of: (i) accuracy of the model size estimation, (ii) total computational cost involved in the training phase, and (iii) classification accuracy. In all the HMMs considered in this paper the emission probability density for each state is a single Gaussian. For the accuracy of the model size estimation, synthetically generated test sets of 3 known HMMs were used. The authors set the number of states allowed from 2 to 10. The selection accuracy ranged from 54% to 100% for standard BIC and MMDL, and from

98% to 100% for pruning BIC and MMDL, with up to 50% less iteration required for the latter.

Classification accuracy was tested on both synthetic and real data. For the synthetic data, the test sets used previously to estimate the accuracy of the model size estimation were used, obtaining 92% to 100% accuracy for standard BIC and MMDL compared to 98% to 100% accuracy for pruning BIC and MMDL, with 35% less iterations for pruning. For classification accuracy on real data, two experiments were conducted. The first involves a 2D shape recognition problem, and uses a data set with four classes each with 12 different shapes. The results obtained are 92.5% for standard BIC, 94.37% for standard MMDL, and 95.21% for pruning BIC and MMDL. The second experiment was conducted on the ORL database, using the method proposed by [Kohir & Desai 1998]. The results are 97.5% for standard BIC and MMDL and 97.63% for pruning BIC and MMDL. The classification accuracies are similar, but the pruning method reduces substantially the number of iterations required.

3.7. A 1d Hmm with 2d-Dct Features and Haar Wavelets

In a following paper [Bicego et al., 2003b], a comparison between DCT coding and wavelet coding is undertaken. The aim is to evaluate the effectiveness of HMMs in modelling faces using these two different forms of image features. Each compresses the relevant image data, but employing different underlying techniques. Also, the suitability of HMM to deal with the JPEG 2000 image compression standard is considered by these authors. They adopt the 1D HMM approach introduced by [Kohir & Desai, 1998]. However, the optimum number of states for the model is selected using the sequential pruning strategy presented in [Bicego et al., 2003a] and described in the preceding section. The same feature extraction used by [Kohir & Desai, 1998] is employed, and both 2D DCT and Haar wavelet coefficients are computed.

These experiments have been conducted on the ORL database, consisting of 40 subjects with 10 sample images of each. The first 5 images are used for training the HMM while the remaining 5 are used in the testing phase. The number of states for each HMM is estimated using the pruning strategy. For feature extraction, a 16×16 pixel sliding window is used, with 50% and 75% overlaps being tested, and in each case the first 4, 8 and 12 DCT or Haar coefficients are retained. The recognition rate scores for 50% overlap are between 97.4% for 4 coefficients to 100% for 12 coefficients, and for 75% overlap between 95.4% for 4 coefficients to 99.6% for 12 coefficients. Slightly better results were obtained for DCT coefficients throughout the experiments. It is worth noting that unlike [Samaria & Harter, 1994] and [Nefian & Hayes, 1998] in the case of [Kohir & Desai, 1998] the method of extracting observation vectors results in better performance for a 50% overlap than for 75% overlap.

A second experiment was performed to prove the effectiveness of HMM in solving the face recognition problem regardless of the coefficients used, by replacing in the proposed system the wavelet coding with a trivial coding represented by the mean of the square window. The results obtained are 84.9% for 50% overlap and 77.8% for 75% overlap.

4. HYBRID APPROACHES BASED ON 1D HMM

From the discussions of the preceding section it can be seen that 1D HMM can perform successfully in face recognition applications. However, the vast majority of early experiments were performed on the ORL database. The images in this dataset only exhibit very small variations in head pose, facial expressions, facial occlusions such as facial hair and glasses, and almost no variations in illumination. For practical applications a face recognition system must be able to handle significant variations in facial appearance in a robust manner. Thus in this next section more challenging face recognition applications are described and further HMM approaches are considered from the literature. Specifically, in this section we consider hybrid approaches based on HMMs used successfully in more challenging applications of face recognition.

There are several core problems that a face recognition system has to solve, specifically those of variations in illumination, variations in facial expressions or partial occlusions of the face, and variations in head pose. Firstly an attempt at solving recognition problems caused by facial occlusions is considered [Martinez, 1999]. The solution adopted by this author was to explore the use of *principle component analysis* (PCA) features to characterize 6 different regions of the face and use 1D HMM to model the relationships between these regions. A second group of researchers [Wallhoff et al., 2001] have tackled the challenging task of recognizing side-profile faces in datasets where only frontal faces were used in the training stage. These authors have used a combination of *artificial neural network* (ANN) techniques combined with 1D HMM to solve this challenging problem.

4.1. Using 1d Hmm with Pca Derived Features

A face recognition system is introduced [Martinez, 1999] for indexing images and videos from a database of faces. The system has to tackle three key problems, identifying frontal faces acquired, (i) under differing illumination conditions, (ii) with varying facial expressions and (iii) with different parts of the face occluded by sunglasses/scarves. Martinez's idea was to divide the face into N different regions analyzing each using PCA techniques and model the relationships between these regions using 1D HMMs.

The problem of different lighting conditions is solved in this paper by training the system with a broad range of illumination variations. To handle facial expressions and occlusions, the face is divided into 6 distinct local areas and local features are matched. This dependence on local rather than global features should minimize the effect of facial expressions and occlusions, which affect only a portion of the overall facial region. Each of these local areas obtained from all the images in the database is projected into a primary eigenspace. Each area is represented in vector form. Figure 6 [Martinez, 1999] shows the local feature extraction process.

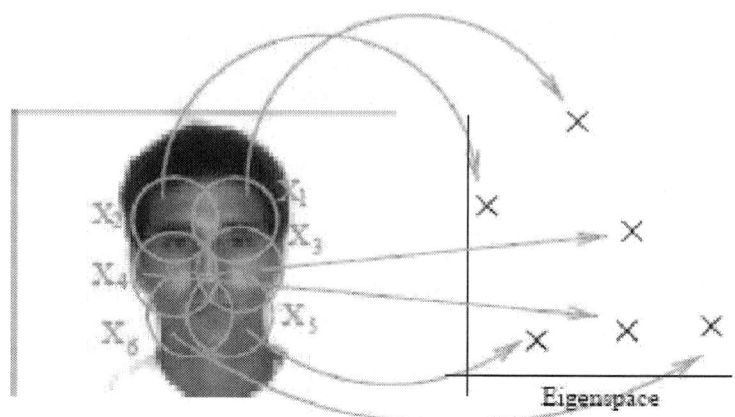

Figure 6. Projection of the 6 different local areas into a global eigenspace Martinez, 1999].

Note that face localization is performed manually in this research and thus cannot be precise enough to guarantee that the extracted local information will always be projected accurately into the eigenspace. Thus information from pixels within and around the selected local area is also extracted, using a rectangular window. By considering these six local areas as hidden states, a 1D HMM was built for each image in the database. However, a more desirable case is to have a single HMM for each person in the database, as opposed to a HMM for each image. To achieve this, all HMMs of the same person were merged together into a single 1D HMM, where the transition probability from one state to another is *1/number of HMMs per person*. In the recognition phase, instead of using the forward-backward algorithm, the authors used the Viterbi algorithm [Rabiner, 1989] to compute the probability of an observation sequence given a model.

Two sets of tests were performed, using pictures and video sequences. The image database[4] - was created by Aleix Martinez and Robert Benavente. It contains over 4,000 colour facial images corresponding to 126 people - 70 men and 56 women. There are 12 images per person, the first 6 frontal view faces with different facial expressions and illumination conditions and the second 6 faces with occlusions (sun-glasses and scarf) and different illumination conditions. These pictures were taken under strictly controlled conditions. No restrictions on appearance including clothing, accessories such as glasses, make-up or hairstyle were imposed on participants. Each person participated in two sessions, separated by 14 days. The same pictures were taken in both sessions. In addition, 30 video sequences were processed consisting of 25 images almost all of them containing a frontal face. Five different tests were run, using 50 people (25 males and 25 females) randomly selected from the database, converted to greyscale images and sampled at half their size, and also using 30 corresponding video sequences. In a first test, all 12 images per person were used in training, and the system was tested with every image by replacing each one of the local features with random noise with mean 0. The recognition rate obtained was 96.83%. For a second test training was with the first six images and testing with the last six images, featuring occlusions. A recognition rate of 98.5% was achieved. In a third test the last six images were used for training and the first six for testing and the resulting recognition rate was 97.1%. A fourth test consisted of training with only two non-occluded images and testing with all the remaining images. A lower recognition rate of 72% was obtained. Finally, the system was trained with all 12 images for each person, and tested with the video sequences, achieving a 93.5% recognition rate.

4.2. Artificial Neural Networks (Ann) in Conjunction with 1d Hmm

[Wallhoff et al., 2001] approached the challenging task of recognizing profile views with previous knowledge from only frontal views, which may prove a challenging task even for humans. The authors use two approaches based on a combination of Artificial Neural Networks (ANN) and a modelling technique based on 1D HMMs: a first approach uses a synthesized profile view, while a second employs a joint parameter-estimation technique. This paper is of particular interest because of its focus on non-frontal faces. In fact these authors are one of the first to address the concept of training the recognition system with conventional frontal faces, but extending the recognition to include faces with only a side-profile view.

The experiments are performed on the MUGSHOT[5] - database containing the images of 1573 cases, where most individuals are typically represented by only two photographs: one showing the frontal view of the person's face and the other showing the person's right hand profile. The database contains pairs of mostly male subjects at several ages and representatives of several ethnic groups, subjects with and without glasses or beards and a wide range of hairstyles. The lighting conditions and the background of the photographs also change. The pictures in the database are stored as 8-bit greyscale images. Prior to applying the main techniques of [Wallhoff et al., 2001] a pre-processing of each image is conducted. Photographs with unusually high distortions, perturbations or underexposure are discarded; all images are manually labelled so that all faces appear in the centre of the image and with a moderate amount of background, and resized to 64x × 64 pixels. Then two sets are defined: a first set consisting of 600 facial image pairs, *frontal* and right-hand *profile*, are used for training the neural network. A second set with 100 facial image pairs is used for testing. The features used for experiments are pixel intensities. In order to obtain the observation vectors, each image which was resized to 64×64 pixels is divided into 64 columns. So from each image 64 observation vectors are extracted. The dimension of the vectors is the number of rows in the image, which is also 64, and these vectors consist of pixel intensities. In the training phase an appropriate neural network is used, estimated by applying the following intuitions: (i) a point in the frontal view will be found in approximately the same row as in the profile view, (ii) considering the right half of the face to be almost bilaterally symmetrical with the left half, only the first 40 columns of the image are used in the input layer to the ANN. Figure 7 taken from [Wallhoff et al., 2001] shows how a frontal view of the face is used to generate the profile view. In the testing phase, a 1D left to right first order HMM is used, allowing self transitions and transitions to the next state only. The models consist of 24 states, plus two non-emitting start and end states.

In the first hybrid approach for face profile recognition there are two training stages. Firstly, a neural network is trained using the first set of 600 images, the frontal image of each individual representing the input and the profile view the output. In this way the neural network is trained to synthesize profiles from the frontal image. In figure 8 [Wallhoff et al., 2001] an example of synthesized profile is shown. In the second training stage, the 100 frontal images are introduced in the neural network and their corresponding profiles are synthesized. Using these profiles, an average profile HMM model is obtained. Then for each testing profile, an HMM model is built using for initialization the average profile model. The Baum-Welch estimation procedure is used for training the HMM.

In a second approach only one training stage is performed, the computation speed being vastly improved as a result. This proceeds as follows: the NN is trained using the frontal images as input; the target outputs are in this case the mean values of each Gaussian mixture used for describing the observations of the corresponding profile image. First, an average profile HMM model is obtained using the 600 training profile images. Using this average model, the mean values for each individual in the training set are computed and used as the target values for the NN to be trained. In the recognition phase, for each frontal face the mean value for profile is returned by the NN. Using this mean and the average profile model, the corresponding HMM is built, then the probability of the test profile image given the HMM model is computed. The recognition rates achieved for the systems proposed in this paper are around 60% for the first approach and up to 49% for the second approach, compared to 70%-80% when humans perform the same recognition task. The approach presented by the authors is very interesting in the context of a mugshot database, where only the two instances, one *frontal* and one *profile* of a face are present. Also the results are quite impressive compared to the human recognition rates reported. However, both ANN and HMM are computationally complex, and using pixel intensities as features also contributes to making this approach very greedy in terms of computing resources.

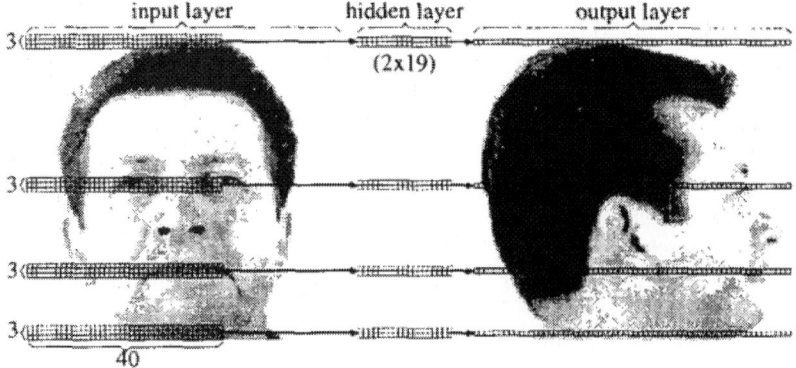

Figure 7. Generation of a profile view from a frontal view [Wallhoff et al., 2001].

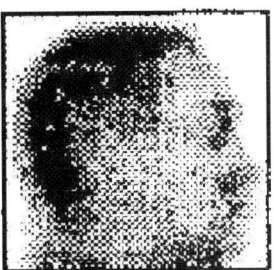

Figure 8. Example of frontal view, generated and real profile [Wallhoff et al., 2001].

5. 2D HMM APPROACHES

In section 3 and section 4 we showed how 1D HMMs might be adapted for use in face recognition applications. But face images are fundamentally 2D signals and it seems intuitive that they would be more effectively processed with a 2D recognition algorithm. Note however that a fully connected 2D extension of HMM exhibits a significant increase in computational complexity making it inefficient and unsuitable for practical face recognition applications [Levin & Pieraccini, 1992]. As a consequence of this complexity of the full 2D HMM approach a number of simpler structures were developed and are discussed in detail in the following sections.

5.1. A First Application of Pseudo 2d Hmm to Facial Recognition

In his PhD thesis, [Samaria, 1994] was the first researcher to use *pseudo-2D* HMMs in face recognition, with pixel intensities as features. In order to obtain a P2D HMM, a one-dimensional HMM is generalized, to give the appearance of a two-dimensional structure, by allowing each state in a one-dimensional global HMM to be a HMM in its own right. In this way, the HMM consists of a top-level set of super states, each of which contains a set of embedded states. The super states may then be used to model the two-dimensional data in one direction, with the embedded HMMs modelling the data along the other direction. This model is appropriate for face images as it exploits the 2D physical structures of a face, namely that a face preserves the same structure of states from top to bottom – forehead, eyes, nose, mouth, chin, and also the same left-to-right structure of states inside each of these super states. An example of state structure for the face model and the non-zero transition probabilities of the P2D HMM are shown in figure 9. Each state in the overall top-to-bottom HMM is assigned to a left-to-right HMM.

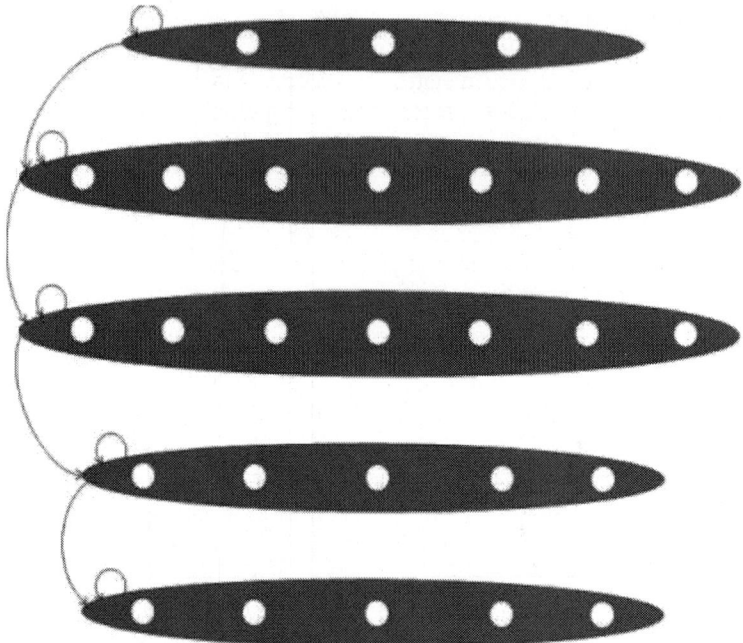

Figure 9. Structure of a P2D HMM.

In order to simplify the implementation of P2D-HMM, the author used an equivalent 1D HMM to replace the P2D-HMM as shown in figure 10. In this case, the shaded states in the 1D HMM represent end-of-line states with two possible transitions: one to the same row of states - *superstate self-transition*- and one to the next row of states - *superstate to superstate transition*. For feature extraction a square window is used sliding from left-to-right and top-to-bottom. Each observation vector contains the intensity level values of the pixels contained by the window, arranged in a column-vector. In order to accommodate the extra end-of-line state, a white frame is added at the end of each line of sampling. Each state is modelled by one Gaussian with mean and standard deviation set, initialized at the beginning of training, to mid-intensity values for normal states and to white with near zero standard deviation for the end-of-line states. The parameters of the model are then iteratively re-estimated using the Baum-Welch algorithm.

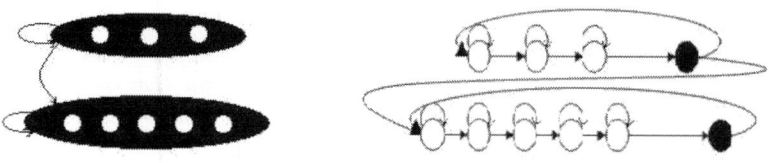

Figure 10. P2D HMM and its equivalent 1D HMM.

Samaria's experiments were carried out on the ORL database. Different topologies and sampling parameters were used for the P2D-HMM: from 4 to 5 superstates and from 2 to 8 embedded states within each superstate. In addition these experiments considered different sizes of sampling windows with different overlaps ranging from 2×2 pixels with 1×1 overlap up to 24×22 (horizontal $\times$ vertical) pixels with 20×13 pixels overlap. The highest error rate of 18% was obtained for a 3-5-5-3 P2D-HMM, using a 10×8 scanning window with an 8×6 overlap, while the smallest error rate of 5.5% was obtained for 3-6-6-6-3 P2D-HMM, with 10×8 (and 12×8) window and 8×6 (and 9×6 respectively) overlap. In the same thesis Samaria also tested the standard *unconstrained* P2D HMM, which does not have an end-of-line state. In this case no attempt is made to enforce the fact that the last frame of a line of observations should be generated by the last state of the superstate. The recognition results for the *unconstrained* P2D HMM are similar to those obtained with constrained P2D-HMM, the error rates ranging from 18% to 6%. We remark that Samaria also obtained a 2% error rate for a 3-7-7-5-3 P2D HMM with 12×8 sampling window and 4×6 overlap, but considering that for only slightly different overlaps (8×6 and 4×4) the error rates were 6% and 8.5% respectively, this particular result appears to be a statistical anomaly. It does serve to remind that these models are based on underlying statistical probabilities and that occasional aberrations can occur.

5.2. Refining Pseudo 2d Hmm with Dct Features

In [Nefian & Hayes, 1999] the authors adapted the P2D-HMM developed by [Kuo & Agazzi, 1994] for optical character recognition analysis, showing how it represented a valid approach for facial recognition and detection. These authors renamed this technique as *embedded* HMM. In order to obtain the observation vectors, a set of overlapping blocks are extracted from the image from left to right and top to bottom as shown in figure 11, the observation vector finally consisting of the 6 lower-frequency 2D-DCT coefficients extracted form each image block. Each state in the embedded HMMs is modelled using a single Gaussian.

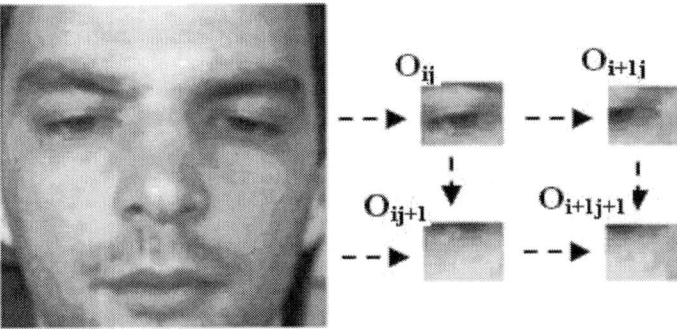

Figure 11. Face image parameterization and blocks extraction.

For face recognition the ORL database was used. The system was trained with half of the database and tested with the other half. The recognition performance of the method presented in this paper is 98%, improving by more than 10% compared with the best results obtained in using 1D HMM in earlier work [Nefian & Hayes, May 1998, October 1998].

This research also considered the problem of face detection. In the testing phase for detection, 288 images of the MIT database were used, representing 16 subjects with different illuminations and head orientations. A set of 40 images representing frontal views of 40 different individuals from the ORL database is used to train one face model. The testing is performed using a doubly embedded Viterbi algorithm described by [Kuo & Agazzi, 1994]. The detection rate of the system described in this paper is 86%. While this version of HMM appears to be relatively efficient in face detection, it is however computationally very complex and slow, particularly when compared with state of art algorithms [Viola & Jones, 2001].

5.3. Improved Initialization of Pseudo 2d Hmm

Also employing a P2D HMM, [Eickler et al., 2000] describe an advanced face recognition system based on the use of standard P2D HMM employing 2D DCT features is presented. The performance of the system is enhanced using improved initialization techniques and mirror images. It is very important to use a good initial model therefore the authors used all faces in the database to build a 'common initial model'. Then for each person in the database a P2D HMM model is refined using this 'common model'. Feature extraction is based on DCT. The image is scanned with a sliding window of size 8×8 from left-to-right and top-to-bottom with an overlap of 6 pixels (75%). The first 15 DCT coefficients are extracted. The use of DCT coefficients allows the system to work directly on images compressed with JPEG standard without a need to decompress these images. The size of the sampling window was chosen as 8×8 because the DCT portion of JPEG image compression is based on this window size.

Tests are performed on the ORL database, described previously, with the first 5 images per person used for training and the remaining 5 for testing. Three sets of experiments are performed in this paper. *First Experiment Set:* the system is tested on different quadratic P2D HMM model topologies (4×4 states to 8×8 states) with 1 to 3 Gaussian mixtures to model the probability density functions. The recognition rates achieved range from 81.5% for 4×4 states with 1 Gaussian to 100% for 8×8 states with 2 and 3 Gaussians. *Second Experiment Set*: the effect of overlap on recognition rates is tested. An overlap of 75% is used for all training while for testing overlaps between 75% and 0% were used, with a 7×7 HMM and from 1 to 3 Gaussian mixtures. The overall result of this experiment is that recognition rates decrease slightly when the overlap is reduced, however, very good recognition rates of 94.5%-99.5% were still obtained even for 0% overlap, compared with 98.5%-100% for 75% overlap. Thus wide variations in overlap have relatively minor effects on overlap for a sophisticated 7x7 HMM model.

Third Experiment Set: comprises an evaluation of the effect of compression artefacts on the recognition rate. Recognition was performed on JPEG compressed images across a range of quality settings ranging from 100 for the best quality to 1 for the highest compression ratio as shown in figure 12[Eickler et al., 2000]. The results are as follows: for compression ratios of up to 7.5 to 1, the recognition rates remain constant around 99.5%±0.5%. For compression ratios over 12.5 to 1, the recognition rates drop below 90%, down to approximately 5% for 19.5 to 1 compression ratio.

There are some additional conclusions we can draw from the work of [Eickler et al., 2000]. Firstly, building an initial HMM model using all faces in the database is an improvement over the intuitive initialization used by [Samaria, 1994] or [Nefian & Hayes, 1999], however this may lead to the dependency of the initial model on the composition of the database. Secondly, these authors obtain excellent results when using JPEG compressed images in the testing phase (overlap 0%), speeding up the recognition process significantly. Note however, in the training stage they use uncompressed images scanned with a 75% overlap and as they have used a very complex HMM model with 49 states the training stage of their approach is resource and time intensive offsetting the benefits of faster recognition speeds.

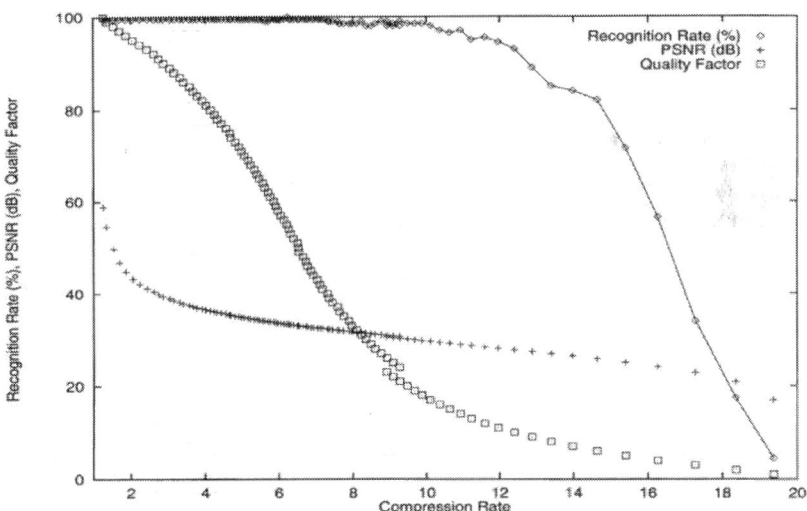

Figure 12. Recognition rates versus compression rates [Eickler et al., 2000].

5.4. Discrete Vs Continuous Modelling of Observation Vectors for P2d Hmm

In another paper on the subject of face recognition using HMM, [Wallhoff et al., 2001] consider if there is a major difference in recognition performance between HMMs where the observation vectors are modelled as continuous or discrete processes. In the continuous case, the observation probability is expressed as a density probability function approximated by a weighted sum of Gaussian mixtures. In the case of a discrete output probability, a discrete set of observation probabilities is available for each state, and input vector. This discrete set is stored as a set of codebook entries. The codebook is typically obtained by k-means clustering of all available training data feature vectors.

The authors used for their experiments 321 subjects selected from the FERET database[6] - . For testing the system, two galleries of images were used: f_a gallery, containing a regular frontal image for each subject, and f_b gallery, containing an alternative frontal image, taken seconds after the corresponding f_a image. First the images are pre-processed, using a semi-automated feature extraction that starts with the manual labelling of the eye and mouth centre-coordinates. The next step is the automatic rotation of the original images so that a line through the eyes is horizontal. After this the face is divided vertically and processing continues on a half-face image. The images are re-sized to the smallest image among the resulting images being 64 × 96 pixels.

For feature extraction, the image is scanned using a rectangular window, with an overlap of 75%. After the DCT coefficients for each block are calculated, a triangular shaped mask is applied and the first 10 coefficients are retained, representing the observation vector. Two sets of experiments were performed, for continuous and discrete outputs. For the case of *continuous output*, the experiments used 8 × 8 and 16 × 16 scanning windows, and 4 × 4 to 7 × 7 state structures for the P2D HMM. Initially only one Gaussian per state was used. The best recognition rate in this case was 95.95%, for 8 × 8 block size and 7 × 7 states for HMM. When the number of Gaussians was increased form 1 to 3, the recognition rate dropped, maybe due to the fact that only one image per person was used in the training phase. In the case of *discrete output values*, identical scanning windows and HMM were used, and two codebook sizes of 300 and 1000 values were used to generate the observation vectors. The highest recognition rate obtained was 98.13%, for 8 × 8 pixels block size, 7 × 7 states HMM, and a codebook size of 1000. In both cases, continuous and discrete, better results were obtained for the smaller size of scanning window.

5.5. Face Retrieval on Large Databases

After using the combination of 2D DCT and P2D HMM for face recognition on small databases, a new HMM-based measure to rank images within a larger database is next presented, [Eickeler, 2002]. The relation of the method presented to confidence measures is pointed out and five different approximations of the confidence measure for the task of database retrieval are evaluated. These experiments were carried out on the C-VIS database, containing the extracted faces of three days of television broadcast resulting in 25000 unlabeled face images. Normal HMM-based face recognition for database retrieval entails building a model for each person in the database. However, in the case of a very large and unlabeled database, that would imply building a model λ_j for each image O_j in the database, which is not only computationally expensive, but results in poor modelling, considering that a robust model for one person requires multiple training images of that person. In this case, calculating the probability of a query image for each built model $P(O\text{query} \mid \lambda j)$ is simply not practical.

A more feasible method for database retrieval is to train a query HMM λ_{query} using the query images O_{query} of the person searched for ω_{query}, but noting that the probability derived by the Forward-Backward algorithm, $P(O_j \mid \lambda_{query})$ cannot be used as ranking measure for the images in the database because inaccuracies in the modelling of the face images have a big influence on the probability. In order to fix this problem, the ranking of the images uses the query model λ_{query} as a representation of the person being searched for and a set of cohort models Λ_{cohort} representing people not being searched for. An easy way to form the cohort is by using former queries or by taking some images form the database. So instead of calculating $P(O_{query} \mid \lambda_j)$, the probability of an image O_j given the person being searched is used:

$$P(O_j \mid \omega_{query}) \propto P(\lambda_{query} \mid O_j) = \frac{P(O_j \mid \lambda_{Query})}{P(O_j \mid \Lambda_{cohort})}$$

(1)

In this research five different confidence measures were used for database retrieval based on this formula. For the confidence measure using normalization, the denominator is replaced:

$$P(O_j \mid \Lambda_{cohort}) = \sum_{\lambda_k \in \Lambda_{cohort}} P(O_j \mid \lambda_k)$$

(2)

Another confidence measure uses one filler (common) model instead of a cohort of HMMs for a group of people. The filler model can be trained on all people of the cohort group. If the denominator is set to a fixed probability, it can be dropped from the formula, in which case the confidence measure will be $P(O_j \mid \lambda_{query})$. The fourth confidence measure is based on the sum of ranking differences between the ranking of the cohort models on the query image and the ranking of the cohort models on each of the database images. Finally, the Levenshtein Distance (the Levenshtein distance between two strings is given by the minimum number of operations needed to transform one string into the other) is considered as an alternative measure for the comparison of the rankings of the cohort models for the query image and the database images.

For the experimental part 14 people with 8 to 16 face images each were used as query images, and also as cohort set. A NN-based face detector was used to detect the inner facial rectangles in the video broadcast and the rectangle of each image is scaled to 66 × 86 pixels. In order to remove the background an ellipsoid mask is applied. A P2D HMM with 5 × 5 states is used. The results of the query are evaluated using precision and recall: precision is the proportion of relevant images among the retrieved images while recall is the proportion of relevant images in the database that are part of the retrieval result. In a first experiment a database retrieval for each person of the query set using the normalization is performed and only the precision is calculated considering the database is unlabeled hence an exact number for each person in unknown. For 12 out of 14 people the precision is constant at 100% for around 40 retrieved images (the number of images per person varies between 20 and 300). In a second experiment all five measures were tested for one person. The results are almost perfect for normalization, a little worse but much faster for the filler model. The 'sum of ranking differences' and Levenshtein Distance measures return relatively good results but are inferior to normalization, while the use of a fixed probability gives significantly worse results than all other measures.

5.6. A Low-Complexity Simplification of the Full-2d-Hmm

An alternative approach to 2D HMM was proposed by [Othman & Aboulnasr, 2000]. These authors propose a *low-complexity* 2D HMM (LC2D HMM) system for face recognition. The aim of this research is to build a full 2D HMM but with reduced complexity. The challenge is to take advantage of a full 2D HMM structure, but without the full complexity implied by an unconstrained 2D model. Their model is implemented in the 2D DCT compressed domain with 8×8 pixel non-overlapping blocks to maintain compatibility with standard JPEG images. The authors claim a computational complexity reduction from N^4 for a fully connected 2D HMM to $2N^2$ for the LC2D HMM, where N is the number of states. Although the accuracy of the system is not better than other approaches, these authors claim that the computational complexity involved is somewhat less than that required for a 1D HMM and significantly less than that of P2D HMM.

The LC2D HMM is based on 2 key assumptions: (i) the active state at the observation block $B_{k,l}$ is dependant only on immediate vertical and horizontal neighbours, $B_{k-1,l}$ and $B_{k,l-1}$;[7] - ; (ii) the active states at the 2 observation blocks in anti-diagonal neighbourhood locations, $B_{k-1,l}$ and $B_{k,l-1}$ are statistically independent given the current state. This assumption allows separating the 3D state transition matrix into two distinct 2D transition matrices, for horizontal and vertical transitions. This decreases the complexity of the model quite significantly. This *low-complexity* model topology and image scanning are illustrated in figure 13.

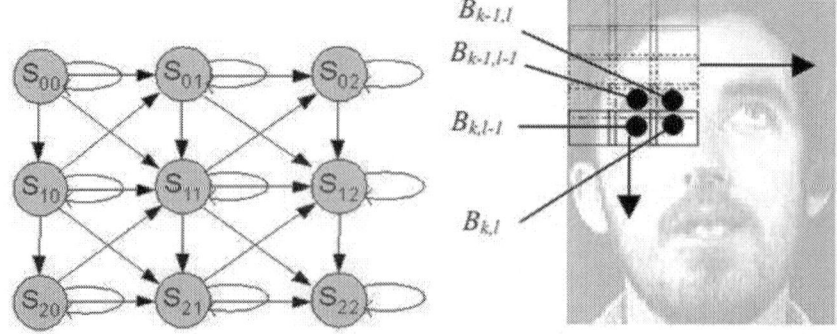

Figure 13. a) Image scanning b) Model topology [Othman & Aboulnasr, 2000]

The authors state that the two assumptions are acceptable for non-overlapped feature blocks, but have less validity for very small sized feature blocks or as the allowable overlap increases. The tests were performed on the ORL database. The model for each person was trained with 9 images, and the remaining image was used in the testing phase. Image scanning is performed in a two dimensional manner, with block size set to 8×8. Only the first 9 DCT coefficients per block were used. Different block overlap values were used to investigate the system performance and the validity of the design assumptions. The recognition rates are around 70% for 0 or 1 pixel overlap, decreasing dramatically down to only 10% for a 6-pixel overlap. This is explained because the assumptions of statistical independence, which are the underlying basis of this model, lose their validity as the overlap increases.

5.7. Refinements of the Low-Complexity Approach

In a subsequent publication by the same authors, [Othman & Aboulnasr, 2001], a hybrid HMM for face recognition is introduced. The proposed system comprises of a LC2D HMM, as described in their earlier work used in combination with a 1D HMM. The LC2D HMM carries out a complete search in the compressed JPEG domain, and a 1D HMM is then applied that searches only in the candidate list provided by the first module.

In the experiments presented in this paper, a 6×2 states model was used for the LC2D HMM, and 4 and 5 state top-to-bottom models were used for the 1D HMM. For the 1D HMM, DCT feature extraction is performed on a horizontal 10×92 scanning window. For the 2D HMM, a 8×8 block size is used for scanning the image, and the first 9 DCT coefficients are retained from each block. No overlap is allowed for the sliding windows. Tests are performed on the ORL database. In a first series of tests the effects of training data size on the model robustness were studied. The accuracy of the system ranges from 48%-58% when trained with only 2 images per person, to almost 95%-100% if trained with 9 images per person. A second series of experiments provides a detailed analysis of the trade-off between recognition accuracy and computational complexity and determines an optimal operating point for this hybrid approach. This appears to be the first research in this field to consider such trade-offs in a detailed study and this methodology should provide a useful approach for other researchers in the future.

In a third paper, [Othman & Aboulnasr, 2003], these authors propose a 2D HMM face recognition system that limits the independence assumptions described in their original work to conditional independence among adjacent observation blocks. In this new model, the active states of the two anti-diagonal observation blocks are statistically independent given the current state and knowledge of the past observations. This translates into a more flexible model, allowing state transitions in the transverse direction as shown in figure 14, taken from, [Othman & Aboulnasr, 2003].

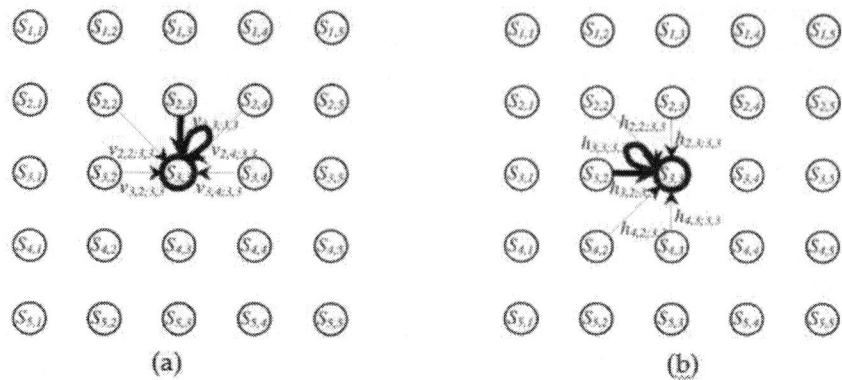

Figure 14. Modified LC2D HMM [Othman & Aboulnasr, 2003]. (a) Vertical transitions to state S3,3 for 5×5 state model (b) Horizontal transitions to state S3,3 also for 5×5 state model.

This modified LC2D HMM face recognition system is examined for different values of the structural parameters, namely number of states per model and number of Gaussian mixtures per state. These tests are again conducted on the ORL database. The images are scanned using 8×8 blocks and the first 9 2D DCT coefficients comprise the observation vector. The HMMs were trained using 9 images per person, and tested using the 10th image. The test is repeated 5 times with different test images and the results are averaged over a total of 200 test images for 40 persons. Test images are not members of the training data set at any time. The results vary from a very low 4% recognition rate for a 7 × 3 HMM with 64 Gaussian mixtures per state, up to 100% for a 7 × 3 HMM with 4 Gaussian mixtures per state. Best results are obtained for 4 and 8 Gaussians per state. The reason for the poor performance for a higher number of Gaussian mixtures is that the model becomes too discriminating and cannot recognize data with any flexibility, outside the original training set. Finally, the reader's attention is drawn to detailed comments by [Yu & Wu, 2007] on the key assumption of conditional independence in the relationship between adjacent blocks. In this communication, [Yu & Wu, 2007] it is shown that this key assumption is entirely unnecessary.

6. MORE RECENT RESEARCH ON HMM IN FACE RECOGNITION

While there have been more recent research which applies HMM techniques to face recognition, most of this work has not refined the underlying methods, but has instead combined known HMM techniques with other face analysis techniques. Some work is worth mentioning, such as that of [Le & Li, 2004] who combined a one-dimensional discrete hidden Markov model (1D-DHMM) with new way of extracting observations and using observation sequences. All subjects in the system share only one HMM that is used as a means to weigh a pair of observations. The Haar wavelet transform is applied to face images to reduce the dimensionality of the observation vectors. Experiments on the AR face database[8] - and the CMU PIE face database[9] - show that the proposed method outperforms PCA, LDA, LFA based approaches tested on the same databases.

Also worth mentioning is the work of [Yujian, 2006]. In this paper, several new analytic formulae for solving the three basic problems of 2-D HMM are provided. Although the complexity of computing these is exponential in the size of data, it is almost the same as that of a 1D HMM for cases where the numbers of rows or columns are a small constant. While this author did not apply these results specifically to facial recognition problem they appear to offer some promise in simplifying the application of a full 2D HMM to the face recognition problem.

Another notable contribution is the work of [Chien & Liao, 2008] which explores a new discriminative training criterion to assure model compactness combined with ability for accurate to discrimination between subjects. Hypothesis testing is employed to maximize the confidence level during model training leading to a *maximum-confidence* model (MC-HMM) for face recognition. From experiments on the FERET[10] - database and GTFD[11] - , the proposed method obtains robust segmentation in the presence of different facial expressions, orientations, and so forth. In comparison with the maximum likelihood and minimum classification error HMMs, the proposed MC-HMM achieves higher recognition accuracies with lower feature dimensions. Notably this work uses more challenging databases than the ORL database.

Finally we conclude this chapter referring to our own recent work in face recognition using EHMM, presented in [Iancu, 2010; Corcoran & Iancu 2011]. This work can be divided in three parts according to our objectives. The tests were performed on a combined database (BioID, Achermann, UMIST) and on the FERET database. The first objective was to build a recognition system applicable on handheld devices with very low computational power. For this we tested the EHMM-based face recognizer for different sizes of the model, different number of Gaussians, picture size, features, and number of pictures per person used for training. The results obtained for very small picture size (32 × 32), with 1 Gaussian per state and on a simplified EHMM are only 58% recognition for only 1 image per person used for training, when we use 5 pictures per person for training the recognition rates go up to 82% [Corcoran & Iancu 2011]. A second objective was to limit the effect of illumination variations on recognition rates. For this three illumination normalization techniques were used and various combinations of these were tested: histogram equalization (HE), contrast limited adaptive histogram equalization (CLAHE) and DCT in logarithm domain (logDCT). The best recognition rates were obtained for a combination of CLAHE and HE (95.71%) and the worst for logDCT (77.86%) on the combined database [Corcoran & Iancu 2011].

A third objective was to build a system robust to head pose variations. For this we tested the face recognition system using frontal, semi-profile and profile views of the subjects. The first set of tests was performed on the combined database. Here the maximum head pose angle is around 30°. We compared recognition rates obtained when building one EHMM model per person versus one EHMM model per picture. The second set of tests was performed on FERET database which has a much bigger variety of head poses. In this case we used one frontal, 2 semi-profiles and 2 profiles for each subject in the training stage and all pictures of each subject in the testing stage. We compared the recognition rates when building 1 model per person versus 2 models per person versus 3 models per person. We obtained better recognition rates for one model per person for the first set of tests where the database has little head pose variation but better recognition rates for 2 models per person for the second set of tests where the database has a very high head pose variation [Iancu, 2010].

7. REVIEW AND CONCLUDING REMARKS

The focus of this chapter is on the use of HMM techniques for face recognition. For this review we have presented a concise yet comprehensive description and review of the most interesting and widely used techniques to apply HMM models in face recognition applications. Although additional papers treating specific aspects of this field can be found in the literature, these are invariably based on one or another of the key techniques presented and reviewed here.

Our goal has been to quickly enable the interested reader to review and understand the state-of-art for HMM models applied to face recognition problems. It is clear that different techniques balance certain trade-offs between computational complexity, speed and accuracy of recognition and overall practicality and ease-of-use. Our hope is that this article will make it easier for new researchers to understand and adopt HMM for face analysis and recognition applications and continue to improve and refine the underlying techniques.

REFERENCES

1. L. E. Baum, T. Petrie, 1966 Statistical inference for probabilistic functions of finite state Markov chains. Annals of Mathematical Statistics, 37 1966.
2. L. E. Baum, T. Petrie, G. Soules, N. Weiss, 1970 A maximization technique occurring in the statistical analysis of probabilistic functions of Markov chains. Annals of Mathematical Statistics, 41 1970.
3. L. E. Baum, 1972 An inequality and associated maximization technique in statistical estimation for probabilistic functions of Markov processes, Inequalities, 3 18 , 1972.
4. M. Bicego, U. Castellani, V. Murino, 2003b Using hidden markov models and wavelets for face recognition. Proceedings of Image Analysis and Processing 2003. 12th International Conference on, 5256 , 2003.
5. M. Bicego, V. Murino, M. Figueiredo, 2003a A sequential pruning strategy for the selection of the number of states in hidden markov models. Pattern Recognition Letters, 24 13951407 , 2003.
6. J. Chien-T, C. Liao-P, 2008 Maximum Confidence Hidden Markov Modeling for Face Recognition. Pattern Analysis and Machine Intelligence, IEEE Transactions on, 30 4 606616 , April 2008

7. P. M. Corcoran, C. Iancu, 2011 Automatic Face Recognition System for Hidden Markov Model Techniques, Face Recognition 2 Intech Publishing, 2011.
8. S. Eickeler, 2002 Face database retrieval using pseudo 2d hidden markov models. Fifth IEEE International Conference on Automatic Face and Gesture Recognition, Proceedings, 5863 , May 2002.
9. S. Eickeler, S. Muller, G. Rigoll, 2000 Recognition of jpeg compressed face images based on statistical methods. Image and Vision Computing Journal, Special Issue on Facial Image Analysis, 18 4 279287 , March 2000.
10. M. Figueiredo, J. Leitao, A. Jain, 1999 On fitting mixture models. Proceedings of the Second International Workshop on Energy Minimization Methods in Computer Vision and Pattern Recognition, Springer-Verlag, 5469 , 1999.
11. C. Iancu, 2010 Face recognition using statistical methods. PhD thesis, NUI Galway, 2010.
12. F. Jelinek, L. R. Bahl, R. L. Mercer, 1975 Design of a linguistic statistical decoder for the recognition of continuous speech. IEEE Transactions on Information Theory, 21 3 250256 , 1975.
13. B. H. Juang, 1984 On the hidden markov model and dynamic time warping for speech recognition-a unified view. AT&T Technical Journal, 63 7 12131243 , September 1984.
14. B. H. Juang, L. R. Rabiner, 2005 Automatic speech recognition- a brief history of the technology development. Elsevier Encyclopedia of Language and Linguistics, Second Edition, 2005.
15. V. V. Kohir, U. B. Desai, 1998 Face recognition using a dct-hmm approach. Applications of Computer Vision, WACV '98, Proceedings, Fourth IEEE Workshop on, 226231 , October 1998.
16. V. V. Kohir, U. B. Desai, 1999 A transform domain face recognition approach. TENCON 99, Proceedings of the IEEE Region 10 Conference, 1 104107 , September 1999.
17. V. V. Kohir, U. B. Desai, 2000 Face recognition. IEEE International Symposium on Circuits and Systems, Geneva, Switzerland, May 2000.
18. S. Kuo, O. Agazzi, 1994 Keyword spotting in poorly printed documents using pseudo 2-d hidden markov models. IEEE Transactions on Pattern Analysis and Machine Intelligence, 16 842848 , August 1994.
19. H. S. Le , H. Li, 2004 Face identification system using single hidden markov model and single sample image per person. IEEE International Joint Conference on Neural Networks, 1 2004.
20. E. Levin, R. Pieraccini, 1992 Dynamic planar warping for optical character recognition. Proceedings ICASSP 1992, San Francisco, 3 149152 , March 1992.
21. S. E. Levinson, L. R. Rabiner, M. M. Sondhi, 1983 An introduction to the application of the theory of probabilistic functions of a markov process to automatic speech recognition. Bell System Technical Journal, 62 4 10351074 , April 1983.
22. A. Martinez, 1999 Face image retrieval using hmms. IEEE Workshop on Content-Based Access of Image and Video Libraries, (CBAIVL '99) Proceedings, 3539 , June 1999.

23. A. V. Nefian, 1999 A hidden markov model based approach for face detection and recognition. PhD Thesis, 1999.

24. A. V. Nefian, I. I. I. M. H. . Hayes, Oct, 1998 Face detection and recognition using hidden markov models. Image Processing, ICIP 98, Proceedings. 1998 International Conference on, 1 141145 , October 1998.

25. A. V. Nefian, I. I. I. M. H. . Hayes, May, 1998 Hidden markov models for face recognition. Acoustics, Speech, and Signal Processing ICASSP '98. Proceedings of the 1998 IEEE International Conference on, 5 27212724 , May 1998.

26. A. V. Nefian, I. I. I. M. H. Hayes, 1999 An embedded hmm-based approach for face detection and recognition. Acoustics, Speech, and Signal Processing, ICASSP '99. Proceedings, IEEE International Conference, 6 35533556 , March 1999.

27. H. Othman, T. Aboulnasr, 2000 Hybrid hidden markov model for face recognition. 4th IEEE Southwest Symposium on Image Analysis and Interpretation, 3440 , April 2000.

28. H. Othman, T. Aboulnasr, 2001 A simplified second-order hmm with application to face recognition. ISCAS 2001 IEEE International Symposium on Circuits and Systems, 2 161164 , May 2001.

29. H. Othman, T. Aboulnasr, 2003 A separable low complexity 2d hmm with application to face recognition. Pattern Analysis and Machine Intelli- gence, IEEE Transactions on, 2003.

30. H. S. Park, S. W. Lee, 1998 A Truly 2D Hidden Markov Model For Off-Line Handwritten Character Recognition. Pattern Recognition, 31 12 18491864 , December 1998.

31. L. R. Rabiner, 1989 A tutorial on hidden markov models and selected applications in speech recognition. Proceedings of IEEE, 77 2 257286 , February 1989.

32. L. R. Rabiner, B. H. Juang, 1986 An introduction to hidden markov models. IEEE ASSP Magazine, 3 1 416 , 1986.

33. F. Samaria, 1994 Face recognition using hidden markov models. Ph.D. thesis, Department of Engineering, Cambridge University, UK, 1994.

34. F. Samaria, F. Fallside, 1993 Face identification and feature extraction using hidden markov models. Image Processing: Theory and Applications, Elsevier, 295298 , 1993.

35. F. Samaria, A. C. Harter, 1994 Parameterization of a stochastic model for human face identification. Applications of Computer Vision, 1994., Proceedings of the Second IEEE Workshop on, 77 138142 , December 1994.

36. P. Viola, M. Jones, 2001 Robust real-time object detection, Technical report 2001/01, Compaq CRL, 2001.

37. F. Wallhoff, S. Eickeler, G. Rigoll, 2001 A comparison of discrete and continuous output modeling techniques for a pseudo-2d hidden markov model face recognition system. International Conference on Image Processing, Proceedings, 2 685688 , October 2001.

38. F. Wallhoff, S. Müller, G. Rigoll, 2001 Hybrid face recognition system for profile views using the mugshot database. IEEE ICCV Workshop on Recognition, Analysis and Tracking of Faces and Gestures in Real-Time Systems, Proceedings, 149156 , July 2001.

39. L. Yu, L. Wu, 2007 Comments on 'a separable low complexity 2d hmm with application to face recognition'. Pattern Analysis and Machine Intelligence, IEEE Transactions on, 29 2 368368 , February 2007.
40. L. Yujian, 2007 An analytic solution for estimating two-dimensional hidden Markov models. Applied Mathematics and Computation, 185 2 810822 , February 2007.

CHAPTER 5

Dimensionality Reduction Techniques for Face Recognition

S S Shylaja, K N Balasubramanya Murthy and S Natarajan

Department of Information Science and Engineering, P E S Institute of Technology, India

1. INTRODUCTION

High level of image content analysis is required for several applications. This is taking more significance as the number of digital images stored is growing exponentially. On the one hand the technology should help store these images, on the other, enable us to develop newer algorithmic models aimed at efficient and quick retrieval of images. The entire captured data may not be applicable for an application and hence deriving a subset of data to achieve objective function is desirable.

Face detection and recognition are preliminary steps to a wide range of applications such as personal identity verification, video-surveillance, facial expression extraction, gender classification, advanced human and computer interaction. A face recognition system would allow user to be identified by simply walking past a surveillance camera. Research has been devoted to facial recognition for years and has brought forward algorithms in an attempt to be as accurate as humans are.

A face recognition system is expected to identify faces present in images and videos automatically. It can operate in either or both of two modes:
- Face verification or authentication,(fig above)
- Face identification or recognition.

Face verification involves a one-to-one match that compares a query face image against a template face image whose identity is being claimed. Face identification involves one-to-many matches that compare a query face image against all the template images in the database to determine the identity of the query face. Another face recognition scenario involves a watch-list check, where a query face is matched to a list of suspects (one-to-few matches). As per Hietmeyer, face recognition is one of the most effective biometric techniques for travel documents and scored higher on several evaluation parameters.

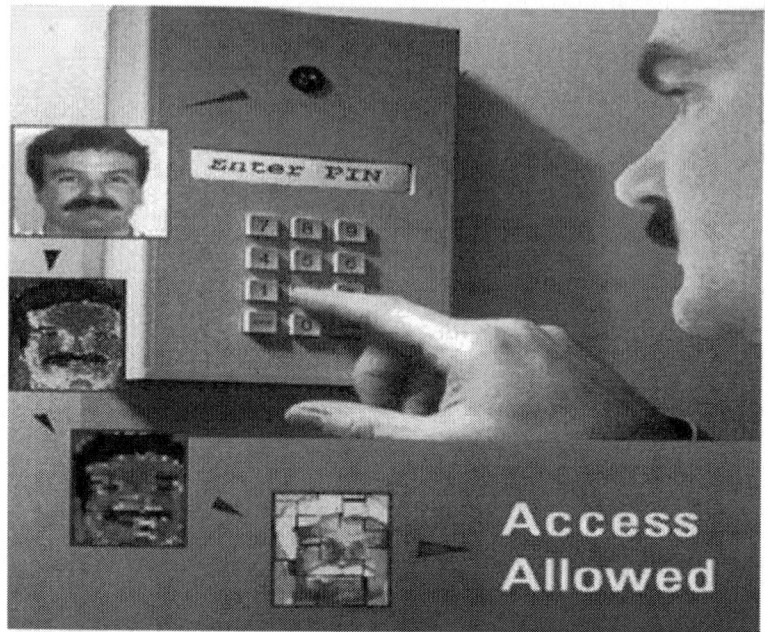

Figure 1. Face Verification System

Computational models of face recognition must address several difficult problems. This difficulty arises from the fact that faces must be represented in a way that best utilizes the available face information to distinguish a particular face from all other faces. The problem of dimensionality reduction arises in face recognition because an m X n face image is reconstructed to form a column vector of mn components, for computational purposes. As the number of images in the data set increases, the complexity of representing data sets increases. Analysis with a large number of variables generally consumes a large amount of memory and computation power.

2. DIMENSIONALITY REDUCTION

Efforts are on for efficient storage and retrieval of images. Considerable progress has happened in face recognition with newer models especially with the development of powerful models of face appearance. These models represent faces as points in high-dimensional image spaces and employ dimensionality reduction to find a more meaningful representation, therefore, addressing the issue of the "curse of dimensionality". Dimension reduction is a process of reducing the number of variables under observation. The need for dimension reduction arises when there is a large number of univariate data points or when the data points themselves are observations of a high dimensional variable. The key observation is that although face images can be regarded as points in a high-dimensional space, they often lie on a manifold (i.e., subspace) of much lower dimensionality, embedded in the high-dimensional image space. The main issue is how to properly define and determine a low-dimensional subspace of face appearance in a high-dimensional image space.

Dimensionality reduction techniques using linear transformations have been very popular in determining the intrinsic dimensionality of the manifold as well as extracting its principal directions. Dimensionality reduction is an effective approach to downsizing data. In statistics, dimension reduction is the process of reducing the number of random variables under consideration, $R^N \rightarrow R^M$ (M<N) and can be divided into feature selection and feature extraction.

Feature selection is choosing a subset of all the features

$$[x_1 \ x_2 \ \dots \ x_n] \qquad \text{Feature selection} \qquad [x_{i1} \ x_{i2} \ \dots \ x_{im}]$$

Feature extraction is creating new features from existing ones

$$[x_1 \ x_2 \ \dots \ x_n] \qquad \text{Feature extraction} \qquad [y_1 \ y_2 \ \dots \ y_m]$$

In either case, the goal is to find a low-dimensional representation of the data while still describing the data with sufficient accuracy.

For reasons of computational and conceptual simplicity, the representation is often sought as a linear transformation of the original data. In other words, each component of the representation is a linear combination of the original variables. Well-known linear transformation methods include principal component analysis, factor analysis, and projection pursuit. Independent component analysis (ICA) is a recently developed method in which the goal is to find a linear representation of nongaussian data so that the components are statistically independent, or as independent as possible. Such a representation seems to capture the essential structure of the data in many applications, including feature extraction and signal separation.

Several techniques exist to tackle the curse of dimensionality out of which some are linear methods and others are nonlinear. PCA, LDA, LPP are some popular linear methods and nonlinear methods include ISOMAP & Eigenmaps. PCA and LDA are the two most widely used subspace learning techniques for face recognition. These methods project the training sample faces to a low dimensional representation space where the recognition is carried out. The main supposition behind this procedure is that the face space (given by the feature vectors) has a lower dimension than the image space (given by the number of pixels in the image), and that the recognition of the faces can be performed in this reduced space. PCA has the advantage of capturing holistic features but ignore the localized features. Fisher faces from LDA technique extracts discriminating features between classes and is found to perform better for large data sets. Its shortcoming is that of Small Sample Space (SSS) problem. LPPs are linear projective maps that arise by solving variational problem that optimally preserves the neighborhood structure of the data set.

In many cases, face images may be visualized as points drawn on a low-dimensional manifold hidden in a high-dimensional ambient space. Specially, we can consider that a sheet of rubber is crumpled into a (high-dimensional) ball. The objective of a dimensionality-reducing mapping is to unfold the sheet and to make its low-dimensional structure explicit. If the sheet is not torn in the process, the mapping is topology-preserving. Moreover, if the rubber is not stretched or compressed, the mapping preserves the metric structure of the original space.

PCA is guaranteed to discover the dimensionality of the manifold and produces a compact representation. Turk and Pentland use Principal Component Analysis to describe face images in terms of a set of basis functions, or "eigenfaces". LDA is a supervised learning algorithm. LDA searches for the project axes on which the data points of different classes are far from each other while requiring data points of the same class to be close to each other. Unlike PCA which encodes information in an orthogonal linear space, LDA encodes discriminating information in a linear separable space using bases are not necessarily orthogonal. It is generally believed that algorithms based on LDA are superior to those based on PCA. However, some recent work shows that, when the training dataset is small, PCA can outperform LDA, and also that PCA is less sensitive to different training datasets.

Recently, a number of research efforts have shown that the face images possibly reside on a nonlinear submanifold. However, both PCA and LDA effectively see only the Euclidean structure. They fail to discover the underlying structure, if the face images lie on a nonlinear submanifold hidden in the image space. Some nonlinear techniques have been proposed to discover the nonlinear structure of the manifold, *e.g.* Isomap, LLE and Laplacian Eigenmap. These nonlinear methods do yield impressive results on some benchmark artificial data sets. However, they yield maps that are defined *only* on the training data points and how to evaluate the maps on novel test data points remains unclear.

3. SINGULAR VALUE DECOMPOSITION (SVD)

Singular value decomposition (SVD) is an important factorization of a rectangular real or complex matrix, with many applications in signal processing and statistics. As applied to face recognition this technique is used to extract the holistic global features of the training set SVD is the best, in the mean-square error sense, linear dimension reduction technique. Being based on the covariance matrix of the variables, it is a second-order method. SVD seeks to reduce the dimension of the data by finding a few orthogonal linear combinations of the original variables with the largest variance.

The basic idea behind SVD is taking a high dimensional, highly variable set of data points and reducing it to a lower dimensional space that exposes the substructure of the original data more clearly and orders it from most variation to the least. What makes SVD practical for pattern recognition applications is that one can simply ignore variation below a particular threshold to massively reduce the data but be assured that the main relationships of interest have been preserved.

Singular value decomposition (SVD) can be looked at from three mutually compatible points of view. On the one hand, we can see it as a method for transforming correlated variables into a set of uncorrelated ones that better expose the various relationships among the original data items. At the same time, SVD is a method for identifying and ordering the dimensions along which data points exhibit the most variation. This ties into the third way of viewing SVD, which is that once we have identified where the most variation is, it's possible to find the best approximation of the original data points using fewer dimensions. Hence, SVD can be seen as a method for data reduction.

As said earlier Singular Value Decomposition is a way of factoring matrices into a series of linear approximations that expose the underlying structure of the matrix. If A is the input matrix, calculating the SVD consists of finding the eigenvalues and eigenvectors of AA^T and A^TA. This yields three matrices U,V & S where the eigenvectors of A^TA make up the columns of V, the eigenvectors of AA^T make up the columns of U. and the singular values in S are square roots of eigenvalues from AA^T or A^TA. The singular values are the diagonal entries of the S matrix and are arranged in descending order. The singular values are always real numbers. If the matrix A is a real matrix, then U and V are also real.

In the factorization, the first principal component is s1, with the largest variance is the linear combination with T T. We have $S_1 = XW_1$, where the p-dimensional coefficient vector solves $W_1 = (W_{11},....,W_{1P})$ where

$$W_1 = \arg\max_{\|w=1\|} Var\{x, w\}$$

(1)

The second PC is the linear combination with the second largest variance and orthogonal to the first PC, and so on. There are as many PCs as the number of the original variables. PCs explain most of the variance, so that the rest can be disregarded with minimal loss of information. Since the variance depends on the scale of the variables, it is customary to first standardize each variable to have mean zero and standard deviation one. After the standardization, the original variables with possibly different units of measurement are all in comparable units.

The mathematical model formulated is given below:
Let A is m' X n' real matrix and $N=A^TA$

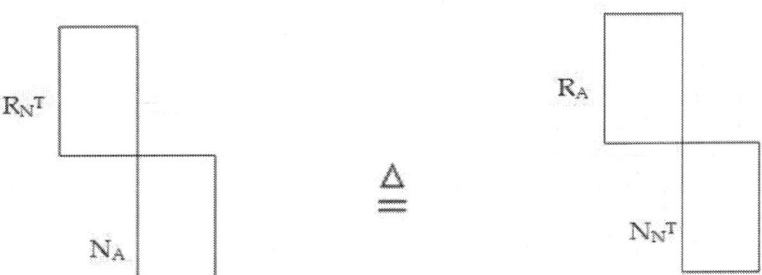

Figure 2. Range and Null space of matrix

R denotes the range space and N denotes the null space of a matrix. Rank of A, A^T, A^TA, AA^T is equal and is denoted by ρ orthonormal basis v_i $1 \leq i \leq \rho$ are sought for $R_A{}^T$ where ρ is the rank of $R_A{}^T$ & u_i $1 \leq i \leq \rho$ for R_A such that,

$$AV_j = S_j U$$

(2)

$$A^T U_j = S_j V_j, \qquad 1 \leq j \leq \rho$$

(3)

Advantages of having such a basis are that geometry becomes easy and gives a decomposition of A into ρ one-ranked matrices. Combining the equations (Eq. 2) & (Eq. 3)

$$A = \sum_j S_j V_j \, U_j{}^T, \qquad 1 \leq j \leq \rho$$

(4)

If V_j is known then, $U_j = (1 / S_j) AV_j$ $|S_j| = \| AV_j \|$ therefore, $s_j \neq 0$, choosing $s_j > 0$,

$$AV_j = S_j U_j$$

(5)

$$A^T A V_j = S_j A^T U_j$$

(6)

$$A^T A V_j = S_j{}^2 V_j$$

(7)

Let $S_j{}^2 = \mu_i$, $N V_j = \mu_i V_i$ is required U_i's as orthonormal eigenvectors of $N=A^TA$ are found and $S_j = \sqrt{\mu_i}$ Where $\mu_i > 0$ are eigen values corresponding to V_j. The resulting U_i span the Eigen subspace. When SVD is applied to the sample set below in figure 3, the corresponding eigen faces obtained are shown in figure 4. The figure is highlighting the holistic features from the given sample set.

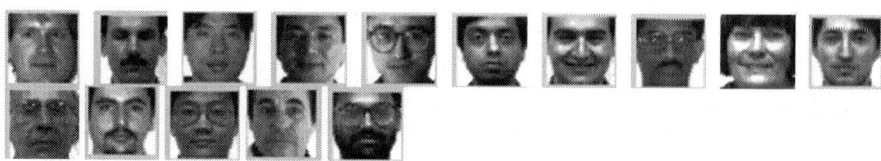

Figure 3. Training set example faces

Figure 4. Eigen faces from SVD

Basis selection from SVD

If A is the face Space, then x vectors are drawn from $[X_1.....X_x] = \Pi_{<1..x>}(A^{-1}UD)$
Where U & D are the unitary and diagonal matrices of SVD of A.

5. LINEAR DISCRIMINANT ANALYSIS

Fisher Linear Discriminant also referred as Linear Discriminant Analysis is a classical pattern recognition method, which was introduced by Fisher (1934). It is a very effective feature extraction method but facing issues for Small Sample Space problem.

The Dimensionality Reduction technique SVD searches for directions in the data that have largest variance and subsequently project the data onto it. In this way, one can obtain a lower dimensional representation of the data, that removes some of the "noisy" directions. There are many difficult issues with how many directions one needs to choose. It is an unsupervised technique and as such does not include label information of the data. For instance, if we imagine 2 clusters in 2 dimensions, one clusters has $y = 1$ and the other $y = ¡1$. The clusters are positioned in parallel and very closely together, such that the variance in the total data-set, ignoring the labels, is in the direction of the clusters. For classification, this would be a terrible projection, because all labels get evenly mixed and will destroy the useful information.

A much more useful projection is orthogonal to the clusters, i.e. in the direction of least overall variance, which would perfectly separate the data-cases (obviously, we would still need to perform classification in this 1-D space).

The conventional solution to misclassification for small sample size problem and large data set with similar faces is the use of PCA into LDA i.e. fisher faces. PCA is used for dimensionality reduction and then LDA is performed on the lower dimensional space. Discriminant analysis often produces models whose accuracy approaches complex modern methods. The target variable may have two or more categories. The following figure 5 shows a plot of the two categories with the two predictors on orthogonal axes:

A transformation function is found that maximizes the ratio of between-class variance to within-class variance as illustrated by this figure 6 produced by Ludwig Schwardt and Johan du Preez:

The transformation seeks to rotate the axes so that when the categories are projected on the new axes, the differences between the groups are maximized. So the question is, how do we utilize the label information in finding informative projections?

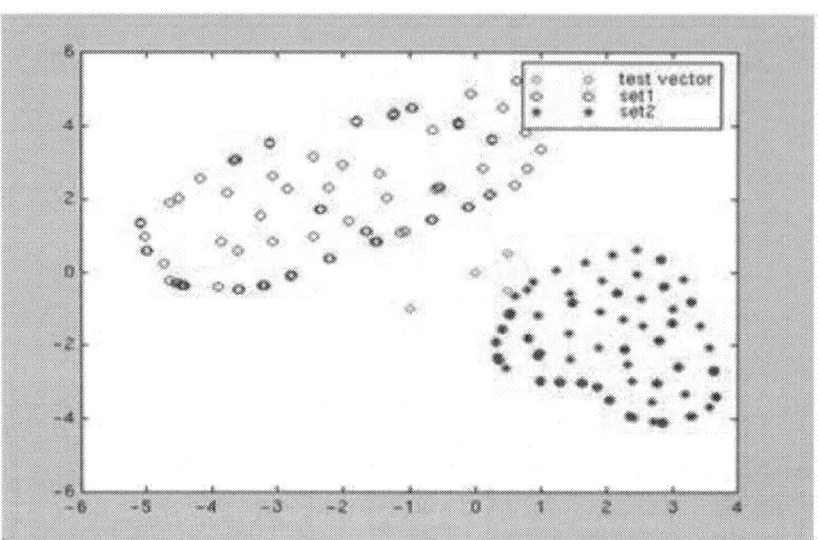

Figure 5. A plot of the two categories with the two predictors on orthogonal axes

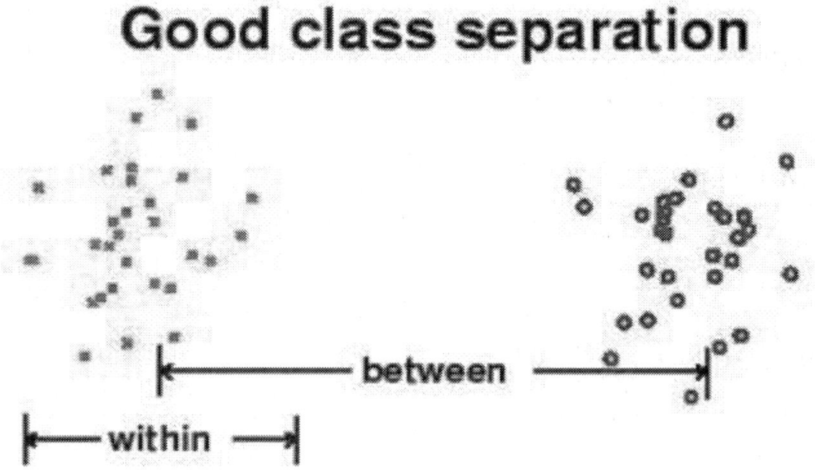

Good class separation

Figure 6. Output of applying transformation function

To that purpose Fisher-LDA considers maximizing the following objective:

$$J(w) = \frac{w^T S_B w}{w^T S_w w}$$

(8)

The second use of the term LDA refers to a discriminative feature transform that is optimal for certain cases [10]. This is what we denote by LDA throughout this paper. In the basic formulation, LDA finds eigenvectors of matrix

$$T = S_w^{-1} \, S_b \tag{9}$$

Here S_b is the between-class covariance matrix, that is, the covariance matrix of class means. S_w denotes the within-class covariance matrix, that is equal to the weighted sum of covariance matrices computed for each class separately. S_w^{-1} captures the compactness of each class, and S_b represents the separation of the class means. Thus T captures both. The eigenvectors corresponding to largest k eigenvalues of T form the rows of the transform matrix w, and new discriminative features d_k are derived from the original ones d simply by

$$d_k = W \, d \tag{10}$$

The straightforward algebraic way of deriving the LDA transform matrix is both a strength and a weakness of the method. Since LDA makes use of only second-order statistical information, covariances, it is optimal for data where each class has a unimodal Gaussian density with well separated means and similar covariances. Large deviations from these assumptions may result in sub-optimal features.

Also the maximum rank of S_b in this formulation is $N_c - 1$ where N_c the number of different classes is. Thus basic LDA cannot produce more than $N_c - 1$ features. This is, however, simple to remedy by projecting the data onto a subspace orthogonal to the computed eigenvectors, and repeating the LDA analysis in this space.

However, the classification performance of traditional LDA is often degraded by the fact that their separability criteria are not directly related to their classification accuracy in the output space. A solution to the problem is to introduce weighting functions into LDA. Object classes that are closer together in the output space, and thus can potentially result in misclassification, should be more heavily weighted in the input space. This idea has been further extended in with the introduction of the fractional-step linear discriminant analysis algorithm (F-LDA), where the dimensionality reduction is implemented in a few small fractional steps allowing for the relevant distances to be more accurately weighted. Although the method has can be applied on low dimensional patterns it cannot be directly applied to high-dimensional patterns, such as those face images due to two factors: (1) the computational difficult of the eigen-decomposition of matrices in the high-dimensional image space; (2) the degenerated scatter matrices caused by the small sample size, which widely exists in the FR tasks where the number of training samples is smaller than the dimensionality of the samples.

The traditional solution to the SSS problem requires the incorporation of a PCA step into the LDA framework. In this approach, PCA is used as a pre-processing step for dimensionality reduction so as to discard the null space of the within-class scatter matrix of the training data set. Then LDA is performed in the lower dimensional PCA subspace. However, it has been shown that the discarded null space may contain significant discriminatory information. To prevent this from happening, solutions without a separate PCA step, called direct LDA (D-LDA) methods have been presented recently. In the D-LDA framework, data are processed directly in the original high-dimensional input space avoiding the loss of significant discriminatory information due to the PCA pre-processing step.

Firstly dimensionality of the original input space is lowered by introducing a new variant of D-LDA that results in a low-dimensional SSS-free subspace where the most discriminatory features are preserved. The variant of D-LDA utilizes a modified Fisher's criterion to avoid a problem resulting from the wage of the zero eigenvalues of the within-class scatter matrix as possible divisors. Also, a weighting function is introduced into the variant of D-LDA, so that a subsequent F-LDA step can be applied to carefully re-orient the SSS-free subspace resulting in a set of optimal discriminant features for face representation.

The DF-LDA is a linear pattern recognition method. Compared with nonlinear models, a linear model is rather robust against noises and most likely will not over fit. Although it has been shown that distribution of face patterns is highly non convex and complex in most cases, linear methods are still able to provide cost effective solutions to the FR tasks through integration with other strategies, such as the principle of "divide and conquer," in which a large and nonlinear problem is divided into a few smaller and local linear sub problems.

Let S_{BTW} and S_{WTH} denote the between- and within-class scatter matrices of the training image set, respectively. LDA-like approaches such as the Fisherface method find a set of basis vectors, denoted by that maximizes the ratio between S_{BTW} and S_{WTH} is

$$\Psi = \arg \max_{\Psi} \frac{\left| \left(\Psi^T S_{BTW} \Psi \right) \right|}{\left| \left(\Psi^T S_{WTH} \Psi \right) \right|}$$

(11)

The maximization process in (3) is not directly linked to the classification error which is the criterion of performance used to measure the success of the Face Recognition procedure. Thus, the weighted between-class scatter matrix can be expressed as:

$$S_{BTW} = \sum_{i}^{C} \Phi_i \Phi_i^{T}$$

(12)

where

$$\Phi_i = \left(L_i/L\right)^{1/2} \sum_{i=1}^{C}\left(w\left(d_{ij}\right)\right)^{1/2} \left(\overline{Z}_i - \overline{Z}_j\right), \overline{Z}_i$$

(13)

is the mean of class Z_i, L_i is the number of elements in Z_i, and

$$d_{i,j} = \left\|\overline{Z}_i - \overline{Z}_j\right\|$$

(14)

is the Euclidean distance between the means of class i and j.

Basis selection from DF-LDA
The set Y vectors are chosen by the equation $[y_1.....y_y] = \Pi_{<1..y>}(U^T S_{TOT} U^T)$ Where S_{TOT} is the sum of between and within class scatter matrices, U is a diagonal matrix from Eigen values and vectors. Fisher faces are shown in figure 7 below.

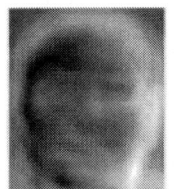 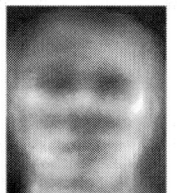  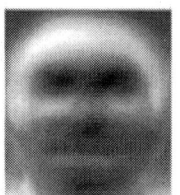

Figure 7. Fisher Faces from DF-LDA

We can clearly see from fisher faces that more pronounced features are highlighted than the rest of the face point like hair, eyebrows etc.

6. LOCALITY PRESERVING PROJECTIONS

Different from Principal Component Analysis (PCA) and Linear Discriminant Analysis (LDA) which effectively see only the Euclidean structure of face space, LPP finds an embedding that preserves local information, and obtains a face subspace that best detects the essential face manifold structure. The Laplacianfaces are the optimal linear approximations to the eigen functions of the Laplace Beltrami operator on the face manifold. In this way, the unwanted variations resulting from changes in lighting, facial expression, and pose may be eliminated or reduced. Theoretical analysis shows that PCA, LDA and LPP can be obtained from different graph models. By using Locality Preserving Projections (LPP), the face images are mapped into a face subspace for analysis.

LPP shares many of the data representation properties of nonlinear techniques such as Laplacian Eigen maps or Locally Linear Embedding. Yet LPP is linear and more crucially is defined everywhere in ambient space rather than just on the training data points. It builds a graph incorporating neighborhood information of the data set. Using the notion of the Laplacian of the graph, transformation matrix is computed which maps the data points to a subspace.

This linear transformation optimally preserves local neighborhood information in a certain sense. The representation map generated by the algorithm may be viewed as a linear discrete approximation to a continuous map that naturally arises from the geometry of the manifold. In the meantime, there has been some interest in the problem of developing low dimensional representations through kernel based techniques for face recognition. These methods can discover the nonlinear structure of the face images. However, they are computationally expensive. Moreover, none of them explicitly considers the structure of the manifold on which the face images possibly reside.

While the Eigen faces method aims to preserve the global structure of the image space, and the Fisher faces method aims to preserve the discriminating information; our Laplacianfaces method aims to preserve the local structure of the image space. In many real world classification problems, the local manifold structure is more important than the global Euclidean structure, especially when nearest neighbor like classifiers are used for classification.

LPP seems to have discriminating power although it is unsupervised. An efficient subspace learning algorithm for face recognition should be able to discover the nonlinear manifold structure of the face space. LPP shares some similar properties to LLE, such as a locality preserving character. However, their objective functions are totally different. LPP is obtained by finding the optimal linear approximations to the eigen functions of the Laplace Beltrami operator on the manifold. LPP is linear, while LLE is nonlinear. Moreover, LPP is defined everywhere, while LLE is defined only on the training data points and it is unclear how to evaluate the maps for new test points. In contrast, LPP may be simply applied to any new data point to locate it in the reduced representation space. LPP seeks to preserve the intrinsic geometry of the data and local structure.

The objective function of LPP is as follows:

$$\min \sum_{ij} \left(y_i - y_j \right)^2 S_{ij}$$

(15)

Where

$$S_{ij} = \begin{cases} \exp\left(\left\| x_i - x_j \right\|^2 / t \right), & \left\| x_i - x_j \right\|^2 < \varepsilon \\ 0 \end{cases}$$

(16)

7. STATISTICAL VIEW OF LPP

LPP can also be obtained from statistical viewpoint. Suppose the data points follow some underlying distribution. Let d be the number of non-zero S_{ij}, and D be a diagonal matrix whose entries are column (or row, since S is symmetric) sums of S, $D_{ii} = \sum_j S_{ji}$. By the Strong Law of Large Numbers, $E(zz^T \mid \| z \| < \varepsilon)$ can be estimated from the sample points as follows:

$$E\left(ZZ^T \mid \|z\| < \varepsilon \right)$$

$$\approx \frac{1}{d} \sum_{\|Z\| < \varepsilon} ZZ^T$$

$$= \frac{1}{d} \sum_{\|x_i - x_j\| < \varepsilon} \left(x_i - x_j \right) \left(x_i - x_j \right)^T$$

$$= \frac{1}{d} \sum_{i,j} \left(x_i - x_j \right) \left(x_i - x_j \right)^T S_{ij}$$

$$= \frac{1}{d} \left(\sum_{i,j} x_i x_i^T S_{ij} + \sum_{i,j} x_j x_j^T S_{ij} - \sum_{i,j} x_i x_j^T S_{ij} - \sum_{i,j} x_j x_i^T S_{ij} \right)$$

$$= \frac{2}{d} \left(\sum_{i,j} x_i x_i^T D_{ii} - \sum_{i,j} x_i x_j^T S_{ij} \right)$$

$$= \frac{2}{d} \left(XDX^T - XSX^T \right)$$

$$= \frac{2}{d} XLX^T \tag{17}$$

where $L = D - S$ is the Laplacian matrix. The *ith* column of matrix X is x_i.

8. THEORETICAL ANALYSIS OF LPP, PCA AND LDA

In this section, we present a theoretical analysis of LPP and its connections to PCA and LDA.

8.1. Connections to Pca

It is worthwhile to point out that XLX^T is the data covariance matrix, if the Laplacian matrix L is

$$\frac{1}{n}I - \frac{1}{n^2}ee^T \tag{18}$$

where n is the number of data points, I is the identity matrix and e is a column vector taking 1 at each entry. In fact, the Laplacian matrix here has the effect of removing the sample mean from the sample vectors.

In this case, the weight matrix S takes $1/n^2$ at each entry, i.e

$$S_{ij} = 1/n^2, \forall i, j \tag{19}$$

$$D_{ii} = \sum_j S_{ji} = 1/n \tag{20}$$

Hence the Laplacian matrix is

$$L = D - S = \frac{1}{n}I - \frac{1}{n^2}ee^T \tag{21}$$

Let m denote the sample mean i.e.

$$m = 1/n \sum_i x_i \tag{22}$$

we have

$$XLX^T = \frac{1}{n}X\left(I - \frac{1}{n^2}ee^T\right)X^T$$

$$= \frac{1}{n}XX^T - \frac{1}{n^2}(X_e)(X_e)^T$$

$$= \frac{1}{n}\sum_i x_i x_i^T - \frac{1}{n^2}(nm)(nm)^T$$

$$= \frac{1}{n}\sum_i (x_i - m)(x_i - m)^T + \frac{1}{n}\sum_i x_i m^T + \frac{1}{n}\sum_i mx_i^T - \frac{1}{n}\sum_i mm^T - mm^T$$

$$= E[(x - m)(x - m)]^T + 2mm^T - 2mm^T$$

$$= E[(x - m)(x - m)]^T$$

(23)

Where $E[(x - m)(x - m)]^T$, is just the covariance matrix of the data set

The above analysis shows that the weight matrix S plays a key role in the LPP algorithm. When we aim at preserving the global structure, we take ε (or k) to be infinity and choose the eigenvectors (of the matrix XLX^T) associated with the largest eigenvalues. Hence the data points are projected along the directions of maximal variance. ε should be sufficiently small to preserve the local structure and choose the Eigen vectors associates with smallest Eigen values.

Hence the data points are projected along the directions preserving locality. It is important to note that, when ε (or k) is sufficiently small, the Laplacian matrix is no longer the data covariance matrix, and hence the directions preserving locality are not the directions of minimal variance. In fact, the directions preserving locality are those minimizing *local* variance.

8.2. Connections to Lda

LDA seeks directions that are efficient for discrimination. The projection is found by solving the generalized Eigen value problem

$$S_B w = \lambda S_W w$$

(24)

where S_B and S_W are between and within class scatter matrices. Suppose there are l classes. The ith class contains ni sample points. Let m(i) denote the average vector of the ith class. Let x(i) denote the random vector associated tothe i^{th} class and) ($i\,j$ x denote the jth sample point in the ith class. We can rewrite the matrix S_W as follows:

$$S_W = \sum_{i=1}^{l}\left(\sum_{j=1}^{n_i} \left(x_j^{(i)} - m^{(i)} \right)\left(x_j^{(i)} - m^{(i)} \right)^T \right)$$

$$= \sum_{i=1}^{l}\left(\sum_{j=1}^{n_i} \left(x_j^{(i)}\left(x_j^{(i)}\right)^T - m^{(i)}\left(m^{(i)}\right)^T - x_j^{(i)}\left(m^{(i)}\right)^T + m^{(i)}\left(m^{(i)}\right) \right.\right.$$

$$= \sum_{i=1}^{l}\left(\sum_{j=1}^{n_i} \left(x_j^{(i)}\left(x_j^{(i)}\right)^T - n_i m^{(i)}\left(m^{(i)}\right)^T \right)\right)$$

$$= \sum_{i=1}^{l}\left(X_i X_i^T - \frac{1}{n_i}\left(x_1^{(i)} + ... + x_{ni}^{(i)}\right)\left(x_1^{(i)} + ... + x_{ni}^{(i)}\right)^T \right)$$

$$= \sum_{i=1}^{l}\left(X_i X_i^T - \frac{1}{n_i} X_i\left(e_i e_i^T\right) X_i^T \right)$$

$$= \sum_{i=1}^{l} X_i L_i X_i^T$$

(25)

Where,

$X_i L_i X_i^T$ is the data covariance matrix of the ith class and $X_i = [\, X_1^{(i)}, X_2^{(i)}, X_3^{(i)}, ... X_{ni}^{(i)} \,]$ is a d x n_i matrix.

$L_i = I - 1/n_i e_i e_i^T$ is a n_i x n_i is a ni x nini x ni matrix where I is the identity matrix and $e_i(1,1,1....1)^T$ is an ni dimensional vector.

To further simplify the above equation, we define

$$W_{ij} = \begin{cases} 1/n_k & \text{if } x_i \text{ and } x_j \text{ both belong to the kth class} \\ 0 & \end{cases}$$

$$\text{otherwise,} \tag{26}$$

It is interesting to note that we could regard the matrix W as the weight matrix of a graph with data points as its nodes. Specifically, W_{ij} is the weight of the edge (x_i, x_j). W reflects the class relationships of the data points. The matrix L is thus called *graph Laplacian*, which plays key role in LPP.

Similarly, we can compute the matrix SB as follows:

$$S_B = \sum_{i=1}^{l} n_i \left(m^{(i)} - m \right) \left(m^{(i)} - m \right)^T$$

$$= \left(\sum_{i=1}^{l} n_i m^{(i)} \left(m^{(i)} \right)^T \right) - m \left(\sum_{i=1}^{l} n_i \left(m^{(i)} \right)^T \right) - \left(\sum_{i=1}^{l} n_i m^{(i)} \right) m^T + \left(\sum_{i=1}^{l} n_i \right) mm^T$$

$$= \sum_{i=1}^{l} \left(\frac{1}{n_i} \left(x_1^{(i)} + \ldots + x_{ni}^{(i)} \right) \left(x_1^{(i)} + \ldots + x_{ni}^{(i)} \right)^T \right) - 2nmm^T + nmm^T$$

$$= \left(\sum_{i=1}^{l} \sum_{j,k=1}^{n_i} \frac{1}{n_i} x_j^{(i)} \left(x_j^{(i)} \right)^T \right) - 2nmm^T + nmm^T$$

$$= XWX^T - 2nmm^T + nmm^T$$

$$= XWX^T - nmm^T \tag{27}$$

$$= XWX^T - X \left(\frac{1}{n} ee^T \right) X^T$$

$$= X \left(W - \frac{1}{n} ee^T \right) X^T$$

$$= X \left(W - I + I - \frac{1}{n} ee^T \right) X^T$$

$$= -XLX^T + X \left(I - \frac{1}{n} ee^T \right) X^T$$

$$= -XLX^T + C$$

where $e = (1,1,...,1)^T$ is a n dimensional vector and $C = X\left(I - \frac{1}{n}ee^T\right)X^T$ is the data covariance matrix. Thus, the generalized eigenvector problem of LDA can be written as follows:

$$S_B w = \lambda S_W w$$

$$\Rightarrow \left(C - XLX^T\right)w = \lambda XLX^T w$$

$$\Rightarrow Cw = (1 + \lambda) XLX^T w$$

$$\Rightarrow XLX^T w = \frac{1}{1+\lambda} Cw$$

(28)

Thus, the projections of LDA can be obtained by solving the following generalized eigenvalue problem,

$$\Rightarrow XLX^T w = \lambda Cw$$

(29)

The optimal projections correspond to the eigenvectors associated with the smallest eigenvalues. If the sample mean of the data set is zero, the covariance matrix is simply XX^T which is close to the matrix XDX^T in the LPP algorithm. Our analysis shows that LDA actually aims to preserve discriminating information and global geometrical structure. Moreover, LDA has a similar form to LPP. However, LDA is supervised while LPP can be performed in either supervised or unsupervised manner.

8.3. Learning Laplacian Faces for Representation

LPP is a general method for manifold learning. It is obtained by finding the optimal linear approximations to the eigenfunctions of the Laplace Betrami operator on the manifold. Therefore, though it is still a linear technique, it seems to recover important aspects of the intrinsic nonlinear manifold structure by preserving local structure. Based on LPP, Laplacianfaces method for face representation is a locality preserving subspace. In the face analysis and recognition problem one is confronted with the difficulty that the matrix XDX^T is sometimes singular. This stems from the fact that sometimes the number of images in the training set (n) is much smaller than the number of pixels in each image (m). In such a case, the rank of XDX^T is at most n, while XDX^T is an $m \times m$ matrix, which implies that XDX^T is singular. To overcome the complication of a singular XDX^T, we first project the image set to a PCA subspace so that the resulting matrix XDX^T is nonsingular. Another consideration of using PCA as preprocessing is for noise reduction. This method, we call*Laplacianfaces*, can learn an optimal subspace for face representation and recognition.

The algorithmic procedure of Laplacianfaces is formally stated below:

1. PCA projection: We project the image set {xi} into the PCA subspace by throwing away the smallest principal components.

2. Constructing the nearest-neighbor graph: Let G denote a graph with n nodes. The *ith* node corresponds to the face image xi. We put an edge between nodes i and j if xi and xj are "close", i.e. xi is among $k$nearest neighbors of xi or xi is among k nearest neighbors of xj. The constructed nearest neighbor graph is an approximation of the local manifold structure. Note that, here we do not use the ε - neighborhood to construct the graph. This is simply because it is often difficult to choose the optimal ε in the real world applications, while k nearest neighbor graph can be constructed more stably. The disadvantage is that the k nearest neighbor search will increase the computational complexity of our algorithm. When the computational complexity is a major concern, one can switch to the ε -neighborhood.

3. Choosing the weights: If node i and j are connected, put

$$S_{ij} = e^{\frac{\left\| x_i - x_j \right\|^2}{t}}$$

(30)

where t is a suitable constant. Otherwise, put $S_{ij} = 0$. The weight matrix S of graph G models the face manifold structure by preserving local structure.

4. Eigenmap: Compute the eigenvectors and eigenvalues for the generalized eigenvector problem:

$$XLX^T w = \lambda XDX^T w$$

(31)

where D is a k-dimensional vector. W is the transformation matrix. This linear mapping best preserves the manifold's estimated intrinsic geometry in a linear sense. The column vectors of W are the so called *Laplacianfaces*.

9. FACE REPRESENTATION USING LAPLACIANFACES

As we described previously, a face image can be represented as a point in image space. A typical image of size $m \times n$ describes a point in $m \times n$-dimensional image space. However, due to the unwanted variations resulting from changes in lighting, facial expression, and pose, the image space might not be an optimal space for visual representation, we have discussed how to learn a locality preserving face subspace which is insensitive to outlier and noise. The images of faces in the training set are used to learn such a locality preserving subspace. The subspace is spanned by a set of eigenvectors of equation (1), i.e. w0, w1, ..., wk-1.

Eigenmaps are obtained from the generalized eigenvector problem as $ALA^T a = \lambda ADA^T a$ where D is a diagonal matrix whose entries are column or row, since W is symmetric sums of W, $D_{ii} = \Sigma j W_{ji}$., L = D -W is the Laplacian matrix is equivalent nonlinear Laplace Beltrami opearator. The ith column of matrix A is xi. Let the column vectors a_0; _ _ _ ; a_{l-1} be the solutions of equation (), ordered according to their eigenvalues, in ascending order Thus, the embedding is as follows: yi = $E^T x_i$; E = (a_0; a_1; _ _ _ ; a_{l-1}) where y_i is a l-dimensional vector, and E is a n x l matrix.. The y_i represent the Laplacian faces.

9.1. Basis Selection From LPP

Locality information can be preserved by the following transformation on A, the input face space $[z_1.... Z_z] = \Pi_{<1..z}(A^T L A)$ Where L =D-W gives the Laplacian matrix. D is the diagonal matrix and W is the weight matrix of the K nearest neighbors clustering.

Basis for the face space is obtained as, $B = [X_1 X_2....X_x Y_1 Y_2......Y_y Z_1 Z_2.....Z_z]$, such that

$$x + y + z = \frac{M}{3}$$

(32)

And

$$x + y + z = \frac{2M}{3}$$

(33)

where M is the dimension of the original face space

9.2. Projection Onto Reduced Subspace

Each face in the training set Φ_i can be represented as a linear combination of these vectors, $U_i \; \varepsilon \; B , 1 \le i \le K$ such that $\Phi_i = \sum_{j=1}^{k} w_j u_j$, where ujuj's are Eigenfaces. These weights are calculated as: $w_j = u_j^T \Phi_i \Omega_i = [w_1 \; w_2...w_k]$ i.e. the orthogonal projection of a face vector on each basis vector.

Figure 8. Laplacian Faces from LPP

10. INDEPENDENT COMPONENT ANALYSIS

Independent component analysis (ICA) is a statistical method, the goal of which is to decompose multivariate data into a linear sum of non-orthogonal basis vectors with coefficients (encoding variables, latent variables, hidden variables) being statistically independent.

ICA generalizes a widely-used subspace analysis method such as principal component analysis (PCA) and factor analysis, allowing latent variables to be non-Gaussian and basis vectors to be non-orthogonal in general. ICA is a density estimation method where a linear model is learned such that the probability distribution of the observed data is best captured, while factor analysis aims at best modeling the covariance structure of the observed data.

The ICA model is a generative model, which means that it describes how the observed data are generated by a process of mixing the components s_i. The independent components are latent variables, meaning that they cannot be directly observed. Also the mixing matrix is assumed to be unknown. All we observe is the random vector X, and we must estimate both A and S using it. This must be done under as general assumptions as possible. The starting point for ICA is the very simple assumption that the components S_i are statistically *independent*. It will be seen below that we must also assume that the independent component must have *nongaussian* distributions. However, in the basic model we do *not* assume these distributions known (if they are known, the problem is considerably simplified.) For simplicity, we are also assuming that the unknown mixing matrix is square, but this assumption can be sometimes relaxed. Then, after estimating the matrix A, we can compute its inverse, say W, and obtain the independent component simply by: s=Wx

ICA is very closely related to the method called *blind source separation* (BSS) or blind signal separation. A "source" means here an original signal, i.e. independent component, like the speaker in a cocktail party problem. "Blind" means that we know very little, if anything, on the mixing matrix, and make little assumptions on the source signals. ICA is one method, perhaps the most widely used, for performing blind source separation.

The task of ICA is to estimate the mixing matrix A or its inverse $W = A^{-1}$ such that elements of the estimate $y = A^{-1}x = Wx$ are as independent as possible. For the sake of simplicity, we often leave out the index t if the time structure does not have to be considered.

PCA makes one important assumption: the probability distribution of input data must be Gaussian. When this assumption holds, covariance matrix contains all the information of (zero-mean) variables. Basically, PCA is only concerned with second-order (variance) statistics. The mentioned assumption need not be true. If we presume that face images have more general distribution of probability density functions along each dimension, the representation problem has more degrees of freedom. In that case PCA would fail because the largest variances would not correspond to meaningful axes of PCA.

$$x_i(t) = a_i1 {}^* s_1(t) + a_i2 {}^* s_2(t) + a_i3 {}^* s_3(t) + a_i4 {}^* s_4(t) \ldots \qquad (34)$$

Here, i = 1:4.
In vector-matrix notation, and dropping index t, this is

$$x = A\, s \qquad (35)$$

$$s = A^{-1}\, x \qquad (36)$$

$$s = W\, x \qquad (37)$$

$$W = A^{-1} \qquad (37)$$

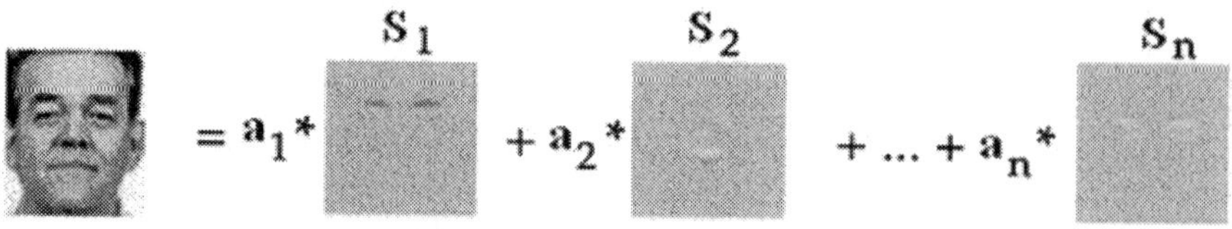

Figure 9. Mixture Matrix forming face

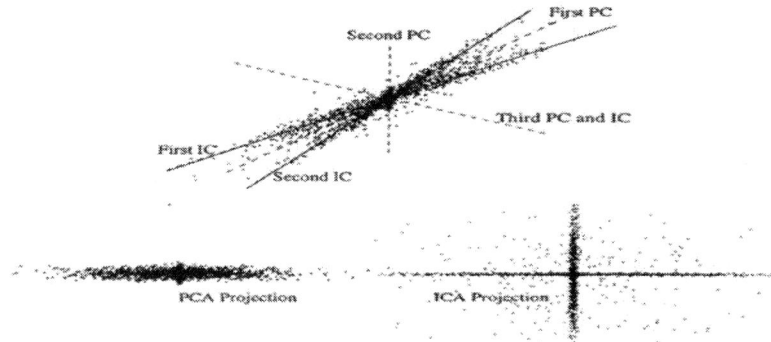

Figure 10. Different Principal Component(PC) directions & PCA vs. ICA Projections

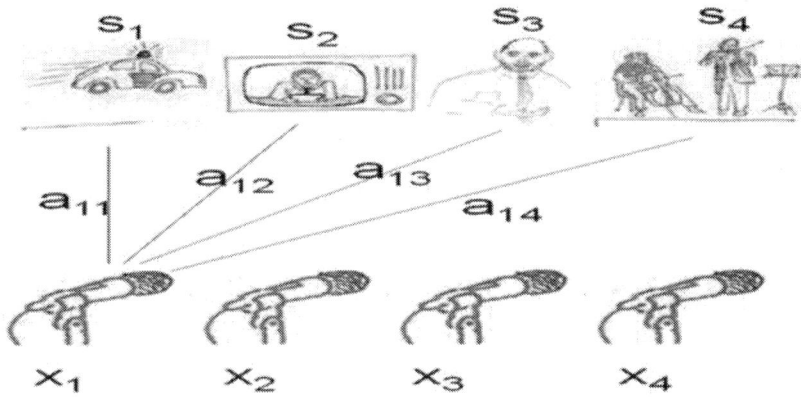

Figure 11. x=As (Blind Source Separation)

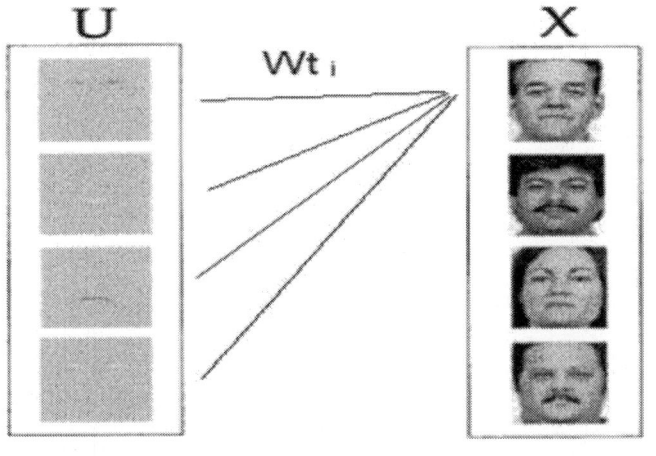

Figure 12. Construction of face from Basis

  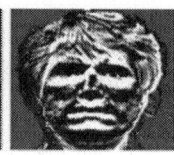  

FIGURE 13. Basis Images from ICA

11. RANDOM PROJECTIONS

There has been a strong trend lately in face processing research away from geometric models towards appearance models. Appearance-based methods employ dimensionality reduction to represent faces more compactly in a low-dimensional subspace which is found by optimizing certain criteria. Recently, Random Projection (RP) has emerged as a powerful method for dimensionality reduction. It represents a computationally simple and efficient method that preserves the structure of the data without introducing significant distortion. D O ($\log n / \varepsilon^2$) dimensional subspace such that the distances between the points are approximately preserved.

Transforms Γ_i to a lower dimension d, with d<<p via the following transformation:

$\Gamma_i = R \ \Gamma_i$ where R is orthonormal and its columns are realizations of independent and identically distributed (i.i.d.) zero-mean normal variables, scaled to have unit length. RP is motivated by the *Johnson-Lindenstrauss* lemma that states that a set of M points in a high dimensional.

Euclidean space can be mapped down onto a $d \geq O$ ($\log n / \varepsilon^2$) dimensional subspace such that the distances between the points are approximately preserved.

The main reason for orthogonalizing the random vectors is to preserve the similarities between the original vectors in the low-dimensional space. In high enough dimensions, however, it is possible to save computation time by avoiding the orthogonalization step without affecting much the quality of the projection matrix. This is due to the fact that, in high-dimensional spaces, there exist a much larger number of almost orthogonal vectors than orthogonal vectors. Thus, high-dimensional vectors having random directions are very likely to be close to orthogonal.

12. MIXTURE OF COMPONENTS

One can use different ratios of feature vectors drawn from SVD, DF-LDA & LPP Techniques. The first step can be normalizing the images in the training set to compensate for the illumination effects. These processed images should be subjected to dimensionality reduction using each of the methods mentioned in the chapter. Basis selection can be carried out using these independent sets of dimension reduced vectors in different proportions aimed at enhancing the efficiency and accuracy of recognition task. Below is a sample example mentioned with two trials, one with for $1/3^{rd}$ dimensionality reduction and another with $2/3^{rd}$ reduction. In each of the trials, several iterations are performed by taking different combinations of the feature vectors. The iterations will converge when the desired precision of recognition rate is obtained.

12.1. Example

12.1.1. Preprocessing

The Face Space: For the recognition task, each m X n I_i image in the training set is transformed into a column vector of mn components. A matrix S (mn X M) is constructed such that $S = [I_1 I_2 \dots I_M]$, where M is number of face images in training set It is found that all N vectors are linearly independent, which implies that the range space of matrix S is the entire region spanned by the columns of S. i.e Range space of S R(S)=[S]

Normalization: Normalize the images,to reduce illumination effects and lighting conditions as,

$$Ai = (\Phi - \mu_i) X \frac{\delta'}{\delta} + \mu'$$

(39)

For i=1,2,3....., M

Where,

$$\mu_i = \frac{1}{N^2} \sum_{j=1}^{N^2} x_j$$

(40)

$$\mu'_i = \frac{1}{M} \sum_{i=1}^{M} \frac{1}{N^2} \sum_{j=1}^{N^2} A_{ij}$$

(41)

$$\delta_i = \sqrt[2]{\frac{1}{N^2 - 1} \sum_{j=1}^{N^2} (x_j - \mu_i)}$$

(42)

$$\delta'_i = \frac{1}{N^2} \sum_{i=1}^{N^2} \sqrt[2]{\frac{1}{N^2 - 1} \sum_{j=1}^{N^2} A_{ji}}$$

(43)

12.1.2. Basis Selection

Recognition Task: Unknow probeface is normalized (ϕ) and projected on to the subspace to get weight for the probe image $w_i = u_j^T \Phi$ Euclidean Distance measure is used in lassification given by $e_r = \min \|\Omega - \Omega_i\|$. And if $e_r < \Theta$ where Θ is a threshold chosen heuristically, then we the probe image is recognized as the image with which it gives the lowest score. If however $e_r > \Theta$ then the probe does not belong to the database.

Deciding on the Threshold: A set of 150 known images other than the ones in the data set is used in the computation of threshold given by $\theta = \mu + \eta\sigma$. Where,

$$\mu = \frac{1}{N} \sum_{i=0}^{N-1} x_i$$

(44)

$$\sigma^2 = \frac{1}{N-1} \sum_{i=0}^{N-1} \left(x_j - \mu \right)^2$$

(45)

$\eta \in I$ is chosen according to level of precision required in the results. $x_i \in \gamma$

The method of choosing right combination of right proportion of feature vectors has been applied on a large database consisting of a variety of still images with illumination, expression variations as well as partially occluded images. The ratio 3:2:5:: SVD:DF-LDA:LPP has yielded highest accuracy in recognition. The example is tried on a total test set of 165 images drawn from YALE dataset and the training set consisting 15 classes having a class count of five images.

An ROC graph is plotted to visualize and analyze the working of face recognition efficiency. It is a two dimensional graph in which TP rate, true positive rate, is plotted on the Y axis and FP rate, false positive rate, is plotted on the X axis. Given a set of test images a two by two contingency table is constructed representing the dispositions of the set of images.

Table 1. Iterations for subspace of dimension M/3

SVD (no. of vectors)	DFLDA (no. of vectors)	LPP (no. of vectors)	EFFICIENCY (in %)
15	5	5	80.00
5	5	15	81.21
8	9	8	81.81
15	5	15	87.27
5	15	5	81.21

Table 2. Comparative results with Iterartion Trial of M/3

Graph No.	True Positive	False Negative	False Positive	True Negative
1	122	28	5	10
2	123	27	4	11
3	125	15	5	10
4	132	18	3	12
5	122	28	3	12

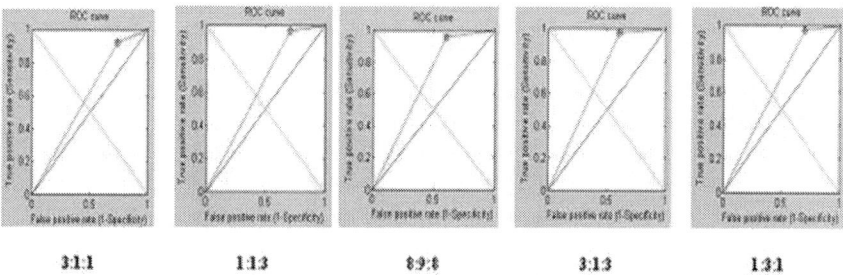

| 3:1:1 | 1:1:3 | 8:9:8 | 3:13 | 13:1 |

Figure 14. Fig. 14. ROC's Indicating the True Positive VS False Positive for M/3

TABLE 3. Iterations for subspace of dimension 2M/3

SVD (no. of vectors)	DFLDA (no. of vectors)	LPP (no. of vectors)	EFFICIENCY (in %)
30	10	10	84.24
10	10	30	85.45
20	15	15	86.67
25	15	10	84.84
15	10	25	92.12

Table 4. Comparative results with Iterartion Trial of M/3

Graph No.	True Positive	False Negative	False Positive	True Negative
1	129	21	5	10
2	132	18	4	11
3	133	17	5	10
4	128	32	4	11
5	140	10	3	12

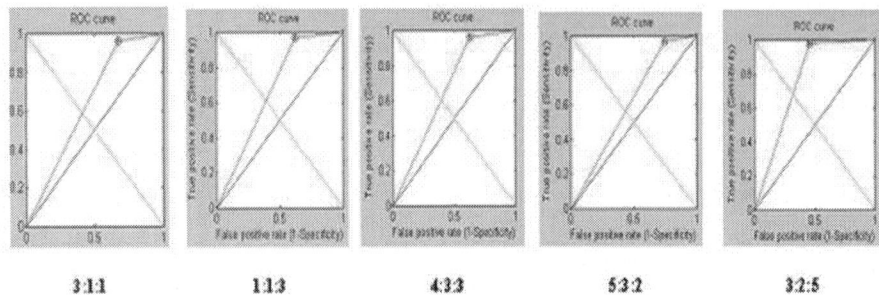

Figure 15. Fig. 15. ROC's Indicating the True Positive VS False Positive for 2M/3

13. CONCLUSION

In this chapter several linear and non linear dimensionality reduction techniques were discussed from the perspective of face recognition. Since the face images contain several characteristic features both global and local, using any one method alone may not yield better recognition accuracy. It may be good to have combinations of the basis vectors from several approaches to achieve higher accuracy. Underlying manifold structure in image space will get face subspace and is possible with LPP, ICA methods. More pronounced features can be drawn from the space in case of LDA based algorithms. Random and PCA projections give appearance models which are holistic in nature.

Future of face recognition can also look at increase in dimension like depth information for recognition purposes. Algorithmic models should aim at addressing scale invariance feature vectors which can hopefully solve recognition task even under extreme variations in images.

The approach to face recognition was motivated by information theory, leading to the idea of basing face recognition on a small set of image features that best approximate the set of known face images, without requiring that they correspond to our intuitive notions of facial parts and features. The approach does provide a practical solution to the problem of face recognition and is relatively simple and has been shown that it can work well in a constrained environment. Anecdotal experimentation with acquired image sets indicates that profile size, complexion, ambient lighting and facial angle play significant parts in the recognition of a particular image.

REFERENCE

1. R. Hietmeyer, Biometric identification promises fast and secure processing of airline passengers.The International Civil Aviation Organization Journal,55 9 1011 .

2. E. Michael, Andreas. Wall, Luis. M. Rechtsteiner, ". Rocha, Value. Singular, And. Decomposition, Component. Principal, A. Analysis", Approach. Practical, Microarray. to, Analysis. Data, Chapter, 2003 91109 ,

3. M. A. Turk, A. Pentland, ". Eigenfaces, Recognition". for, of. Journal, Neuro-science. Cognitive, vol, 3, 7186 , 1991

4. Xiaofei He, Partha Niyogi, " Locality Preserving Projections", IN the proceedings of Advances in Neural Information Processing Systems, 2003.

5. Juwei Lu, K.N. Plataniotis, and A.N. Venetsanopoulos, "Face Recognition Using LDA Based Algorithms", IEEE Transactions on Neural Networks, VOL. 2003 1 195200 ,.

6. Yu, H., Yang, J.,"A Direct LDA #algorithm for High-Dimensional Data with Application to Face Recognition. Pattern Recognition, 34 34 20672070 ., 2001

7. Roweis. Sam, K. Lawrence, ". Saul, Dimensionality. Nonlinear, by. Reduction, Linear. Locally, In. Embedding,", proceedings. the, Science. of, Journal, 290 23232326 , 2000.

8. B. Joshua, Tenenbaum, Silva. Vin de, C. John, Langford,", Langford,"A Global Geometric Framework for Nonlinear Dimensionality Reduction", In the proceedings of Science Journal, 2000290 23192323 ,.

9. Xiaofei He, Shuicheng Yan, Yuxiao Hu, Partha Niyogi, and Hong-Jiang Zhang, "Face Recognition Using Laplacianfaces", In the proceedings of IEEE Transactions on Pattern Analysis and Machine Intelligence 2005 3 328340 ,

10. M. W. Berry, S. T. Dumais, G. W. O'Brian, 1995 Using Linear Algebra for Intelligent Information Retrieval., In the proceedings of SIAM Review Journal, 37(4).573595 , 1995.

11. Liu, S. Chen, Discriminant common vecotors versus neighbourhood components analysis and laplacianfaces: A comparative study in small sample size problem. Image and Vision Computing, 3 24 249262 2006

12. Navin Goela, George Bebisa, Ara Nefianb "Face Recognition Experiments with Random Projection"

13. Muhammad Imran Razzak, Muhammad Khurram Khan, Khaled Alghathbar, Rubiyah Yousaf, "Face Recognition using Layered Linear Discriminant Analysis and Small Subspace", In 10 10th IEEE International Conference on Computer and Information Technology (CIT 2010), 14071412 , 2010.

14. MaxWelling, "Fisher Linear Discriminant Analysis", Classnotes in Machine Learning.

15. Kari Torkkola, "Discriminative Features for Text Document Classification", In the proceedings of International Conference on Pattern Recognition, pp. 472475 .

16. Shermina.J, "Application of Locality Preserving Projections in Face Recognition", In the proceedings of International Journal of Advanced Computer Science and Applications 2010 3 8285 September

17. Deng Cai, Xiaofei He, Jiawei Han, Hong-Jiang Zhang," Orthogonal Laplacianfaces for Face Recognition ", In the proceedings of IEEE Transactions on Image Processing,pp.1-7 2010

18. S.Sakthivel, R.Lakshmipathi, "Enhancing Face Recognition Using Improved Dimensionality Reduction And Feature Extraction Algorithms-An Evaluation With ORL Database", International Journal of Engineering Science and Technology 2010 22882295 ,.

19. Ruba Soundar Kathavarayan,, Murugesan Karuppasamy, "Preserving Global and Local Features for Robust Face Recognition under Various Noisy Environments", In the proceedings of International Journal of Image Processing (IJIP) Volume(3), Issue(6), pp. 328340 .

20. Nain. Neeta, Gour. Prashant, Agarwal. Nitish, P. Rakesh, Subhash. Talawar, ". Chandra, Recognition. Face, P. C. A. using, L. D. A. with, Value. Singular, S. V. D. Decomposition, . D. L. D. using, 2008 Proceedings of the World Congress on Engineering Vol I, 978-9-88986-719-5 14 ,.

21. Tat-Jun Chin, Konrad Schindler, David Suter, "Incremental Kernel SVD for Face Recognition with Image Sets", In the proceedings of Seventh IEEE International Conference on Automatic Face and Gesture Recognition pp.461-466, 2006.

22. Stewart. Marian, Bartlett, R. Javier, Terrence. J. Movellan, ". Sejnowski, Recognition. Face, Independent. by, Analysis". I. E. E. E. Component, on. Transactions, Networks. Neural, 2002 6 , 14501464 ,.

23. K.J. Karande, S.N. Talbar, 'Independent Component Analysis of Edge Information for Face Recognition', In the Proceedings of International Journal of Image Processing 3 2009 3 120-130,.

24. M. Belkin, P. Niyogi, "Using Manifold Structure for Partially Labeled Classification", In the Proceedings of Conference on Advances in Neural Information Processing System, 2002.

25. W. Zhao, R. Chellappa, P.J. Phillips, 'Subspace Linear Discriminant Analysis for Face Recognition', Technical Report CAR-TR-914, Center for Automation Research, University of Maryland, 1999.

26. P. C. Yuen and J. H. Lai, "Independent Component Analysis of Face Images,", In the proceedings of IEEE Workshop Biologically Motivated Computer Vision, Seoul, Korea, 2000.

27. M. Martinez, A. C. Kak, ". P. C. A. versus, L. D. A,", the. In, of. I. E. E. E. proceedings, on. Transaction, Analysis. Pattern, Intelligence. Machine, vol, 2001 2 228233 ,.

28. T. Shakunaga and K. Shigenari, "Decomposed Eigenface for Face Recognition under Various Lighting Conditions," In the proceedings of IEEE Conference on Computer Vision and Pattern Recognition, 2001.
29. Zhonglong Zheng, Fan Yang, Wenan Tan, Jiong Jia and Jie Yang, "Gabor feature-based face recognition using supervised locality preserving projection", In the proceedings of Signal Processing, Vol. 87, 10 2007 24732483 ,.

CHAPTER 6

Efficiency of Recognition Methods for Single Sample per Person Based Face Recognition

Miloš Oravec, Pavlovičová, Ján Mazanec, Ľuboš Omelina, Matej Féder and Jozef Ban

Faculty of Electrical Engineering and Information Technology Slovak University of Technology in Bratislava, Slovakia

1. INTRODUCTION

Even for the present-day computer technology, the biometric recognition of human face is a difficult task and continually evolving concept in the area of biometric recognition. The area of face recognition is well-described today in many papers and books, e.g. (Delac et al., 2008), (Li & Jain, 2005), (Oravec et al., 2010). The idea that two-dimensional still-image face recognition in controlled environment is already a solved task is generally accepted and several benchmarks evaluating recognition results were done in this area (e.g. Face Recognition Vendor Tests, FRVT 2000, 2002, 2006, http://www.frvt.org/). Nevertheless, many tasks have to be solved, such as recognition in unconstrained environment, recognition of non-frontal images, single sample per person problem, etc.

This chapter deals with single sample per person face recognition (also called one sample per person problem). This topic is related to small sample size problem in pattern recognition. Although there are also advantages of single sample – fast and easy creation of a face database and modest requirements for storage, face recognition methods usually fail to work if only one training sample per person is available.

In this chapter, we concentrate on the following items:

- Mapping the state-of-the-art of single sample face recognition approaches after year 2006 (the period till 2006 is covered by the detailed survey (Tan et al., 2006)).
- Generating new face patterns in order to enlarge the database containing single samples per subject only.

Such approaches can include modifications of original face samples using e.g. noise, mean filtering, suitable image transform (forward transform, then neglecting some coefficients and image reconstruction by inverse transform), or generating synthetic samples by ASM (active shape method) and AAM (active appearance method).

• Comparing recognition efficiency using single and multiple samples per subject.

We illustrate the influence of number of training samples per subject to recognition efficiency for selected methods. We use PCA (principal component analysis), MLP (multilayer perceptron), RBF (radial basis function) network, kernel methods and LBP (local binary patterns). We compare results using single and multiple training samples per person for images taken from FERET database. For our experiments, we selected large image set from FERET database.

• Highlighting other relevant important facts related to single sample recognition.

We analyze some relevant facts that can influence further development in this area. We also outline possible directions for further research.

2. FACE RECOGNITION BASED ON A SINGLE SAMPLE PER PERSON

2.1. General Remarks

Generally, we can divide the face recognition methods into three groups (Tan et al., 2006): holistic methods, local methods and hybrid methods.

Holistic methods like PCA (eigenfaces), LDA (fisherfaces) or SVM need principally more image samples per person in the training phase. To solve the one sample problem there are basically two ways how to deal with it:

• To extend the classical methods to be trained from single sample more efficiently – e.g. 2D-PCA (Yang et al., 2004), (PC)2A (Wu & Zhou, 2002), E(PC)2A (Chen et al., 2004a), SPCA (Zhang, et al., 2005), APCA (Chen & Lovell, 2004), FLDA (Chen, et al., 2004b), Gabor+PCA+WSCM (Xie & Lam, 2006).

• To enlarge the training set by new representations or generating new views.

• Local methods can be divided into 2 groups:

• Local feature based, which mostly work with some type of graph spread over the face regions with corners in important face features – face recognition is formulated as a problem of graph matching. These methods deal with the one sample problem better than the typical holistic methods (Tan et al., 2006). EBGM (Elastic Bunch Graph Matching) or DCP (directional corner points) are examples of this type of methods.

- Local appearance-based methods extract information from defined local regions. The features are extracted by known methods for texture classification (Gabor wavelets, LBP, etc.) and the feature space is reduced by known methods like PCA or LDA.

An excellent introduction to the single sample problem and survey of related methods mapping state-of-the-art till 2006 is described and discussed in (Tan et al., 2006).

2.2. State-of-the-Art in Single Sample Per Person Face Recognition From 2006

After year 2006, new approaches were proposed. They are based mainly on enhancement of various conventional methods.

Principal Component Analysis (PCA) is still one of the most popular methods used to deal with one sample problem. Despite of its popularity, calculating of representative covariance matrix from one sample is very difficult task. In contrast to conventional application of PCA, 2DPCA (Yang et al. 2004) is based on two dimensional matrices, where the image does not need to be previously transformed into a 1D vector.

In (Que et al., 2008) a new face recognition algorithm MW(2D)2PCA was proposed. Modular Weighted (2D)2PCA (MW(2D)2PCA) is based on the study of (2D)2PCA. Weighting method (W) emphasizes the different influence of different eigenvectors and image blocking method (M) can extract detailed information of face image more effectively. Modularization of image into several blocks according to face elements provides more detailed information of face and assigns this approach rather to local appearance than holistic methods. The best recognition rate achieved by this method was 74.14%.

Similar approach, that deals with the single sample problem from human perception point of view, was proposed in (Zhan et al., 2009) where modularized image was processed by 2D DCT to extract features, instead of (2D)2PCA. Gabor filters can be applied even to the image divided into several areas to reduce illumination impact as it is shown in (Nguyen & Bai, 2009).

Standard way to solve single sample problem is to use local facial representations. Conventional procedure in local methods is face image partitioning into several segments. In (Akbari et al., 2010), an algorithm based on single image per person, with input images segmented into 7 partitions was proposed. The moment feature vectors of a definite order for all images are extracted and distance measure is used to recognize the person.

Another way to get better results of recognition is a fusion of more biometrics. In (Ma et al., 2009) a new multi-modal biometrics fusion approach was presented. They used face and palmprint biometrics and combined the normalized Gaborface and Gaborpalm images at the pixel level. They presented a kernel PCA plus RBF classifier (KPRC) to classify the fused images. Using both face and palmprint samples, the average recognition results were improved from 42.60% and 52.36% (single-modal biometrics) to 87.01% (multi-modal biometrics).

In (Xie & Lam, 2006) novel Gabor-based kernel principal component analysis with doubly nonlinear mapping for human face recognition was proposed. The algorithm is evaluated using 4 databases: Yale, AR, ORL and YaleB database. The best of the proposed variations of the algorithm GW+DKPCA get very good results even under varying lighting, expression and perspective conditions.

(Kanan & Faez, 2010) presents a new approach for face representation and recognition based on Adaptively Weighted Sub-Gabor Array (AWSGA). The proposed algorithm utilizes a local Gabor array to represent faces partitioned into sub-patterns. It employs an adaptively weighting scheme to weight the Sub-Gabor features extracted from local areas based on the importance of the information they contain and their similarities to the corresponding local areas in the general face image. Experiments on AR and Yale databases show, that the proposed method significantly outperforms eigenfaces and modular eigenfaces in most of the benchmark scenarios under both ideal conditions and varying expressions and lighting conditions and this method achieves better results under partial occlusion conditions than the local probabilistic approach.

A novel feature extraction method named uniform pursuit (UP) was proposed in (Deng et al., 2010). A standardized procedure on the large-scale FERET and FRGC databases was applied to evaluate the one sample problem. Experimental results show that the robustness, accuracy and efficiency of the proposed UP method can compete successfully with the state-of-the-art one sample based methods.

In (Qiao et al., 2010), a new graph-based semi-supervised dimensionality reduction algorithm called sparsity preserving discriminant analysis (SPDA) based on SDA was developed. Experiments on AR, PIE and YaleB databases show that proposed method outperforms the SDA method.

Solution for single sample problem based on Fisherface method on generic dataset was presented in (Majumdar & Ward, 2008). The method was also extended to multiscale transform domains like wavelet, curvelet and contourlet. Results on Faces94 and the AT&T database show, that this approach outperforms SPCA and Eigenface Selection methods. Best results came from the Pseudo-fisherface method in the wavelet domain.

In (Gao et al., 2008), a method based on singular value decomposition (SVD) was used to evaluate the within-class scatter matrix so that the FLDA could be applied for face recognition with only one sample image in training set. The experiments on FERET, UMIST, ORL and Yale databases show, that the proposed method outperforms other state-of-the-art methods like E(PC)2A, SVD perturbation and different FLDA implementations.

A novel local appearance feature extraction method based on multi-resolution Dual Tree Complex Wavelet Transform (DT-CWT) was presented in (Priya & Rajesh, 2010). Experiments with ORL and Yale databases show, that this method and its block-based modification get very good results under illumination, perspective and expression variations conditions compared to PCA and global DT-CWT, while keeping low computational complexity.

In (Tan & Triggs, 2010) original LBP method used for face recognition was extended. More efficient preprocessing was proposed to eliminate illumination variances using LTP (local ternary patterns) – generalization and enhancement of the original LBP texture descriptor. By replacing the local histogram with a distance transform based similarity metrics the performance of the LBP/LTP face recognition was further improved. Experiments under difficult lighting conditions with Face Recognition Grand Challenge, Extended Yale-B, and CMU PIE databases provide results comparable to up to date methods.

Another extension of the LBP algorithm was presented in (Lei et al., 2008). The face image is first decomposed by multi-scale and multi-orientation Gabor filters. Local binary pattern analysis is then applied on the derived Gabor magnitude responses. Using FERET database with 1 image per person in the gallery, the method achieved results outperforming LBP, PCA and FLDA. To improve the recognition accuracy, it helps to add some synthetic samples of subject to the learning process. Standard procedures to create synthetic samples are the parallel deformation method (generate novel views of a single face image under different poses) (Tan et al., 2006), modification by noise or filtering original images. In (Xu & Yang, 2009) the feature extraction technique called Local Graph Embedding Discriminant Analysis(LGEDA) was proposed, where the imitated images were generated using a mean filter.

In (Su et al., 2010) an Adaptive Generic Learning (AGL) method was described. To better distinguish the persons with single face sample, a generic discriminant model was adopted. As a specific implementation of the AGL, a Coupled Linear Representation (CLR) algorithm was proposed to infer, based on the generic training set, the within-class scatter matrix and the class mean of each person given its single enrolled sample. Thus, the traditional Fisher's Linear Discriminant (FLD) can be applied to one sample problem task. Experiments are taken on images from FERET, XM2VTS, CAS-PEAL databases and a private passport database. The results show, that the Adaptive Gabor-FLD outperforms other methods like E(PC)2A, LBP and other FLD implementations. The proposed method is related to methods using virtual sample generation although it does not explicitly generate any virtual sample.

3. FACE RECOGNITION METHODS

We use various methods in order to deeply explore the behavior of face recognition methods for single sample problem and to compare the methods using multiple face samples - both real-world samples and virtually generated samples. Used methods are briefly introduced in this subchapter.

3.1. Methods Based On Principal Component Analysis - Pca (Pca, 2d Pca And Kpca)

3.1.1. Principal Component Analysis – Pca

One of the most successful techniques used in face recognition is principal component analysis (PCA). The method based on PCA is named eigenface and was pioneered by Turk and Pentland (Turk & Pentland, 1991). In this method, each input image must be transformed into one dimensional image vector and set of these vectors forms input matrix. So the main idea behind PCA is that each n-dimensional face image can be represented as a linearly weighted sum of a set of orthonormal basis vectors.

This standard statistical method can be used for feature extraction. Principal component analysis reduces the dimension of input data by a linear projection that maximizes the scatter of all projected samples (Bishop, 1995).

For classification of projected samples Euclidean distance or other metrics can be used. Mahalanobis Cosine (MahCosine) is defined as the cosine of the angle between the image vectors that were projected into the PCA feature space and were further normalized by the variance estimates (Beveridge et al., 2003).

3.1.2. Two-Dimensional Pca – 2d Pca

PCA is well-known feature extraction method mostly used as a baseline method for comparison purpose. Several extensions of PCA have been proposed. A major problem of using PCA lies in computation of covariance matrix what is computationally expensive. This computation can be significantly reduced by computing PCA features for columns (or rows) without previous matrix-to-vector conversion. This approach is also called two dimensional PCA (Yang et al., 2004). Main idea behind 2D PCA is the projection of image columns (rows) onto covariance matrix computed as the average of covariance matrices of each column for all training images. Let AA be an m by n image matrix and average image $\overline{A}$ defined as $A = 1/M \sum_k A_k$, where M is number of all k training images. Then covariance matrix can be calculated by

$$G = \frac{1}{M} \sum_{k=1}^{M} \sum_{i=1}^{m} (A_k^{(i)} - \overline{A}^{(i)})^T (A_k^{(i)} - \overline{A}^{(i)})$$

(1)

Equation (1) reveals that the image covariance matrix can be obtained from the outer product of column (row) vectors of images, assuming the training images have zero mean.

For that reason, we claim that original 2D PCA works in the column direction of images. Result of feature extraction is then a matrix instead of a vector. Feature matrix has the same number of columns (rows) as width (height) of face image.

The extraction of image features is computationally more efficient using 2D PCA than PCA since the size of the image covariance matrix is quite small compared to the size of a covariance matrix in PCA (by using Turk & Pentlands optimization it depends on number of training images). 2D PCA is not only more efficient than PCA but it is possible to reach even higher recognition accuracy (Yang et al., 2004).

Despite its better efficiency, 2D PCA has also one disadvantage because it needs more coefficients for image representation than PCA. Because the size of the image covariance matrix for 2D PCA is equal to the width of images, which is quite small compared to the size of a covariance matrix in PCA, 2D PCA evaluates the image covariance matrix more accurately and computes the corresponding eigenvectors more efficiently than PCA.

3.1.3. Kernel Pca – Kpca

PCA is a linear algorithm that is not able to work with nonlinear data. Kernel PCA (Müller et al., 2001) is a method computing a nonlinear form of PCA. Instead of directly doing nonlinear PCA, it implicitly computes linear PCA in high-dimensional feature space that is in non-linear relation to input space.

3.2. Support Vector Machine - Svm

Support vector machines (SVM) (Asano, 2006; Hsu et al., 2003; Müller et al., 2001; Boser et al, 1992) are based on the concept of decision planes that define optimal boundaries. Its fundamental idea is very simple: the boundary is located to achieve the largest possible distance for the vectors of different sets. Example of this is shown in the Fig. 1. This figure illustrates linearly separable problem. In the case of linearly nonseparable problem, kernel methods are used. The concept of kernel method is a transformation of the vector space into a higher dimensional space.

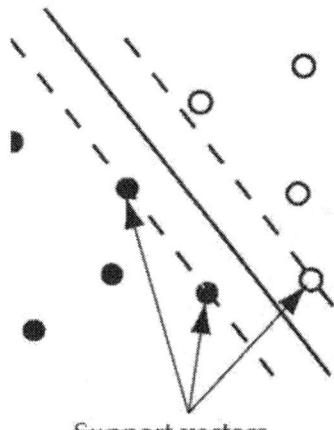

Support vectors

Figure 1. Optimal boundary of support vector machine

The kernel function is defined as follows:

$$K(\mathbf{x}, \mathbf{x}') = \Phi(\mathbf{x})^T \Phi(\mathbf{x}')$$

(2)

Kernel function is equivalent to the distance between x and x' measured in the higher dimensional space transformed by a nonlinear mapping Φ.

3.3. Methods Based on Neural Networks (Mlp, Rbf Network)

Neural network (Bishop, 1995; Haykin, 1994; Oravec et al., 1998) is a massive parallel processor that is inspired by biological nervous systems. Neural network is able to learn and to adapt its free parameters (connections between neurons known as synaptic weights are adjusted during the learning process).

3.3.1. Multilayer Perceptron

Multilayer perceptron (MLP) (Bishop, 1995; Haykin, 1994; Oravec et al., 1998) is a layered feedforward network consisting of input, hidden and output layers.

Multilayer perceptron operates with functional and error signals. The functional signal propagates forward starting at the network input and ending at the network output as an output signal. The error signal originates at output neurons during the learning and propagates backward. MLP is trained by backpropagation algorithm.

MLP represents nested sigmoidal scheme (Haykin, 1994), its form for single output neuron is

$$F(\mathbf{x}, \mathbf{w}) = \varphi\left(\sum_j w_{oj}\varphi\left(\sum_k w_{jk}\varphi\left(\ldots\varphi\left(\sum_i w_{li}x_i\right)\ldots\right)\right)\right)$$

(3)

where $\varphi(\cdot)$ is a sigmoidal activation function, w_{oj} is the synaptic weight from neuron j in the last hidden layer to the single output neuron o, and so on for the other synaptic weights, x_i is the i-th element of the input vector $\mathbf{x}$. The weight vector $\mathbf{w}$ denotes the entire set of synaptic weights ordered by layer, then neurons in a layer, and then number in a neuron.

3.3.2. Radial Basis Function Network

Radial basis function network (RBF) (Oravec et al., 1998; Hlaváčková, 1993) is a feedforward network consisting of input, one hidden and output layer. Input layer distributes input vectors into the network, hidden layer represents RBFs h_i. Linear output neurons compute linear combinations of their inputs. RBF network topology is shown in Fig. 2.

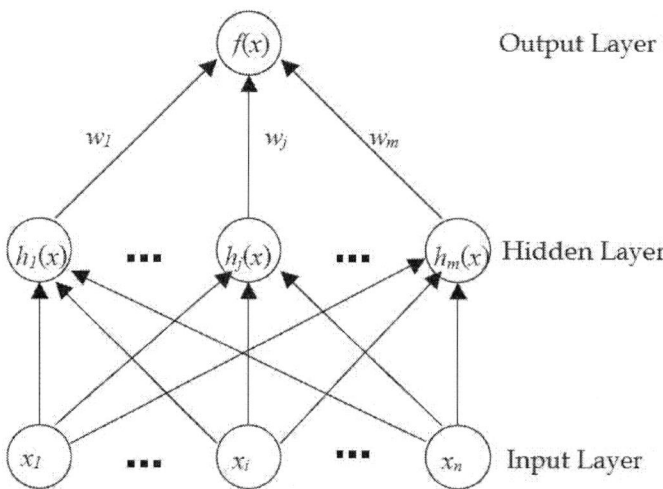

Figure 2. RBF network topology

RBF network is trained in three steps:
1.   Determination of centers of the hidden neurons
2.   Computation of additional parameters of RBFs
3.   Computation of output layer weights.

RBF network from Fig. 2 can be described as follows (Mark, 1996):

$$f(\mathbf{x}) = w_0 + \sum_{i=1}^{m} w_i h_i(\mathbf{x})$$

(4)

where x is the input of RB activation function h_i and w_i are weights. Output of network is a linear combination of RBFs.

3.4. Local Binary Patterns – LBP

Local binary patterns (LBP) were first described in (Ojala et al., 1996). It is a computationally efficient descriptor to capture the micro-structural properties and was proposed for texture classification. The operator labels the pixels of an image by thresholding the 3x3-neighbourhood of each pixel with the center value and considering the result as a binary number. Later the LBP operator has been extended to use circle neighborhoods of different sizes - the pixel values are bilinearly interpolated (Fig. 3).

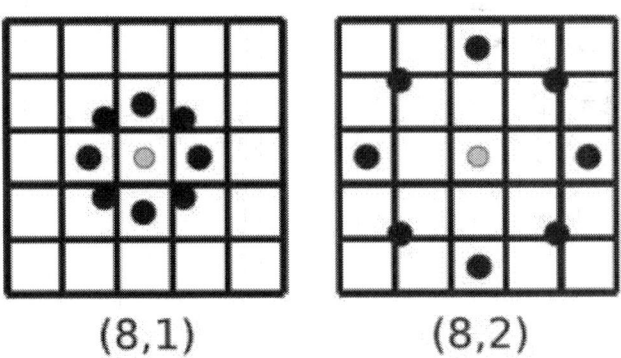

(8,1)   (8,2)

Figure 3. The extended LBP operator with circular neighborhood

Another extension uses just uniform patterns. A local binary pattern is called uniform if it contains at most two bitwise transitions from 0 to 1 or vice versa when the binary string is considered circular. For example, 00000000, 00011110 and 10000011 are uniform patterns. Such patterns represent important features on the image like corners or edges. Uniform patterns account for most of the pattern in images (Ojala et al., 1996).

A system using LBP for face recognition is proposed in (Ahonen et al., 2004, 2006). Image is divided into non-overlapping regions. In each region a histogram of uniform LBP patterns is computed, the histograms are concatenated into one histogram (see Fig. 4 for illustration), which represents features extracted from the image in 3 levels (pixel, region and whole image).

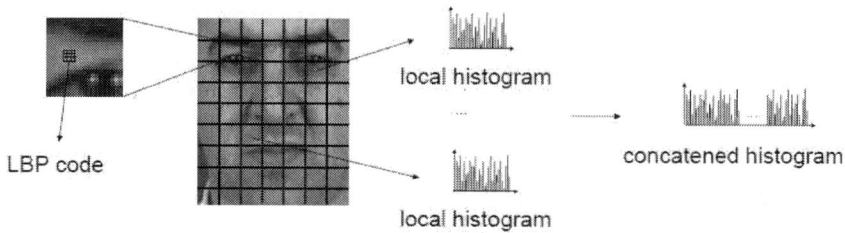

Figure 4. Description of face using concatenated LBP histogram (image taken from (Marcel et al., 2007))

The χ2 metric is used as the distance metric for comparing the histograms:

$$\chi^2(\mathbf{S},\mathbf{M}) = \sum_{r,i} \frac{(S(i) - M(i))^2}{S(i) + M(i)} \tag{5}$$

where S and M are the histograms to be compared and i is the i-th bin of histogram.

4. FACE DATABASE

We used images selected from FERET image database (Phillips et al., 1998). FERET face images database is de facto standard database in face recognition research. It is a complex and large database, which contains more than 14126 images of 1199 subjects of dimensions 256 x 384 pixels. Images differ in head position, lighting conditions, beard, glasses, hairstyle, expression and age of subjects.

We worked with grayscale images from Gray FERET (FERET Database, 2001). We selected image set containing total 665 images from 82 subjects. It consists of all available subjects from whole FERET database that have more than 4 frontal images containing also corresponding eyes coordinates (i.e. largest possible set fulfilling these conditions from FERET database was chosen). The used image sets are visualized in Fig. 5.

The images were preprocessed. Our preprocessing consists of
- geometric normalization (aligning according to eye coordinates)
- histogram equalization
- masking (cropping an ellipse around the face)
  resizing to 65x75pix

Fig. 6 shows an example of the original image and the image after preprocessing.

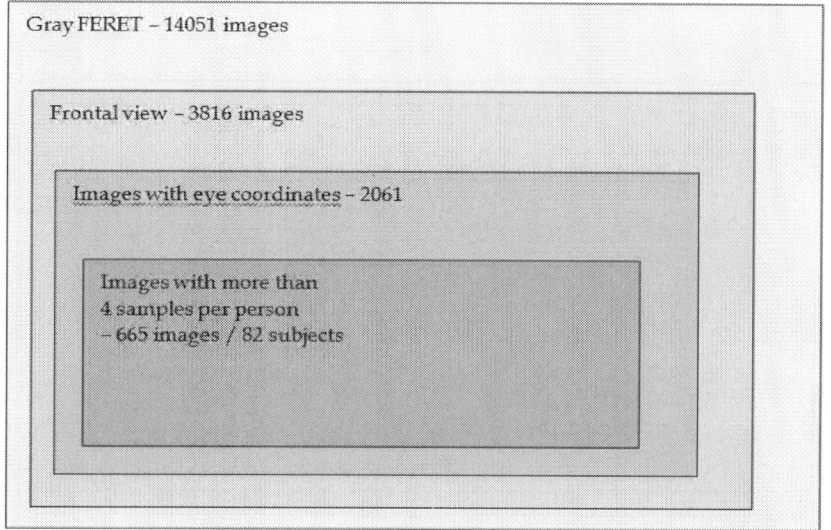

Figure 5. Visualization of subset of images from FERET used in our experiments

Figure 6. Original images and corresponding images after preprocessing

5. SIMULATION RESULTS FOR 1 - 4 ORIGINAL TRAINING IMAGES PER SUBJECT

In our experiments with original training images we compared the efficiency of several algorithms in scenario with 1 (single sample problem), 2, 3 and 4 images/subject. We carefully selected algorithms generally considered to play the major role in today face recognition research. Also standard PCA was included for comparison purposes. All these methods are briefly reviewed in subchapter 3. Face recognition methods.

Table 1. Results for different training sets (dependence of face recognition accuracy in % with regard to number of samples per subject in the training set)

	1_train	2_train	3_train	4_train
PCA	72.26	82.43	86.60	89.43
2D-PCA	73.41	81.90	86.36	89.02
KPCA	73.97	82.28	87.83	91.62
PCA+SVM	79.97	91.52	95.76	97.18
SVM	66.32	68.76	91.73	95.05
MLP	61.69	72.86	83.40	85.86
RBF	66.41	85.26	93.16	96.79
LBP-5x5	83.02	89.37	92.06	94.21
LBP 7x7	85.29	91.47	94.45	95.99
LBP 7x7w	85.81	92.91	95.05	96.59

In each test with different number of images in training set we made 4 runs with different selection of the images into the training set: original one with choosing the first images alphabetically by name and 3 additional training/testing collections with randomly shuffled images. The final test results are the average from these 4 values.

Our results are summarized in Table 1 and in Fig. 7. All figures and tables in this chapter contain values whose meaning is recognition accuracy in % achieved on test sets. The notation *n_train* means *n images (samples) per subject in training sets.*

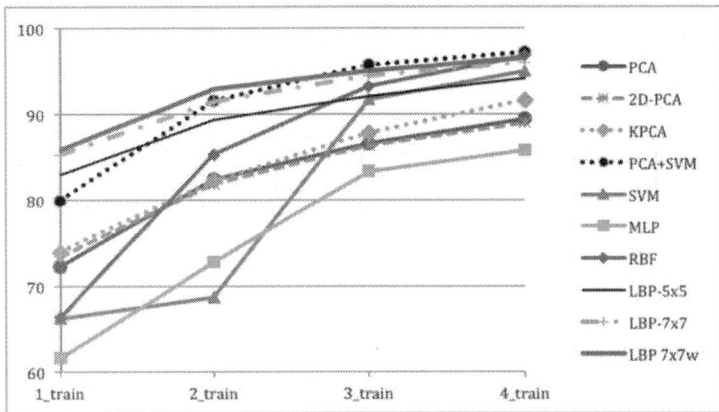

Figure 7. Graphic comparison of the results for different training sets (dependence of face recognition accuracy in % with regard to number of samples per subject in the training set)

Presented results are summarized as follows:

Neural networks and SVM

For single sample per person training sets, methods based on neural networks (RBF network and MLP) and also SVM achieved less favorable results (below 70%). The extension of the training sets by second sample per person slightly increased face recognition test results for MLP and SVM methods. For RBF network, the second sample improved the result to the value above 85%. Impact of adding third sample per person into training sets caused a significant improvement of test results (for RBF and SVM above 90% accuracy was achieved). Adding more than four samples per person into training sets has only a minimal effect on increasing the face recognition results and has a negative impact on the computational and time complexity. The larger training sets the better recognition results were achieved.

PCA-based methods

PCA with Euclidean distance metric as a reference method shows that more images per subject in training set lead to more accurate recognition results, improving from 68% with 1 img./subj. to 89% with 4 img./subj. Although there was reported that 2D PCA can reach higher accuracy in term of precision, PCA slightly overcome 2D PCA in our experiments. However, 2D PCA still has big advantage in comparison to PCA which lies in faster training time due to using smaller covariance matrix. As it is shown in (Li-wei et al., 2005), 2D PCA is equal to block-based PCA and it means that it uses only several parts of covariance matrix used in PCA. In other words we lose information from rest of covariance matrix that can lead to worse recognition rates. KPCA achieved slightly better results compared to 2D PCA (KPCA is included for comparison purposes here and it will not be used further within this chapter).

PCA+SVM

PCA+SVM method is a two-stage setup including both feature extraction and classification. Features are first efficiently extracted by PCA with optimal truncating the vectors from the transform matrix. The parameters for the selection of the transformation vectors are based on our previous research (Oravec et al., 2010). The classification stage is performed by SVM. SVM model is created with the best parameters found using cross-validation on the training set. PCA+SVM has very good recognition rate even with 1 img./subj. and with 3 and 4 img./subj. it outperforms all other methods in our tests reaching 97% recognition rate with 4 img./subj.

LBP

In our experiments, we used local binary patterns method for face recognition in 3 different modifications. The image is divided into 5x5 or 7x7 blocks from which the concatenated histogram is computed. The "LBP 7x7w" modification adds also weighting of the histogram with different weights according to corresponding image regions. This weighting has been proposed in (Ahonen et al., 2004).

Results for all LBP methods are the best in our tests and were outperformed only slightly with PCA+SVM method with 3 and 4 img./subj. The main characteristic of LBP is that the recognition results are very good even for 1 img./subj.. From the graph in Fig. 7 we see that the recognition rates for the three LBP methods go parallel with each other. The LBP is starting with 83% reaching 94% accuracy with 4 img./subj. LBP 7x7 is approximately 1.5% better than the 5x5modification and the LBP 7x7w more than 2% better reaching almost 97% accuracy with 4 img./subj.

Within this chapter, we work with images of size 65x75 pixels after preprocessing. In Table 2, results for image size 130x150pix (FERET default standard) are shown for illustration. Generally, larger size of images can yield slightly better recognition rates.

Table 2. Recognition rates for different LBP modifications

Method	1_train		Method	1_train
LBP-5x5	83.02		LBP 7x7 130x150pix	84,82
LBP 7x7	85.29		LBP 7x7w 130x150pix	86,66
LBP 7x7w	85.81		LBP 10x10 130x150pix	87,48
			LBP 10x10w 130x150pix	88,34

6. SIMULATION RESULTS FOR TRAINING SETS ENLARGED BY GENERATING NEW SAMPLES

In the previous subchapter, we presented recognition results for methods trained by 1 img./subj. in training sets. We also presented the comparison to results for 2, 3 and 4 img./subj. in training sets, while 2nd, 3rd and 4th images were the *original* images, i.e. the images were *real*, taken from the*original* face database.

Herein we consider different situation: only 1 original sample is available and we try to enhance recognition accuracy by generating new samples to the training sets in artificial manner. Thus, we try to enlarge the training sets by generating new (virtual, artificial) samples. We propose to generate new samples by modifying single available original image in different ways – this is why we will use the term *image modification* (or *modified image*). Natural continuation of such approach leads to generating synthetic face images.

In our tests we use different modifications of available single per person images: adding noise, applying wavelet transform and performing geometric transformation.

6.1. Modifications of Face Images by Adding Gaussian Noise

Noise in face images can seriously affect the performance of face recognition systems (Oravec et al., 2010). Each image capturing generates digital or analog noise of diverse intensity. The noise is also generated while transmitting and copying analog images. Noise generation is a natural property for image scanning systems. Herein we use noise for generating modified samples of original image. In our modifications, we use Gaussian (Truax, 1999) noise.

Gaussian noise was generated using Gaussian distribution function

$$p(x) = \frac{1}{\sqrt{2\pi\sigma^2}} e^{-\frac{(x-\mu)^2}{2\sigma^2}}$$

(6)

where μ is the mean value of the required distribution and σ^2 is a variance (Truax, 1999; Chiodo, 2006).

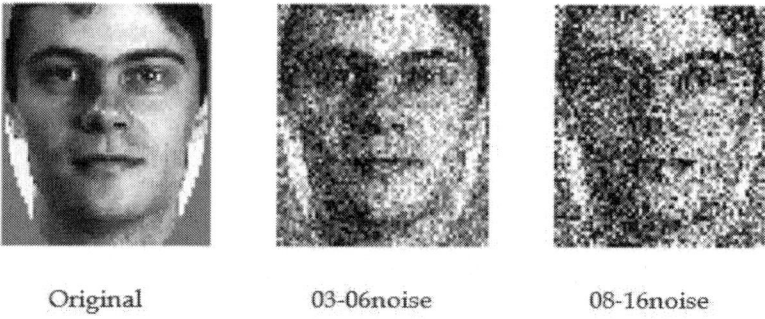

| Original | 03-06noise | 08-16noise |

Figure 8. Examples of images modified by Gaussian noise

Gaussian noise was applied on each image with zero mean and in two random intervals of variance. Examples of images degraded by Gaussian noise can be seen in Fig. 8. The labels *03-06noise* and *08-16noise* mean that the variance of Gaussian noise is random between values 0.03 - 0.06 and 0.08 - 0.16, respectively. The same notation is used also in presented graphs and tables (Tab. 3a and 3b, Fig.9a and 9b). Noise parameters settings for our simulations were determined empirically. Training sets were created by noise modification of samples added to the original one (*1+1noise, 1+2noise* and *1+3noise*).

Presented results for noise modifications shown in Tab. 3a, 3b and Fig. 9a, 9b are summarized as follows:

Neural networks and SVM

The improvement for RBF, MLP and SVM is clearly visible. In both noise modifications (*03-06noise* and *08-16noise*), the most significant increase in accuracy of test results is achieved by RBF network (about 80% for 1+3 training sets). Similarly to the tests in subchapter 5, adding more samples into training sets has a constant effect on the recognition results.

Table 3. Results for generating new face samples (modifications of original face samples by Gaussian noise – lower variance)

	1_train	1+1_train 03-06noise	1+2_train 03-06noise	1+3_train 03-06noise
PCA	72.26	72.16	72.16	72.16
2D-PCA	73.41	73.18	73.24	73.24
PCA+SVM	79.97	78.73	79.03	78.43
SVM	66.32	67.44	74.76	74.84
MLP	61.69	65.99	68.39	71.48
RBF	66.41	79.19	79.73	79.87
LBP-5x5	83.02	83.02	83.02	83.02
LBP-7x7	85.29	85.29	85.29	85.29
LBP-7x7w	85.81	85.81	85.81	85.81

Table 4. Results for generating new face samples (modifications of original face samples by Gaussian noise – higher variance)

	1_train	1+1_train 08-16noise	1+2_train 08-16noise	1+3_train 08-16noise
PCA	72.26	72.16	72.16	72.16
2D-PCA	73.41	73.24	73.13	73.07
PCA+SVM	79.97	78.90	77.83	77.79
SVM	66.32	74.84	74.87	74.81
MLP	61.69	64.83	65.57	65.63
RBF	66.41	77.1	79.3	80.07
LBP-5x5	83.02	83.02	83.02	83.02
LBP-7x7	85.29	85.29	85.29	85.29
LBP-7x7w	85.81	85.81	85.81	85.81

PCA-based methods

The results of PCA and 2D PCA methods are only slightly affected when adding additional images with different amount of noise to the training set. The results with the noise images added are approximately 1% worse than the original recognition rate with 1 img./subj. Reason for this effect can be probably found in the fact that the transformation matrix computed from the training sample with added noise represents the variances in the space worse than after computing it from original images only. Adding samples to training set is also very uneconomical from the point of view of PCA methods since the time needed to compute the transform matrix grows.

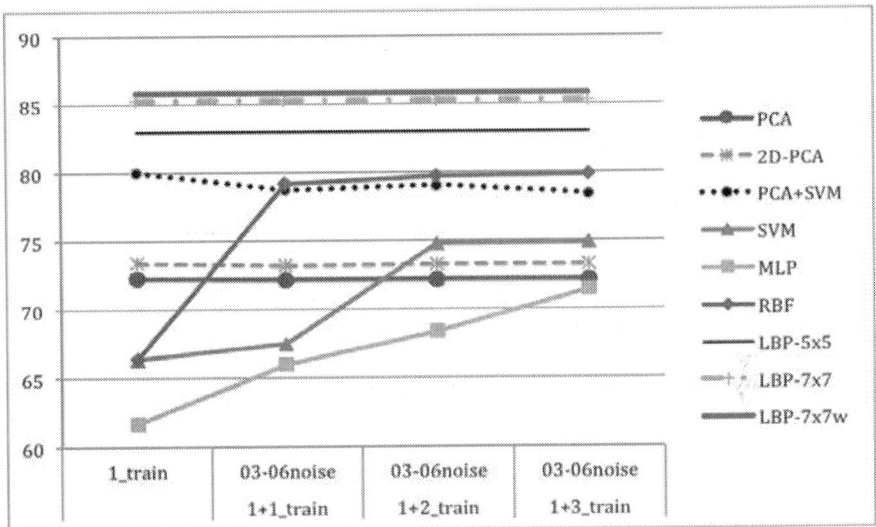

Figure 9. Graphic comparison of the results for generating new face samples (modifications of original face samples by Gaussian noise – lower variance)

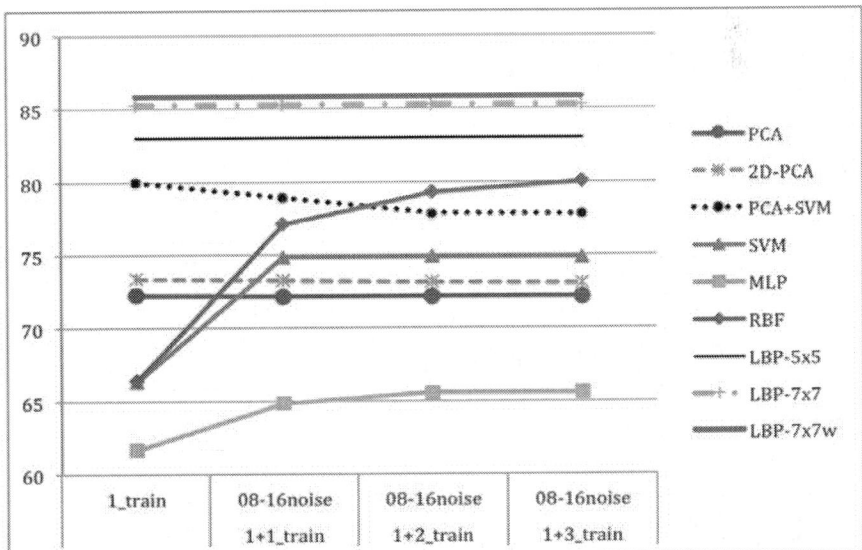

Figure 10. Graphic comparison of the results for generating new face samples (modifications of original face samples by Gaussian noise – higher variance)

PCA+SVM

The effect observed with PCA can be observed also with PCA+SVM method. Adding the noise images to the training set leads to worse results than with the original training set for about 1% for every scenario. The SVM classification model is influenced by the features extracted from noisy samples, but this accuracy drop is not dramatic.

LBP

The results of LBP methods are not influenced with the noisy samples at all. This has two reasons:

- By LBP method no model or transformation is calculated from the training images, so there cannot be such global effect to the recognition results as with PCA or SVM.
- The histograms of LBP patterns in noisy images change rapidly so the distance between the noisy image and the original image of the same person is higher than the distance between two original images of different persons. The consequence is that the minimal distances between the testing and training images do not change and the results are the same as without the noisy images in training set. See Table 4. for illustration of the distances between original and noisy images.

Table 5. Distances between the LBP 7x7 histograms for original and noisy train images compared with the same and different subject (see Fig. 10 for illustration of the images compared)

	Testimg1_subj1	Testimg1_subj2
Trainimg1_subj1	21.45	22.52
Trainimg1_subj1_noise	28.81	30.90

Figure 11. Illustration of images used in comparison in Table 4

6.2 Modifications of Face Images Based on Wavelets

Discrete wavelet transform DWT (Puyati et al., 2006; Sluciak & Vargic 2008) (notation wavelets is used in our tables and charts) is defined as follows:

$$DWT(j,k) = \frac{1}{\sqrt{2^j}} \int_{-\infty}^{\infty} f(x)\psi\left(\frac{x}{2} - k\right)dx$$

(7)

where j is the power of binary scaling, k is a constant of the filter and
function ψ is a basic wavelet, f(x) is a function which is to be transformed.

Our modifications of face images were done by three steps:
1. Forward transform of image by DWT
2. Setting horizontal, diagonal and vertical details in frequency spectrum
3. Image reconstruction by inverse DWT

We used two types of wavelets: Reverse biorthogonal 2.4 (Vargic &
Procháska, 2005) and Symlets 4 (Puyati et al., 2006) (Fig. 11.). These wavelets
were chosen empirically – our aim was to produce slight change in the
expression of a face. The training sets were created similarly to those with the
noise modification (1+1, 1+2 and 1+3), see subchapter 6.1. An example of new
samples is shown inFig. 12.

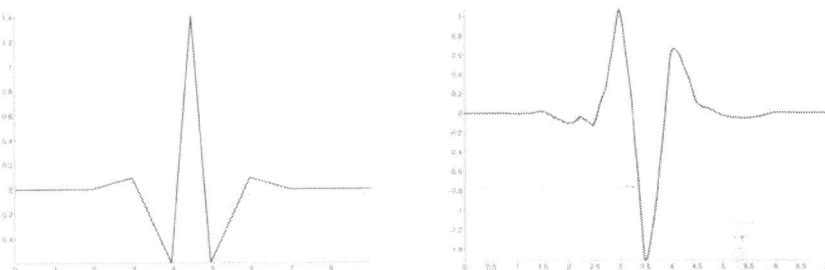

Figure 12. Wavelet function ψ: Reverse biorthogonal 2.4, Symlets 4

 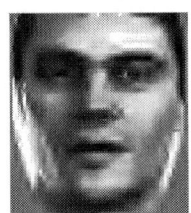 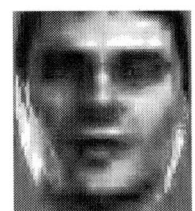 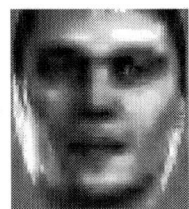

Figure 13. Original image and three types of images modified by wavelet
transform

Presented results for wavelet modifications (shown in Table 5 and in Fig.
13) are summarized as follows:

Neural networks and SVM

Experiment with wavelet transform demonstrated improvement of one sample
per person face recognition using neural network methods – RBF network and

Table 6. Results for generating new face samples (modifications of original face samples by wavelet transform)

	1_train	1+1_train wavelets	1+2_train wavelets	1+3_train wavelets
PCA	72.26	72.16	73.07	73.47
2D-PCA	73.41	73.30	73.18	73.36
PCA+SVM	79.97	78.43	65.27	50.47
SVM	66.32	28.04	24.71	23.61
MLP	61.69	68.92	73.65	74.73
RBF	66.41	77.49	77.70	79.12

MLP. These methods confirmed increase of recognition rate with extending the training sets with images modified by wavelet transform. Improvement above 10% was achieved for RBF network with adding three samples per person (1+3_train) into training sets. On the other hand, SVM method achieved very low face recognition accuracy.

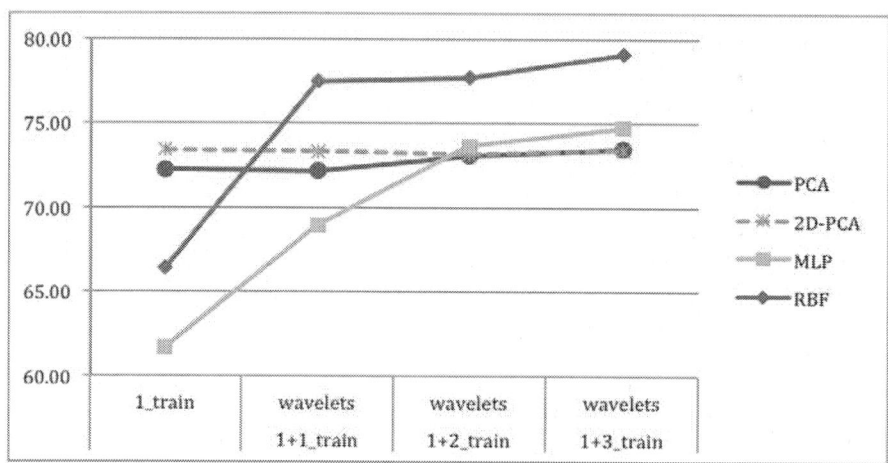

Figure 14. Graphic comparison of the results for generating new face samples (modifications of original face samples by wavelet transform)

PCA-based methods

Experiments with extending the training set with images modified by wavelet transform show that there is only a small influence to the results of PCA and 2D PCA methods. The accuracy increases when adding the images with stronger wavelet modification. The accuracy of recognition results is only 1% higher than original 1img./subj. The modified images do not cause any significant change in recognition and so there is almost no gain by adding new sample.

PCA+SVM

In contrast to PCA, the effect of decreasing accuracy can be seen when also SVM is involved. When 3 images modified by wavelets are added to the training set, the recognition result is almost 30% worse than using the original image only. In this case, only 50% accuracy can be obtained.

LBP

The effect of the wavelet modifications to the LBP histogram is similar to that with the noise images, so the LBP results stay the same as with the original training set.

6.3. Modifications of Face Images Based on Geometry

One of the most successful approaches to samples generation is that based on geometric transformation. The idea is to learn some suitable manifolds and extend training set by new synthetic poses or expressions based on original image (Wen et al., 2003). Because generation of new samples is based on facial features and their position on the face, these features need to be localized at first.

After the all facial features are properly localized and represented by contour and middle points, the next step is to generate target expressions. Because the change of an expression involves moving detected feature points, there is a need to change texture information as well. Real expressions and direction of movements during the expression depends on strength of muscles contractions. We divided each face image into triangles according to direction of these contractions. Face features localization process and dividing into triangles (also called triangulation) is fully automated (unlike usual manual method described in (C.-kai Yang & Chiang, 2007)) using active shape models (Milborrow, 2008). Using active shape models produces very precise positions of facial features and facial boundaries. Result of triangulation is facial graph containing only triangles among detected points determining facial features.

Making use of rule based system, similar to system described in (Yang & Chiang, 2007), we generated different expressions from each training sample by moving location of points in the facial graph. Texture in each triangle containing moved points is then interpolated from original according new coordinates. This procedure with different rules creates new "smile" and "sad" expressions (Fig. 14) and represents more sophisticated approach to generating additional training samples.

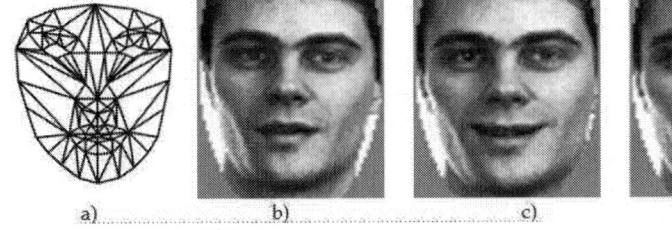

a)   b)   c)   d)

Figure 15. Example of image modified by geometric transformation: a) example of triangular division of face, b) original face image, c) synthetic smile expression, d) synthetic sad expression

Table 7. Results for generating new face samples (modifications of original face samples by geometric transformation)

	1_train	1+1_train SMILE	1+1_train SAD	1+2_train SMILE+SAD
PCA	72.26	73.07	73.13	73.87
2D-PCA	73.41	73.58	73.24	73.87
PCA+SVM	79.97	80.19	80.32	75.09
SVM	66.32	48.26	47.63	28.82
MLP	61.69	71.62	69.60	75.61
RBF	66.41	76.50	75.69	72.96
LBP-5x5	83.02	82.98	83.10	83.23
LBP-7x7	85.29	85.33	85.33	85.51
LBP-7x7w	85.81	85.89	85.98	87.22

Simulation results for geometric modifications are summarized in Table 6 and Fig. 15. Only results for SMILE expression were included in the graph since it helps to improve recognition. It agrees with the fact that the face database contains more faces with smiles than sad faces. In this way it is also possible to present results consistent with other graphs – 1, 2 and 3 samples per face.

The results are summarized as follows:

Neural networks and SVM

Both RBF network and MLP achieved better recognition accuracy using SMILE face expression images (the increase compared with one sample per person about 10%). Tests with extending the training set by SMILE+SAD face expression images were most effective for MLP method (75.61%). For SVM method, these new samples caused the drop of recognition rate about 25%, similar to the wavelet transform.

PCA-based methods

Geometric transformation results show comparable influence as those of PCA and 2D PCA using wavelet modifications. The accuracy increases when adding samples with SMILE expression. The accuracy of recognition results is only 1% higher than original 1 img./subj. The modified images do not cause any significant change in recognition. An improvement could be expected when more face expressions is taken in account.

PCA+SVM

Adding one image modified by geometry into the training set (either SAD or SMILE modification) improved the recognition rate for only about 0.2-0.3% (adding SMILE transformation helps slightly more). Surprisingly, when both transformed images were added to the training set, the recognition rate drops almost 5%.

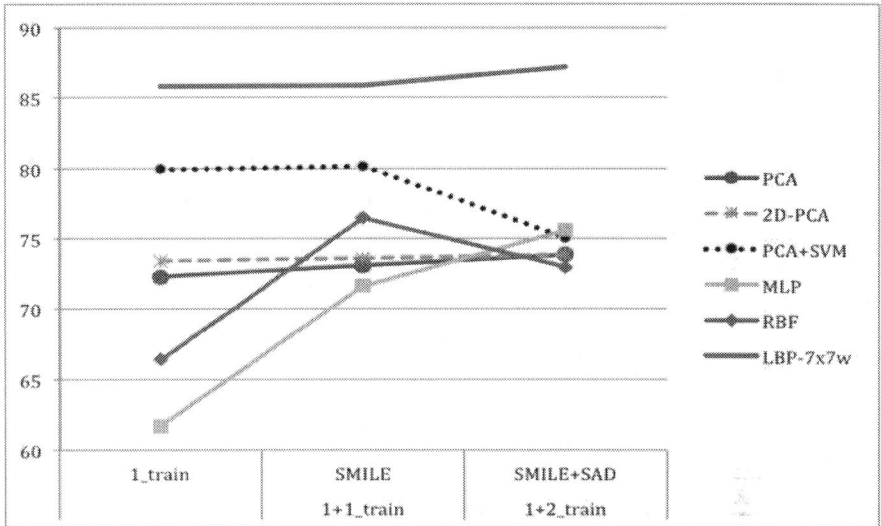

Figure 16. Graphic comparison of the results for generating new face samples (modifications of original face samples by geometric transformation)

LBP
As expected, adding transformed images with artificial change of expression (SAD and SMILE emotion) to the training set improves recognition. LBP method reaches better results because the system is more resistant against change in expression. Better results are reached when both transformed images (SAD+SMILE) are used. When also the images in the test set are transformed (for every sample also distances for SAD and SMILE transformation are computed), the results are even better, yielding 87.22% accuracy for LBP 7x7w method with 1 img./subj.

6.4. Comments and Summary for Methods That are Influenced Significantly by Enlarging Training Sets by Adding Modified Samples

This subchapter deals with methods for which extending the training set by modified images influences recognition results significantly (compared to recognition using multiple original images). The modifications of images described above (noise, wavelets and geometric tranformations) may be most helpful to neural networks. The comparison of recognition results for original training sets and extended training sets for RBF network and MLP is shown in Fig. 16. In Fig. 16 (and similarly in Fig. 17), the horizontal axis represents the number of images per person in training sets: the meaning for method using original images is 1, 2,3 and 4 original images in the training set; the meaning for modified images is 1 original image, 1 original plus 1, 2, or 3 modified images. For RBF network, above 10% improvement using modified images was

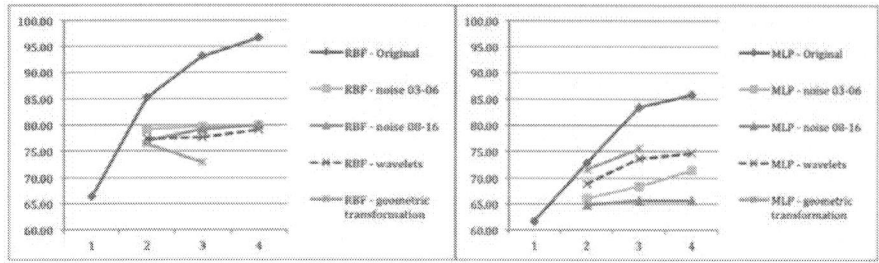

Figure 17. Comparison of the results obtained using original and modified training images for RBF network and MLP (generated samples improve recognition)

achieved. For MLP, geometric transformation was the most successful modification of face images (75.61%).

Figure 17. shows the negative effects of adding newly generated samples into training sets. This effect is clearly visible for PCA+SVM and SVM, when training sets are extended by wavelet transform and geometric transformation.

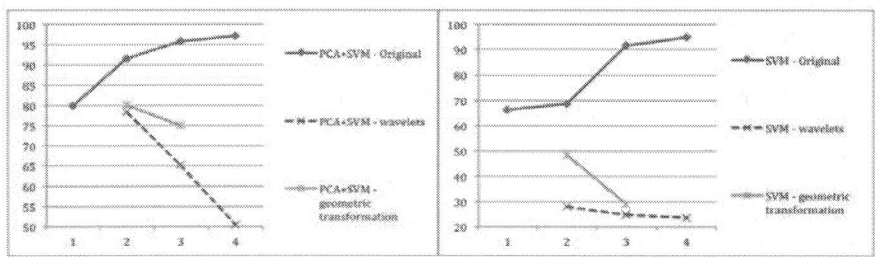

Figure 18. Comparison of the results obtained using original and modified training images for PCA+SVM and SVM (generated samples degrade recognition)

7. CONCLUSION

In this chapter, we considered relevant issues related to one sample per person problem in the area of face recognition. We focused mainly on recognition efficiency of several methods working with single and multiple samples per subject. We researched techniques for enlargement of the training set by new (artificial, virtual or nearly synthetic) samples, in order to improve recognition accuracy. Such samples can be generated in many ways – we concentrated on modifications of the original samples by noise, wavelets and geometric transformation. We proposed methods for modifying expression of a subject by geometric transformation and by wavelet transform. We examined the impact of these extensions on various methods (PCA, 2D PCA, SVM, PCA+SVM, MLP, RBF and LBP variants).

Methods such as PCA+SVM or LBP achieved recognition results above 80% for single sample per person in the training set. For these methods, adding new samples (modified images) did not help significantly. On the other hand, the utilization of the extended training sets for neural networks (MLP and RBF network) always increased the face recognition rate. This confirms that an appropriate extension of the input data set enhances the learning process and the recognition accuracy. Adding more than three new samples per person into the training sets has almost no influence on the recognition rate and has a negative impact on the computational and time complexity. The SVM method improved recognition accuracy only for extension of the training set by noise modification of images.

Experimental results for PCA and 2D PCA show only negligible influence of adding modified samples. We can conclude that the use of modified samples for PCA and 2D PCA has no added value, especially when samples are modified by Gaussian noise only.

PCA+SVM (two-stage method with PCA for feature extraction and SVM for classification) achieved very good results even for 1 img./subj. Adding any modified images to the training set did not improve the recognition rates, but the results were still one of the best from the compared methods.

Our experiments show that LBP is one of the most efficient state-of-the-art methods in face recognition. Adding noise and wavelet modified images to the training set does not have any effect on the recognition rates of LBP – unlike other methods that use the training sample to compute models or transformation matrices. This is caused by the nature of the method, where the histogram of LBP patterns of the noisy image differs too much from the original images. This can be also a disadvantage, when the images in the test set are corrupted with noise. On the other hand, adding images with transformed face expression helps and the system is more resistant to expression change in the images.

LBP for face recognition has obvious advantages such as state-of-the-art recognition rates even with 1 img./subj. in the training set, no need to train models or transformation matrices and good computational efficiency. But there is still potential to improve the results by possible modifications and optimization, which can be researched further: selection of LBP patterns, different preprocessing or modifications of LBP operator. The geometric transformation of images (emotional expression or head pose) and generating synthetic samples seem to be good ways how to improve the results. Further research is needed, since a simple extension of the training set with modified images does not always help.

We are currently working on a more sophisticated geometric transformation to cover more facial expressions. Although the results in section 6.3 show only a small improvement (with the exception of MLP where the improvement was significant), we suppose there is great potential of using samples with synthetic expression. The triangular model of face enables to extend the generation algorithm by other possibilities like generation of samples with different poses and illumination conditions. In the future, we also plan to publish modules generating new samples (with different expressions, poses and illumination) for our universal biometric system BioSandbox[1] - (used in our experiments).

Modification of images using wavelet transform has also large potential to generate new samples. One way to create new samples by wavelet transform is a fusion of two face images, where a new image is generated by applying the wavelet transform on two original images, followed by suitable manipulations of coefficients in a transformed space and finally merging images by inverse transform.

Using mean filter (Xu, J. & Yang, J., 2009) is another simple way of creating modified images. By using mean filter with different kernels (2x2, 3x3…15x15), we achieved results close to the modifications by wavelet transform.

Evaluating face recognition in single sample image per subject conditions reflects the real-world scenario. Also other effects such as various occlusions or lighting variation need to be taken into account when trying to reflect real conditions. We also need to test our methods using face databases that contain samples with these variations. Face databases such as ORL or AR could be used for this purpose.

For authentication and identification purposes, face recognition with 1 img./subj. only may not be enough, because its accuracy does not necessarily reach the required level. Therefore face recognition methods can be combined with different biometrics to form a multimodal system with much better characteristics than each of the biometrics itself (Ross & Jain, 2004).

8. ACKNOWLEDGEMENTS

Research described in this paper was done within the grants 1/0214/10 and 1/0961/11 of the Slovak Grant Agency VEGA. Portions of the research in this paper use the FERET database of facial images collected under the FERET program, sponsored by the DOD Counterdrug Technology Development Program Office. We would like to thank to our colleague Radoslav Vargic for valuable consultation regarding practical use of wavelets. We also thank to our student Ján Režnák for preparation of KPCA results.

REFERENCE

1. T. Ahonen, A. Hadid, M. Pietikäinen, 2006 Face Description with Local Binary Patterns: Application to Face Recognition. IEEE Transactions on Pattern Analysis and Machine Intelligence, 28 12 20372041 , TPAMI.2006.244

2. T. Ahonen, A. Hadid, M. Pietikäinen, 2004 Face Recognition with Local Binary Patterns, Proceedings of Computer Vision- ECCV, 978-3-54021-981-1 Prague, Czech Republic, May, 2004

3. R. Akbari, M. K. Bahaghighat, J. Mohammadi, 2010 Legendre Moments for Face Identification Based on Single Image per Person, 2nd Int. Conference

on Signal Processing Systems (ICSPS), 978-1-42446-892-8 Dalian, August 2010

A.   Asano, 2006 Pattern Information Processing, (lecture 2006 Autumn semester), Hiroshima University, Japan, Available from <http://laskin.mis.hiroshima-u.ac.jp/Kougi/06a/ PIP/PIP12pr.pdf>

4.   R. Beveridge, D. Bolme, M. Teixeira, B. Draper, 2003 The CSU Face Identification Evaluation System User's Guide, Version 5.0, Technical Report., Colorado State University, May 2003, Available from <http://www.cs.colostate.edu/evalfacerec/ algorithms/version5/faceIdUsersGuide.pdf>

5.   C. M. Bishop, 1995 Neural Networks for Pattern Recognition, Oxford University Press, Inc., 0-19853-864-2 York

6.   B. Boser, I. Guyon, V. Vapnik, 1992 A Training Algorithm for Optimal Margin Classifiers, Proceedings of the Fifth Annual Workshop on Computational Learning Theory, Pittsburgh, PA, USA, July 1992

7.   K. Delac, M. Grgic, M. S. Bartlett, 2008 Eds. Recent Advances in Face Recognition, IN-TECH, Vienna, Retrieved from <http://intechweb.org/book.php?id=101>

8.   W. Deng, J. Hu, J. Guo, W. Cai, D. Feng, 2010 Robust, Aaccurate and Efficient Face Recognition From a Single Training Image:A Uniform Pursuit Approach. Pattern Recognition, 43 5 May, 2010, 17481762 , 0031-3203

9.   FERET Database. 2001 Available from: <http://www.itl.nist.gov/iad/humanid/feret/, NIST>

10. Q. X. Gao, L. Zhang, D. Zhang, 2008 Face Recognition Using FLDA With Single Training Image Per Person, Applied Mathematics and Computation, 205 2 726734 , 0096-3003

11. S. Haykin, 1994 Neural Networks- A Comprehensive Foundation, New York: Macmillan College Publishing Company, 0-02352-781-7

12. K. Hlaváčková, R. Neruda, 1993 Radial Basis Function Networks. Neural Network World, 3 1 93102

13. C. W. Hsu, C. C. Chang, C. J. Lin, 2003 A Practical Guide to Support Vector Classification. Dept. of Computer Science and Inf. Engineering, National Taiwan University, April 15, 2010, Available from: <http://www.csie.ntu.edu.tw/~cjlin>

14. S. C. Chen, B. C. Lovell, 2004 Illumination and Expression Invariant Face Recognition with One Sample Image, 17th Int. Conference on Pattern Recognition (ICPR' 04)-1 0-76952-128-2 UK, Aug. 2004

15. S. C. Chen, D. Q. Zhang, Z. H. Zhou, 2004 Enhanced (PC)2A For Face Recognition With One Training Image Per Person. Pattern Recognition Letters, 25 10 11731181 , 0167-8655

16. S. C. Chen, J. Liu, Z. H. Zhou, 2004 Making FLDA Applicable to Face Recognition with One Sample Per Person. Pattern Recognition, 37 (7), July 2004, 15531555 , 0031-3203

17. K. Chiodo, 2006 Normal Distribution. NIST/SEMATECH e-Handbook of Statistical   Methods,   Retrieved   from <http://www.itl.nist.gov/div898/handbook/eda/section3/ eda3661.htm>

18. H. R. Kanan, K. Faez, 2010 Recognizing Faces Using Adaptively Weighted Sub-Gabor Array From a Single Sample Image Per Enrolled Subject. Image and Vision Computing, 28 1 Jan. 2010, 438448 , 0262-8856

19. Z. Lei, S. Liao, R. He, M. Pietikäinen, S. Z. Li, 2008 Gabor Volume Based Local Binary Pattern for Face Representation and Recognition, 8th IEEE Int. Conf. on Automatic Face & Gesture Recognition 2008 FG '08, Amsterdam, 978-1-42442-153-4

20. S. Z. Li, A. K. Jain, 2005 Eds. Handbook of Face Recognition, Springer, 038740595 New York

21. W. Li-Wei, W. Xiao, C. Ming, F. Ju-Fu, 2005 Is Two-dimensional PCA a New Technique ? Acta Automatica, 31 (5), 782787 , 0254-4156

22. W. J. S. Li, Y. F. Yao, C. Lan, S. Q. Gao, H. Tang, X. Y. Jing, 2009 Multimodal Biometrics Pixel Level Fusion and KPCA-RBF Feature Classification for Single Sample Recognition Problem, NSFC, 978-1-42444-131-0

A. Majumdar, R. K. Ward, 2008 Pseudo-fisherface Method For Single Image Per Person Face Recognition, ICASSP 2008, Las Vegas, NV, 978-1-42441-483-3

23. S. Marcel, Y. Rodriguez, G. Heusch, 2007 On the Recent Use of Local Binary Patterns For Face Authentication. International Journal on Image and Video Processing Special Issue on Facial Image Processing, 2007. IDIAP-RR 0634

24. J. L. O. Mark, 1996 Introduction to Radial Basis Function Networks. Centre for Cognitive Science, University of Edinburgh, April 1996

25. S. Milborrow, 2008 Locating Facial Features With an Extended Active Shape Model. Proc. of the 10th European Conf. on Computer Vision: Part IV, Marseille, France, Springer-Verlag, Nov. 2010, Retrieved from from http://www.springerlink.com/index/ 5t8hjm7j02qx6184.pdf

26. K. R. Müller, S. Mika, G. Rätsch, K. Tsuda, B. Schölkopf, 2001 An Introduction to Kernel-Based Learning Algorithms. IEEE Trans. on Neural Networks, 12 2 March 200), 181201 , 1045-9227

27. H. Nguyen, L. Bai, 2009 Local Gabor Binary Pattern Whitened PCA: A Novel Approach for Face Recognition from Single Image Per Person. Advances in Biometrics, 5558 269278 , 0030-2974 November 11, 2010, Retrieved from http://www.springerlink.com/index/t011q303142j7772.pdf

28. T. Ojala, M. Pietikäinen, D. Harwood, 1996 A Comparative Study of Texture Measures with Classification Based on Feature Distributions. Pattern Recognition, 29 5159 , 0031-3203

29. M. Oravec, J. Mazanec, J. Pavlovicova, P. Eiben, F. Lehocki, 2010 Face Recognition in Ideal and Noisy Conditions Using Support Vector Machines, PCA and LDA, In: Face Recognition, Milos Oravec (Ed.), 978-9-53307-060-5 INTECH, available from: http://sciyo.com/articles/show/title/face-recognition-in-ideal-and-noisy-conditions-using-support-vector-machines-pca-and-lda

30. M. Oravec, J. Polec, S. Marchevský, 1998 Neural Networks for Digital Signal Processing (in Slovak), Bratislava, Slovakia, 8-09675-039-9

31. P. J. Phillips, H. Wechsler, J. Huang, P. Rauss, 1998 The FERET Database and Evaluation Procedure For Face Recognition Algorithms, Image and Vision Computing, 16 5 295306 , 0262-8856

32. K. J. Priya, R. S. Rajesh, 2010 Dual Tree Complex Wavelet Transform Based Face Recognition with Single View, The Int. Conference on Computing, Communications and Information Technology Applications (CCITA-2010), 1994-460819944608

33. W. Puyati, S. Walairacht, A. Walairacht, 2006 PCA in Wavelet Domain For Face Recognition, The 8th International Conference Advanced Communication Technology, 455 Phoenix Park, 8-95519-129-4

34. L. Qiao, S. Chen, X. Tan, 2010 Sparsity preserving discriminant analysis for single training image face recognition. Pattern Recognition Letters 31 422429 , 0167-8655

35. D. Que, B. Chen, J. Hu, Y. Ax, 2008 A Novel Single Training Sample Face Recognition Algorithm Based on Modular Weighted (2D)2 PCA, 9th Int. Conference on Signal Processing, 2008. ICSP 2008, 15521555 , Beijing

36. D. Que, B. Chen, J. Hu, Y. Ax, 2008 A Novel Single Training Sample Face Recognition Algorithm Based on Modular Weighted (2D)2 PCA, 9th Int. Conf. on Signal Processing, ICSP 2008, 15521555 , 978-1-42442-178-7

37. O. Sluciak, R. Vargic, 2008 An Audio Watermarking Method Based on Wavelet Patchwork Algorithm, Proceedings of IWSSIP 2008, June 2008, Bratislava, Slovak Republic, 117120 , 978-8-02272-856-0

38. Y. Su, S. Shan, X. Chen, W. Gao, 2010 Adaptive Generic Learning for Face Recognition from a Single Sample Per Person, Proceedings of Int. Conference on Computer Vision and Pattern Recognition, CVPR2010, 2699 2706

39. X. Tan, B. Triggs, 2010 Enhanced Local Texture Feature Sets for Face Recognition under Difficult Lighting Conditions. IEEE Transactions on Image Processing, 19 6), 16351650 , 1057-7149

40. X. Tan, S. Chen, Z. H. Zhou, F. Zhang, 2006 Face recognition from a single image per person: A survey. Pattern Recognition, 39 9), 17251745 . 0031-3203

41. B. Truax, 1999) Ed. Gaussian Noise In: Handbook for Acoustic Ecology, Available on: http://www.sfu.ca/sonic-studio/handbook/Gaussian_Noise.html Cambridge Street Publishing

42. M. Turk, A. Pentland, 1991 Eigenfaces for Recognition. Journal of Cognitive Neuroscience, 3 1 Win. 1991, 7186 , Retrieved September 25, 2010, from http://portal.acm.org/citation.cfm?id=1326894#

43. R. Vargic, J. Procháska, 2005 An Adaptation of Shape Adaptive Wavelet Transform for Image Coding, EURASIP2005, Smolenice, Slovakia, June 2005

44. G. Wen, S. Shiguang, C. Xiujuan, F. Xiaowei, 2003 Virtual Face Image Generation for Illumination and Pose Insensitive Face Recognition, Proceedings of IEEE Int. Conference on Acoustics, Speech, and Signal Processing, ICASSP '03, pp. IV-7769 , 0-78037-663-3

45. J. Wu, Z. H. Zhou, 2002 Face Recognition with One Training Image Per Person. Pattern Recognition Letters, 23 14) 17111719 , 0167-8655

46. X. Xie, K. M. Lam, 2006 Gabor-Based Kernel PCA With Doubly Nonlinear Mapping for Face Recognition With a Single Face Image. IEEE Trans. on Image Processing, 15 9 1057-7149

47. J. Xu, J. Yang, 2009 Local Graph Embedding Discriminant Analysis for Face Recognition, School of Computer Science & Technology, Nanjing University of Science & Technology, Nanjing 210094, China, 2009

48. C. K. Yang, W. T. Chiang, 2007 An Interactive Facial Expression Generation System. Multimedia Tools and Applications, 40 1 4160

49. J. Yang, D. Zhang, A. F. Frangi, J. Y. Yang, 2004 Two-dimensional PCA: a New Approach to Appearance-Based Face Representation And Recognition. IEEE Transactions on Pattern Analysis and Machine Intelligence, 26(1), 131137 , 0162-8828

50. C. Zhan, W. Li, P. Ogunbona, 2009 Face Recognition from Single Sample Based on Human Face Perception. 24th International Conference Image and Vision Computing New Zealand, Wellington, 5661 , 978-1-42444-698-8

51. D. Zhang, S. Chen, Z. H. Zhou, 2005 A New Face Recognition Method Based on SVD Perturbation for Single Example Image Per Person. Applied Mathematics and Computation, 163 2 895907

CHAPTER 8

Face Recognition in the Presence of Expressions

Xia Han[1], Moi Hoon Yap[2], Ian Palmer[3]

[1]Centre for Visual Computing, University of Bradford, Bradford, UK
[2]School of Computing, Mathematics, and Digital Technology, Manchester Metropolitan University (MMU), Manchester, UK
[3]School of Computing, Informatics and Media, University of Bradford, Bradford, UK.

ABSTRACT

The purpose of this study is to enhance the algorithms towards the development of an efficient three dimensional face recognition system in the presence of expressions. The overall aim is to analyse patterns of expressions based on techniques relating to feature distances compare to the benchmarks. To investigate how the use of distances can help the recognition process, a feature set of diagonal distance patterns, were determined and extracted to distinguish face models. The significant finding is that, to solve the problem arising from data with facial expressions, the feature sets of the expression-invariant and expression-variant regions were determined and described by geodesic distances and Euclidean distances. By using regression models, the correlations between expressions and neutral feature sets were identified. The results of the study have indicated that our proposed analysis methods of facial expressions, was capable of undertaking face recognition using a minimum set of features improving efficiency and computation.

Keywords: Face Recognition; Feature Distances; Expressions; Regression Analysis

1. INTRODUCTION

The 3D face recognition techniques have drawn people's attention. Many researchers have moved towards 3D human face recognition techniques. 3D face recognition is based on 3D images in which either the shape (3D surface) of the human face is used individually or the combination with the texture (2D intensity image) is used for the purpose of recognition. The 3D images are specifically captured and generated by a 3D camera, which produces point clouds, triangle meshes or polygonal meshes and texture information to represent 3D facial images.

In a face recognition system based on either 2D or 3D images, there are several procedures to be carried out. The initial step is the face detection. This involves the extraction of a face image from a large scene or image or video sequences. The following procedure is for face alignment, which involves aligning the face image with a certain coordinate system and aids accurate results. These procedures can be accomplished by several key tasks: Firstly, model and representation of facial surfaces. In an efficient way, it aims to reduce the computational complexity and help to increase the good performance in the recognition part; Secondly, extraction of 3D facial features. Feature extraction is to localise a set of feature points associated with the main facial characteristics, eyes, nose and mouth. At present the most popular methods for facial feature extraction are point clouds, facial profiles and surface curvature-based features etc. [1]. Thirdly, algorithm and methodology used for 2D/3D facial surface data in an efficient way.

Our study aims to cover state-of-the-art 3D face recognition techniques, including the technical background, ideas, methodologies, and concepts. More specifically, there are a few issues we will address in the study: Explore the existing face recognition systems dealing with expressions and relevant researchers as well as the current state-of-the-art in this area; Investigate the challenges in existing face recognition systems; Identify the problems in the 3D face recognition, i.e. recognising faces in the presence of expressions; Propose relevant methods to enhance it in the development of an efficient 3D face recognition dealing with expressions; Investigate the effects of our previous curvature-based method [2] in face recognition under expression variations.

2. BACKGROUND

Face recognition under expression variations is required in some applications and over last two decades many computer vision researchers have been attracted to focus on the problems of recognising faces under expression variations. Significant progress [3-6] has been made. Many studies of 3D face recognition differentiate in terms of techniques. Lu et al. [7] match 2.5D facial images in the presence of expression variations and pose variations to a stored 3D face model with neutral expression by fitting the two types of deformable models, expression-specific and expression generic, generated from a small number of subjects to a given test image which is formulated as a minimisation of a cost function. Basically, mapping a deformable model to a given test image involves two transformations, rigid and non-rigid transformation. Therefore, the cost function is formed by translation vector and rotation matrix for rigid transformation, and a set of weights α. Kakadiaris et al. [8] address main challenges concerning the 3D face expressions recognition; Accuracy gain: For using 3D facial expression recognition independently or combined with other modalities, a significant accuracy gain of the 3D system with respect to 2D face recognition system must be produced in order to justify the introduction of a 3D system; Efficiency: 3D acquisition captures and creates larger data files per subject which causes significant storage requirements and slow processing. The data preprocessing for efficient data must be addressed. Automation: A system designed for the applications must be able to function fully automatically. Testing database: Larger and widely accepted databases for testing the performance of 3D facial expression recognition system should be produced. Kakadiaris have addressed the majority of the challenges by utilising a deformable model and mapping the 3D geometry information onto a 2D regular grid. The advantage of this is combining the 3D data with the computational efficiency of 2D data. They achieve the recognition rate of 97%. Lee et al. [9] propose an expression-invariant face recognition method. They extract the facial feature vector and obtain the facial expression state by the facial expression recogniser from the input image. The two main strategies for expression transformations are direct and indirect transformations. Direct transformation transforms a facial image with an arbitrary expression into its corresponding neutral face, whereas indirect transformation obtain relative expression parameters, shape difference and appearance ratio by model translation. By transforming them into its corresponding neutral facial expression vector using direct or indirect facial expressions transformation, they compare the recognition rate of each proposed method based on three different distance-based matching methods, nearest neighbor classifier (NN), LDA and generalised discriminant analysis (GDA). They achieve the highest recognition rate 96.67% based on NN, LDA using indirect expression transformation. Al-Osaimi et al. [10] introduce a new definition called the shape residue between the non-neutral and the estimated neutral 3D face models and present a method for decomposing an unseen 3D facial image under facial expressions to neutral face estimates and expression residues based on PCA. The residues are used for expression classification while the neutral face estimates are used for expression robust face recognition. In a result, 6% increase in the recognition performance is achieved when the decomposition method is employed. Li et al. [11-13] use low-level geometric features to create sparse representation models collected and ranked by the feature pooling and ranking scheme in order to achieve satisfactory recognition rate. They intentionally discard the expression-variant features, which are considered as higher-ranked. The recognition rate 94.68% is achieved.

Expressions seem to occur in a real word scenarios as even subtle expression variations can be captured into the 3D acquisition system. It has been claimed that face expressions can affect the accuracy and the performance of face recognition systems since the geometry of the face significantly changes as a result of facial expression. In general, the six significant expressions, happiness, anger, disgust, surprise, sad and fear, which make an adverse effect on face recognition. The adverse influence of face expression on face recognition is listed by Bronstein et al. [14] and needs to be solved no matter what dimensions face representation is being used (2D or 3D). However, its nonlinear nature and a lack of an associated mathematical model make the problem of face expression hard to deal with. There is no doubt that some progress has been made to solve this problem existing in 3D face recognition, but there are still some challenges remaining at this stage. For instance, Bronstein et al. [15,16] assume facial scans are isometric surfaces, which are not stretched by expressions so as to produce an expression invariant facial surface representation for recognising faces. However, there is one constraint in that they only considered frontal face scans and assumed the mouth to be closed in all expressions, which is not considered realistic.

Another issue with expressions is that there are less reliable invariants when faces carry heavy expressions. In addition, another issue is how to optimise the combination of small rigid facial regions for matching in order to reduce the effect of expressions. Using rigid facial regions can improve the performance on a database with expression variations [17,18]. The selection of rigid regions, however, is based on the optimal extraction and combination.

Researchers' attempts to reduce the computational cost have left another unresolved issue. Achieving less computational cost for real world applications has become a big challenge. Current studies intentionally use additional 2D texture information with an attempt to deal with expressions, which makes impact on computational time. In general, more information trained in 3D face data leads to more computational cost and time. Some algorithms can work on the verification process with a time cost of about 10 seconds on a normal PC [18], whereas efficient face matching with less computational cost is still a problem when dealing with a large gallery with thousands of faces. In addition, modeling relations between expressions and the neutral by expressionvariant features and combining with expression-invariant features still remains a research question.

3. 3D FACE DATABASES

The development of face recognition systems somehow relies on face image databases for the purpose of comparative evaluations of the systems. With the techniques of 3D face capture rapidly developing, currently more and more face databases on 3D have been built and oriented to different experimentation purposes: automatic face recognition, gender classification and facial expressions. In our proposed face recognition methods, two public databases will be utilised and compared in order to compare to benchmarks. From literature review, there are two popular databases among available public databases, GavabDB [12] and BU3D-FE [13]. The GavabDB contains 427 3D facial surface images corresponding to 61 individuals (45 males and 16 females), and there are nine different images per person. The other database BU-3DFE is considered as a facial expression database. In brief, each one of 100 subjects (56 female, 44 male) is instructed to perform seven universal expressions, i.e. neutral, happiness, surprise, fear, sadness, disgust and anger. The facial data contains about 13,000 - 21,000 polygons. In addition, 83 feature points on each facial model are located, refer to [13].

4. PROPOSED METHOD

4.1. Extraction of Shape Features

Facial feature extraction is important in many face-related applications [19]. Our 3D feature is a set of expression-invariants and expression-variants representations. This is achieved by analysing the facial region's sensiviety to the expressions and representing by a set of geometric descriptors. To eliminate the computational time from the preprocessing step, such as face alignment and normalisation, we introduce a distance-based feature to avoid the preprocessing.

Having considered computational time, we intentionally discard the texture information in our proposed method, in a similar way to the other chapters. Hence, the resulting performance is completely reliant on the features extracted from the shape of the image. The shape feature analysis based on the triangle meshes of faces, reflecting the facial skin wave, represents the intrinsic facial surface structure associated with the specific facial expressions. Motivated by this idea, we propose a set of novel distance-based geometric descriptors based on Euclidean and geodesic distances. Instead of focusing on the entire face we investigate face models by segmenting into two regions: the top face region, as shown in **Figure 1**(a), including eyes, eyebrows and nose; the low region, as shown in **Figure 1**(b), including the mouth only. The top region is used because it is comparatively insensitive to expression variations [4]. More specifically, the nose region suffers less from the effect of expression variations. However, the nose region alone is not sufficiently discriminatory for the purpose of face recognition because it represents only a small fraction of the face. Thus, we introduced more information, such as eyebrows and eyes, to be used to perform face recognition under expression variations. Expressionvariant regions are introduced and evaluated in our proposed method, found in the low region. Due to the significant change of the mouth region from laugh to neutral, we consider the low region as being sensitive to expression variations [17].

Based on the analysis of the expression-variant and expression-invariant regions, there are a few concerns about selecting the feature points. Firstly, selecting meaningful and significant positions as feature points, such as eye corners and top of the eyebrows, and the most distinctive position, the nose tip, is one concern. By taking this into consideration, accuracy of the extracted features can be improved. Secondly, for the insensitive region, another concern is that feature points are chosen to ensure minimum variations under change of expressions. For example, compared to the eye regions, the cheek region produces visible variations caused by expressions. Thus, we avoid locating feature points in the cheek region. Thirdly, for the expression-sensitive region, distinct positions along the outer mouth contour are taken into consideration rather than the areas including undesirable feature points, such as the chin regions. Fourthly, due to the missing data in the dark region resulting from scanning, such as the inner contour of mouths when laughing, we exclude those areas when selecting feature points. Fifthly, we determine the number of feature points by analysing the face representation efficiency and computation requirements. More features make face representation more accurate; however, they cause more computational time. Thus, choosing the number of feature points is based on the requirement that fewer features represent faces efficiently.

Taking advantage of 83 feature points on annotated face models in the BU3D-FE database [13], twelve significant positions are selected as a set of feature points. More specifically, they are top of the eyebrow, top of the upper eyelid, lowest point of the lower eyelid, the outer eye corner, the inner eye corner, the nose tip, the left and right edges of nose wings for insensitive regions; and mouth corners, mid-upper and mid-lower lips in expression-variant regions, as shown in **Figure 2**. The explanation and illustration of the set of distance-based features is shown in **Table 1** and **Figure 3** respectively.

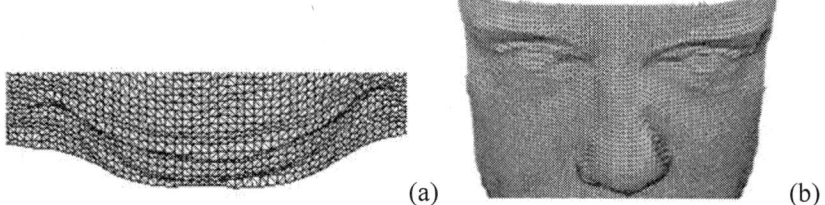
(a)   (b)

Figure 1. Illustration of expression-sensitive and expression-insensitive regions on a sample subject. (a) The upper face: expression-insensitive region; (b) The lower face: expression-sensitive region.

Not only is this set of features utilised in our method, but also contour shape features are considered, as shown in **Figure 3**. The contour shape features describe face elements, i.e. contour shape of eyebrows, eyes, nose and mouth. We select a set of control points on annotated face models to form the contours. Specifically, the contour shape features are represented by the vector $x = \left(x_1, y_1, z_1, x_2, y_2, z_2, \cdots, x_n y_n z_n \right)$ where x_i, y_i and z_i are coordinates of controlpoints.

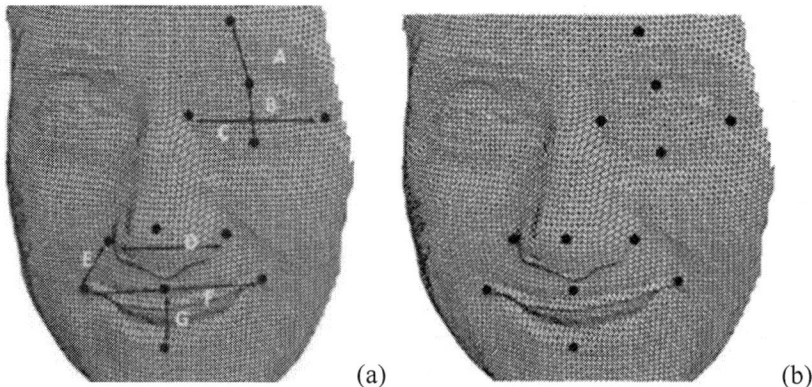
(a)   (b)

Figure 2. (a) Localisation of twelve landmarks; (b) Labelled of distance-based features.

Table 1. Seven distance-based features definition.

Distance index	Distance name	Distance definition
A	Eyebrow height	Distance between uppermost eyebrow and mid-upper eyelid
B	Eye height	Distance between mid-upper eyelid and mid-lower eyelid
C	Eye width	Distance between outer eye corner and inner eye corner
D	Nose width	Distance between left edge and right edge of nose wings
E	Lip stretching	Distance between nose tip and left lip corner
F	Mouth opening	Distance between left lip corner and right lip corner
G	Mouth height	Distance between mid-upper lip and mid-lower lip

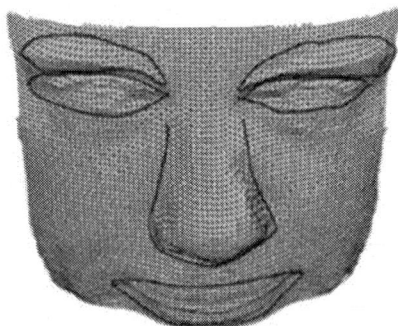

Figure 3. Illustration of the contour shape features describe face elements, i.e. contour shape of eyebrows, eyes, nose and mouth.

We utilise two intrinsic geometric descriptors, namely geodesic distances and Euclidean distances, to represent the facial models by describing the set of distance-based features and contours. The intrinsic geometric descriptors are independent of the chosen coordinate system.

We can observe that not all the distance vectors are invariant to expression variations. For example, the mouth opening (G) in **Figure 2**(b) shows a change of a distance vector caused by an open mouth, when the face of the same individual changes from neutral to a laugh expression. However, there are some certain distance vectors that remain stable under expression variations, for example, the eye width (C) in **Figure 2**(b). Thus, for the distance-based features, as we mentioned, we consider A, B, C and D distances in the top region as expression-invariants since they are insensitive under expression variations, whereas E, F and G are considered as expression-variants. We utilise Euclidean distances as the geometric descriptor to represent the set of distancebased features.

4.2. Geometric Descriptors

4.2.1. Euclidean Distance
In general, the distance between point p and point q in Euclidean n-space is

$$d_{e(p,q)} = \sqrt{\sum_{i=1}^{n}\left|p_i - q_i\right|^2}$$

(1)

where n is the dimension of Euclidean space. Specifically in this case, the Euclidean distance between points p and q in three dimensional Euclideanspace is

$$d_{e(p,q)} = \sqrt{\left(p_1 - q_1\right)^2 + \left(p_2 - q_2\right)^2 + \left(p_3 - q_3\right)^2}$$

(2)

The face model in the database comprises a triangle mesh, which is discretely defined to be a set of connected point clouds. The geodesic distance computation of triangle meshes is initialised by one or more isolated points on the mesh and the distance is propagated from them. More specifically, the geodesic distance of discrete meshes is considered as a finite set of Euclidean distances between pair-wise involved vertices. Thus, another geometric descriptor, geodesic distance, is utilised in our feature sets. Geodesic distance is capable of representing the contour shape features on the discrete meshes. Similarly, the selected contour features contain expression-invariants and expression-variants. The eye contour and the mouth contour are sensitive to expression variations. However, the eyebrow contour and the nose contour comparatively remain stable during expression variations.

4.2.2. Geodesic Distance
On a triangle mesh, the geodesic distance with respect to a point turns out to be a piecewise function, where in each segment the distance is given by the Euclidean distance function. Thus the geodesic distance computation is initialised by one or more isolated points on the mesh and the distance is propagated from them.

$$d_{g(p,q)} = \sum_{i=1}^{n} d_{e(p_i,q_i)}$$

(3)

where $d_{g(p,q)}$ is geodesic distance of a contour, $d_{e(p_i,q_i)}$ is the Euclidean distance between two points and n is the number of control points of each contour.

Depending on the different facial feature extraction methods, the slight influence of face sizes and different scales of the faces can be eliminated, either by normalisation or by preprocessing before the recognition process. Thus, in addition to the geometric descriptors derived from the face models, we also consider defining two distance-based features to avoid the face alignment process and the normalisation process.

We introduce a distance-based feature for normalising the set of seven distance-based features, which is considered as a stable expression-invariant feature, as shown in **Figure 4**(a). In order to be consistent with the geometric descriptor used for the features, we utilise Euclidean distance to represent the feature, named N_1 Thus seven normalized Euclidean distances are derived by the ratios of seven Euclidean distances to N_1. Similarly, since geodesic distances are not scale-invariant, the next step is to normalise each geodesic distance by another distance-based feature [20], the eyes-to-nose distance, as shown in **Figure 4**(b), i.e. N_2, sum of geodesic distances between the nose tip and the two inner eye corners. This guarantees invariance with respect to scaling and facial sizes under expression variations. Thus, for the set of contour shape features, this stable expression-invariant feature, N_2 is represented by geodesic distance descriptor to ensure its consistency with the descriptor for the contour shape features. Deriving six normalised geodesic distances is accomplished by the ratios of the six geodesic distances to N_2.

Thus, the geometric descriptors for the whole set of features are comprised of two sets of ratios. Meanwhile, the attributes of the ratio-based geometric descriptors that are unique to each face model are investigated and proved before carrying out the next step. Compared to the commonly used descriptors in [14,21], our geometric descriptors benefit from fast computation due to their simplicity. In the next section, we adopt regression models to learn the relationship between pair-wise expressions based on the combination of these thirteen ratio-based geometric descriptors.

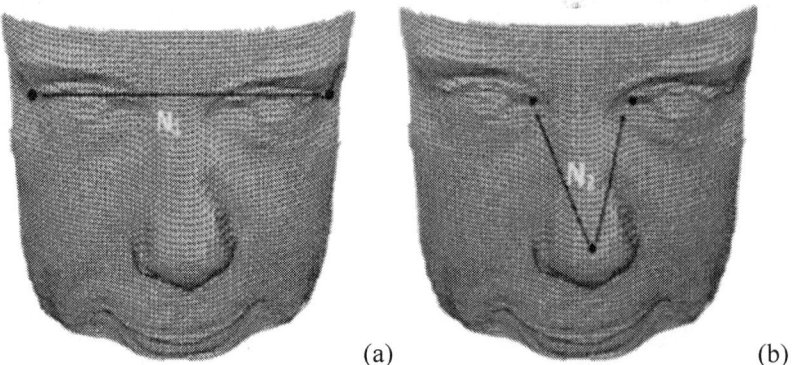

(a)   (b)

Figure 4. Illustration of two distances for normalisation of two feature sets. (a) For distance-based feature set; (b) For the contour shape feature set.

4.3. Regression Analysis Models

The regression analysis model is utilised for analysing the variables and modelling the relationship between them. Recently, regression analysis has been imported and applied in the face recognition area [22,23]. In this chapter, we evaluate two types of regression model: partial least square regression and multiple linear regressions. Specifically, we employ them to learn the correlation between pair-wise expressions and predict the 3D face neutral shape information for dealing with the problem of matching faces under expression variations.

4.3.1. Partial Least Square Regression

In this section, we will introduce a commonly used regression model that will be used to train and predict the feature set when the face models are neutral. Owing to the multiple dimensions of the involved variables, i.e. the total number of the ratio-based geometric descriptors, we will use a subspace regression model based on latent variables, named partial least square (PLS) [24].

X refers to a vector with the independent variables (predictors) and Y refers to a related vector of the dependent variables (responses).

$$X = TP + E,$$
$$Y = UQ + F,$$
$$U = BT + \varepsilon, \tag{4}$$

X is a matrix of predictors and Y is a matrix of responses. E and F are the error terms. There is a linear relation between T and U given by a set of coefficients B. A number of variants of PLS exist for estimating the T, P and Q.

The goal of PLS is to predict Y from X using a common structure of reduced dimensionality. For this purpose, PLS introduces some latent variables:

$$\{T_i\} = T_1, \cdots, T_k,$$
$$\{U_i\} = U_1, \cdots, U_k, \tag{5}$$

T and U preserve the most relevant information of the interaction model between X and Y.

4.3.2. Multi Linear Regression

To compare with the performance of PLS, we utilise the multiple linear analysis regression (MLR) [25] method to model and learn the relationship between neutral and non-neutral facial geometric descriptors and performed recognition rate. A regression model relates Y to a function of X and β.

$$Y \approx f(X, \beta) \tag{6}$$

Given a data set $\{y_i, x_{i1}, \cdots, x_{ip}\}_{i=1}^{n}$ of n statistical units, a linear regression model assumes that the relationship between the dependent variable y_i and the p-vector of regressors x_i is approximately linear. This approximate relationship is modelled through a term ε_i, an unobserved random variable that adds noise to the linear relationship between the dependent variables and responses. Thus, in detail,the model is of the form:

$$y_i = \beta_1 x_{i1} + \cdots + \beta_p x_{ip} + \varepsilon_i = x_i'\beta + \varepsilon_i,$$
$$i = 1, \cdots, n \tag{7}$$

where β_i' denotes the transpose, so that $x_i'\beta$ is the inner product between vector x_i and β. Commonly these n equations are integrated together and written in vector form as:

$$y = X\beta + \varepsilon \tag{8}$$

where:

$$y = \begin{pmatrix} y_1 \\ y_2 \\ \vdots \\ y_n \end{pmatrix}, X = \begin{pmatrix} x_1' \\ x_2' \\ \vdots \\ x_n' \end{pmatrix} = \begin{pmatrix} x_{11} & & x_{1p} \\ x_{21} & \cdots & x_{2p} \\ \vdots & \ddots & \vdots \\ x_{n1} & \cdots & x_{np} \end{pmatrix}$$

$$\beta = \begin{pmatrix} \beta_1 \\ \vdots \\ \beta_p \end{pmatrix}, \varepsilon = \begin{pmatrix} \varepsilon_1 \\ \varepsilon_2 \\ \vdots \\ \varepsilon_n \end{pmatrix} \tag{9}$$

Multi linear regression is based on linear regression but dealing with multiple X and Y.

5. RESULTS AND ANALYSIS

We have established a set of ratio-based geometric descriptors of face models that serves as input to the regression analysis model for simulating the relation between non-neutral and neutral faces. Relying on the ratios-based geometric descriptors, the effects of head rotation, translation and different scales can be eliminated, even without face alignment and normalisation in the preprocessing stage. Our proposed method is carried out on the BU3D-FE database [13]. In order to evaluate the two regression models, we set up a framework allowing for investigating improvement of expression-variants. To enhance the significance of the regression models, a Neural Network (NN) approach is employed for comparison. The comparison results are listed in **Table 2**.

From **Table 2**, we can observe that PLS, NN and MLR improve the rates by introducing expression-variants; however, PLS somehow places more emphasis on the expression-invariants in comparison to variants. Another conclusion is that employing variants indeed enhances the significance of expression-invariants. To further investigate the optimised performance of PLS, we set up another experiment allowing for PLS modelling multiple relationships between four intensities of six expressions and neutral, and vice versa. The results are shown in Figures 5 and 6, respectively.

Figure 5 shows the results of modelling intensity 1 angry to other expressions and neutral. The blue bars present the rate of expression-invariants and the red bars present the accuracy rate of the combination of expression-variants and expression-invariants. The improvement varies in four intensities. However, on average the results further confirm the power of the expressionvariants. In **Figure 6**, the diagram shows the results of modelling neutral to six other random intensity expressions. The blue bars present the accuracy rate of expression-invariants and the red bars present the accuracy rate of the combination. Thus the results confirm that ratio-based expression-variants can improve the performance. The recognition experiments and results will be carried out and described in the following.

Table 2. Comparison of multiple regression models.

Laugh matching to neutral	PLS	NN	MLR
Expression-invariants (4)	71.63%	72.12%	65.53%
Expression-variants (3)	55.12%	54.66%	61.78%
Combination (7)	64.78%	60.73%	69.01%

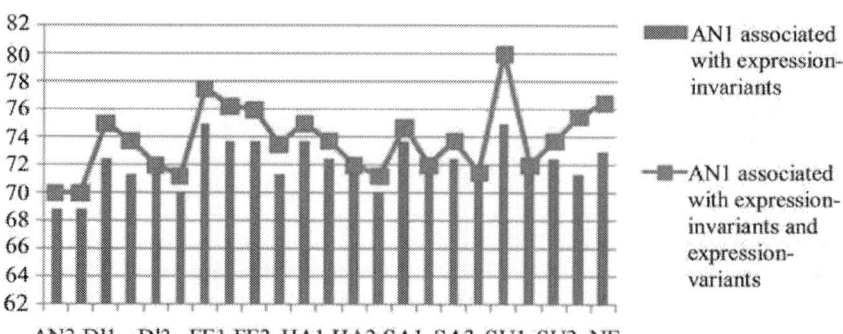

Figure 5. Illustration of the improvement with additional expression-variants employed under intensity 1 of angry expression.

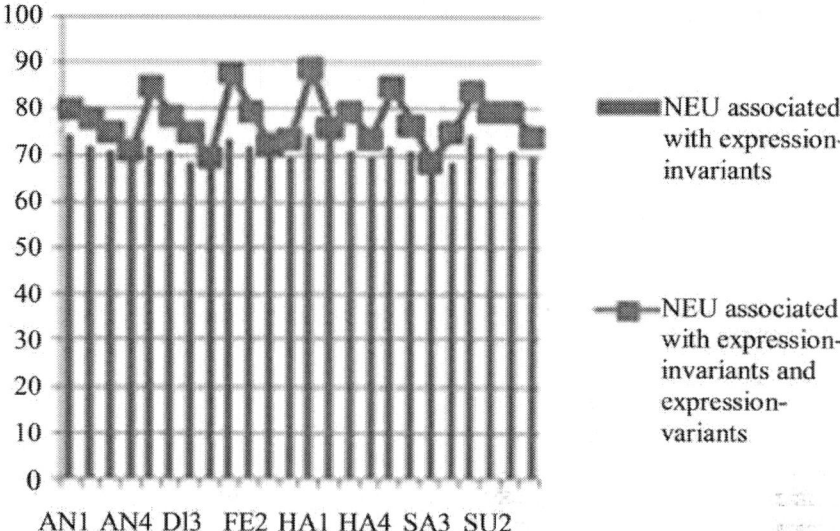

AN1 AN4 Dl3 FE2 HA1 HA4 SA3 SU2

Figure 6. Illustration of the improvement with additional expression-variants employed under neutral.

In summary, for the purpose of recognising faces in our experiments, we choose PLS for modelling relationships and use the combination of ratio-based expression-variants and expression-invariants as the feature sets.

In this experiment, we present the experimental results for recognising faces under expression variations. In particular, the experiment is implemented on the BU-3DFE database. In the experiments, we divide these 100 subjects into two sets: the training set with 80 subjects; and the testing set with 20 subjects. The experiments ensure that any subject used for training does not appear in the testing set because random partition is based on the subjects rather than the individual expressions. Four intensities of each expression are ensured to be involved in this experiment. For each iteration, the PLS regression analysis model is reset and retrained from the initial state.

In **Table 3**, we report the recognition rate of matching six expressions with four intensities to neutral. The promising recognition accuracy is achieved even without the texture information. Happy and surprise expressions achieve the highest accuracy rate of 89%, however, sad, angry, disgust and fear perform less accurately than the other three expressions. This is because happy and surprise expressions are comparatively clean and simple; their measurement allows for simplicity; some noises are produced when other expressions occur. Interestingly, the accentuated intensity of expressions achieves the lowest recognition rate and the recognition rate ascends along with a descending intensity. However, as we can see, the facial expressions have significantly degraded the performance of the face recognition in the absence of expression variations since the neutral face recognition accuracy rate achieves the highest at 90.3%.

Table 3. Neutral face recognition rates (RR) using PLS regression analysis model.

Intensity	RR (%)	Intensity	RR (%)
Happy01	80	Angry01	85
Happy02	75	Angry02	73.75
Happy03	70	Angry03	70
Happy04	70	Angry04	68.75
Disgust01	88	Fear01	89
Disgust02	73.75	Fear02	76.25
Disgust03	72.5	Fear03	72.5
Disgust04	73.75	Fear04	73.75
Sad01	85	Surprise01	84
Sad02	73.75	Surprise02	73.75
Sad03	68.75	Surprise03	72.5
Sad04	70	Surprise04	71.25
Neutral	90.3		

Comparison with Benchmark Algorithms

The final recognition rates of our proposed 3D face recognition method under expressions will be listed and discussed. Some researchers have developed and achieved their results of face recognition under expressions. Despite our results are not the best, we further discuss the feature sets, database, speed and the size of face models of our method in comparison with the current existing 3D face recognition methods, as shown in **Table 4**. The highest accuracy rate of 97% is generated by Alyuz et al. [4] using the nose, the forehead, the eyes and the central face, which covers the whole face with overlapping. In their work, face registration and face alignment are employed. These techniques cost computational time, although they play important roles in recognition rates. The point sets in the determined regions and the curvature related to the corresponding regions are used for dissimilarity calculation. The accuracy rate of 81.7% is achieved by Gervei et al. [26] using a least feature set with 40 dimensions. Compared to other methods, our proposed method use a minimum feature set to achieve a promising accuracy rate.

To further investigate the effect of the curvature descriptors of the T shape profile [2], incorporating curvature information is analysed, as shown in **Table 4**.

It has been proved that the proposed method is promising in dealing with expressions. Thus, we combine the distance-based feature sets with the curvature descriptors of the T shape profiles for face recognition under expressions. The results are listed in **Table 4**. We notice that the recognition rates are slightly higher with the curvature descriptors of the T shape profiles. Based on the results, it is observed that the use of the T shape profiles in the expression-invariant regions makes it possible to improve recognition performance.

Table 4. Our method of face recognition under expressions in comparison with other existing methods.

Authors	Database	Feature sets	Recognition rate	Vertices
Alyuz [4]	Bosphorus	Eye, nose, central face, forehead	97.28%	35,000
Smeets [5]	BU-3DFE	Whole face and nose region based on distances	94.5%	8000
Li [27]	FRGC v2.0	Whole face deformation model	91.9%	4500 after resample
Our proposed method	BU-3DFE	59 features	91.1%	8000
Our proposed method	BU-3DFE	14 features	90.3%	8000
Wang [6]	FRGC v2.0	Whole face	89.7%	8000
Our proposed method	GavabDB	59 features	89.6%	6000 after cropping face
Our proposed method	GavabDB	14 features	89.47%	6000 after cropping face
Chang [17]	Collective database of 546 subjects	Overlapping regions around the nose	83.5%	
Mahoor [28]	GavabDB	Contour lines	82%	13,000
Gervei [26]	GavabDB	Whole face through PCA down to 40 dimension	81.7%	13,000

The design criteria of the methods are based on the accuracy rate and speed (computational time). Regarding to various circumstances face recognition methods apply, there is no simple answer to the question of which method would fit best. In addition, the value of methods varies significantly with each application requirements and circumstances. Based on the design criteria, in some cases, fast processors are facilitated for face recognition techniques and the recognition rate can be considered as the top priority. The computational time can be sacrificed due to fast processors for high accuracy rate. However, the accuracy rate could be trade-off for the fast computation if the face recognition techniques are installed on a portable or wireless device. Therefore, in some circumstances, accuracy rate and speed can be trade off. In summary, our proposed methods are suitable for the cases essentially requiring fast computation.

6. CONCLUSIONS

The fundamental face recognition technique in the presence of expression variations was based on a correlation-learning model which generated a non-neutral model with a neutral model in order to match non-neutral faces to neutral faces. One of the issues pointed out by the researcher was reducing the computational cost. Rather than taking a large number of features into account, we intended to use the limited feature set representing contour information and face structure information. We built a novel framework of learning the correlation between various expressions and neutral with the limited feature set. Thus, there were two main advantages of our proposed method: training the correlations between expressions and neutral provide the flexibility of extending the feature sets; using the minimum feature set extracted from the facial structure information and the contour information explicitly to represent face models. Furthermore, incorporation of our previous T shape method leads to better performance.

There are two main requirements, however, needed to be investigated for improving the recognition rate in the future work. One is employing more geometric descriptors to represent the natural structure of the face model, for example area, curvature and weights allocated to descriptors. By introducing a set of weights to those geometric descriptors, the framework will be able to manipulate the weights and render them reliable for the task. The other one is to define the feature set to each particular expression in order to obtain the corresponding natures of each expression. These possible improvements are considered as the tasks in the future work for the purpose of strengthening the power of the framework.

REFERENCES

1. Y. Wang, C. Chua and Y. Ho, "Facial Appearance Detection and Face Recognition from 2D and 3D Images," Pattern Recognition Letters, Vol. 23, No. 10, 2001, pp. 1191-1202.
2. X. Han, H. Ugail and I. Palmer, "Method of Characterising 3D Faces Using Gaussian Curvature," Chinese Conference on Pattern Recognition, Nanjing, 4-6 November 2009, pp. 528-532.
3. B. Amberg, R. Knothe and T. Vetter, "Expression Invariant 3D Face Recognition with a Morphable Model," IEEE International Conference on Automatic Face and Gesture Recognition, 2008, pp. 1-6.
4. N. Alyuz, B. Gokberk, H. Dibeklioglu and L. Akarun, "Component-Based Registration with Curvature Descriptors for Expression Insensitive 3D Face Recognition," IEEE International Conference on Automatic Face & Gesture Recognition, Amsterdam, 17-19 September 2008, pp. 1-6.
5. D. Smeets, T. Fabry, J. Hermans, D. Vandermeulen and P. Suetens, "Fusion of an Isometric Deformation Modeling Approach Using Spectral Decomposition and a RegionBased Approach Using ICP for Expression-Invariant 3D Face Recognition," The 20th International Conference on Pattern Recognition, Istanbul, 23-26 August 2010, pp. 1172-1175.
6. Y. Wang, G. Pan and Z. Wu, "3D Face Recognition in the Presence of Expression: A Guidance-Based Constraint Deformation Approach," IEEE Conference on Computer Vision and Pattern Recognition, 2007, pp. 1180-1187.
7. X. Lu and A. K. Jain, "Deformation Modeling for Robust 3D Face Matching," IEEE Computer Society Conference on Computer Vision and Pattern Recognition, Vol. 2, 2006, pp. 1377-1383.
8. I. A. Kakadiaris, G. Passalis, G. Toderici, M. N. Murtuza, Y. Lu, N. Karampatziakis and T. Theoharis, "Three-Dimensional Face Recognition in the Presence of Facial Expressions: An Annotated Deformable Model Approach," IEEE Transactions on Pattern Analysis and Machine Intelligence, Vol. 29, No. 4, 2007, pp. 640-649.
9. H. Lee and D. Kim, "Expression-Invariant Face Recognition by Facial Expression Transformations," Pattern Recognition Letters, Vol. 29, No. 13, 2008, pp. 1797-1805.
10. F. R. Al-Osaimi, M. Bennamoun and A. Mian, "On Decomposing an Unseen 3D Face into Neutral Face and Expression Deformations," Advances in Biometrics, Vol. 5558, 2009, pp. 22-31.
11. X. Li, T. Jia and H. Zhang, "Expression-Insensitive 3D Face Recognition Using Sparse Representation," IEEE Conference on Computer Vision and Pattern Recognition, 2009, pp. 2575-2582.

12. A. B. Moreno and A. Sanchez, "GavabDB: A 3D Face Database," Proceedings 2nd COST Workshop on Biometrics on the Internet: Fundamentals, Advances and Applications, Vigo, 25-26 March 2004, pp. 77-82.

13. L. J. Yin, X. Z. Wei, Y. Sun, J. Wang and M. J. Rosato, "A 3D Facial Expression Database for Facial Behavior Research," International Conference on Automatic Face and Gesture Recognition, Southampton, 2-6 April 2006, pp. 211-216.

14. A. M. Bronstein, M. M. Bronstein and R. Kimmel, "Robust Expression-Invariant Face Recognition from Partially Missing Data," European Conference on Computer Vision, Vol. 3953, 2006, pp. 396-408.

15. A. M. Bronstein, M. M. Bronstein and R. Kimmel, "ThreeDimensional Face Recognition," International Journal of Computer Vision, Vol. 64, No. 1, 2005, pp. 5-30.

16. A. M. Bronstein, M. M. Bronstein and R. Kimmel, "Expression-invariant 3D Face Recognition," International Conference on Audioand Video-based Biometric Person Authentication, Guildford, 9-11 June 2003, pp. 62-69.

17. K. I. Chang, K. W. Bowyer and P. J. Flynn, "Multiple Nose Region Matching for 3D Face Recognition under Varying Facial Expression," IEEE Transaction on Pattern Analysis and Machine Intelligence, Vol. 28, No. 10, 2006, pp. 1695-1700.

18. T. Faltemier, K. W. Bowyer and P. Flynn, "A Region Ensemble for 3D Face Recognition," IEEE Transactions on Information Forensics and Security, Vol. 3, No. 1, 2008, pp. 62-73.

19. K. Fatimah, N. Khalid and A. Lili, "3D Face Recognition Using Multiple Features for Local Depth Information," International Journal of Computer Science and Network Security, Vol. 9, No. 1, 2009, pp. 27-32.

20. S. Berretti, A. Del Bimbo and P. Pala, "3D Face Recognition Using Isogeodesic Stripes," IEEE Transactions on Pattern Analysis and Machine Intelligence, Vol. 32, No. 12, 2010, pp. 2162-2177.

21. C. Samir, A. Srivastava and M. Daoudi, "3D Face Recognition Using Shapes of Facial Curves," IEEE Transactions on Pattern Analysis and Machine Intelligence, Vol. 28, No. 11, 2006, pp. 1858-1863.

22. X. Chai, S. Shan, X. Chen and W. Gao, "Locally Linear Regression for Pose-Invariant Face Recognition," IEEE Transactions on Image Processing, Vol. 16, No. 7, 2007, pp. 1716-1725.

23. I. Naseem, R. Togneri and M. Bennamoun, "Linear Regression for Face Recognition," IEEE Transactions on Pattern Analysis and Machine Intelligence, Vol. 32, No. 11, 2010, pp. 2106-2112.

24. R. D. Tobias, "An Introduction to Partial Least Squares Regression," SUGI Proceedings, Orlando 2-5 April 1995, 1995, pp. 1-8.

25. M. Tranmer and M. Elliot, "Multiple Linear Regression," The Cathie Marsh Centre for Census and Survey Research (CCSR) 2008.
26. O. Gervei, A. Ayatollahi and N. Gervei, "3D Face Recognition Using Modified PCA Methods," World Academy of Science, Engineering and Technology, No. 39, 2010, pp. 264-267.
27. X. Li and F. Da, "3D Face Recognition by Deforming the Normal Face," International Conference on Pattern Recognition, Istanbul, 23-26 August 2010, pp. 3975-3978.
28. M. H. Mahoor and M. Abdel-Mottaleb, "3D Face Recognition Based on 3D Ridge Lines in Range Data," IEEE International Conference on Image Processing, Vol. 1, 2007, pp. 137-140

CHAPTER 9

Novel Solution Based on Face Recognition to Address Identity Theft and Cheating in Online Examination Systems

Ayham Fayyoumi[1], Anis Zarrad[2]

[1]Department of Information Systems, Al Imam Mohammad IBN Saud Islamic University (IMSIU), Riyadh, Saudi Arabia
[2]Department of Computer Science and Information Systems, Prince Sultan University (PSU), Riyadh, Saudi Arabia

ABSTRACT

The main objective of this research is to provide a solution for online exam systems by using face recognition to authenticate learners for attending an online exam. More importantly, the system continuously (with short time intervals), checks for learner identity during the whole exam period to ensure that the learner who started the exam is the same one who continued until the end and prevent the possibility of cheating by looking at adjacent PC or reading from an external paper. The system will issue an early warning to the learners if suspicious behavior has been noticed by the system. The proposed system has been presented to eight e-learning instructors and experts in addition to 32 students to gather feedback and to study the impact and the benefit of such system in e-learning environment.

Keywords: Online Exam, Face Recognition, Authentication, Exam Cheating

1. INTRODUCTION

Nowadays, e-learning systems have become vital components in the education and training domains. Several countries are attempting to overcome the Knowledge Divide. Through education and training, countries are able to develop the skills of their citizens, consequently bridging the Knowledge Divide within the country and with more developed ones. Success in the Knowledge

Economy relies heavily on a qualified and skilled population, thus effective education and training systems are required.

Simultaneously, Information and Communication Technologies (ICTs) continue to grow at a rapid pace and have changed the way people live, work, and learn. The integration of ICT tools in education and training has created new ways of delivering, accessing, and processing useful knowledge, as well as has provided support to knowledge sharing between different actors and to lifelong learning [1] . In addition, technological development and the growth of the Internet have resulted in the emergence of e-learning as an important learning approach. E-learning provides innovative methods for educating people. Moreover, the e-learning market is expanding because of its many advantages over traditional education. E-learning is also highly flexible, scalable, employs a rapid learning method, less expensive, and proven to be effective compared with traditional education. In particular, the following are the three main drivers for the increasing global importance of e-learning:

- Movement toward a knowledge-based economy;
- Paradigm shift in education delivery;
- Technological developments and Internet growth.

The development of e-learning and online assessment systems is increasing rapidly, both globally and locally, with many universities and corporations investing significant capital in e-learning programs and initiatives. This growth is also reflected in the report by Ambient Insight, which was published in 2010, indicating that the e-learning market has reached US$ 27.1 billion in 2009 and will surpass $49.6 billion by 2014 [2] . The growth of the e-learning industry requires new services to ensure reliability and effectiveness of its systems, especially during the examinations process, by addressing the issue of cheating in online examinations and identity theft.

E-learning is prospering on global and local levels. In Saudi Arabia, the government is focusing to the education sector in general and to e-learning in particular in responding to the increasing number of male and female students enrolled in educational institutions. Many universities in Saudi Arabia have already implemented e-learning systems and are offering distance learning courses and degrees. Thus, ensuring the reliability of e-learning systems, especially during examinations, is highly critical. Online examination cheating and identity theft should be considered, while the privacy of learners' data and more importantly, their images is guaranteed.

2. E-LEARNING SYSTEMS

The Web is easy to use, easy to update, and is available worldwide. The Web is the driver of the knowledge economy and is therefore a natural vehicle for learning [3]. Hall and Snider defined e-learning as synonymous to all computer-related applications, tools, and processes of learning and teaching [4].

E-learning offers flexible learning anytime and anywhere. The increasing speed of the Internet, the growth of the World Wide Web, and the emergence of high-speed computers contribute to the availability of e-learning 24/7 and worldwide. Moreover, e-learners can access materials at any time and place convenient for them. E-learning has other advantages for learners, teachers, and instructional developers. Learners, for example, will benefit from their interaction with other students and with their instructors, and they can study at their preferred pace. Learners can also select the material they want or be directed to the content that meets their level of knowledge, interest, and needs. Furthermore, learners become responsible for their learning. Meanwhile, teachers can develop materials using online resources, and then publish them in many different ways such as text, images, video, audio, simulations, and games. Teachers will subsequently gain satisfaction through quality student participation. Finally, developers can develop detailed and standardized courses. E-learning allows developers to design the course and use it for multiple times by using learning objects.

E-learning provides much enrichment to the many models, markets, interest groups, and different degrees of satisfaction in the educational process. E-learning technologies offer a potential for high-quality formative assessment. Online assessment provides dynamic visuals, sound, user interactivity, adaptation to individual learners, and almost real-time score reporting, thus expanding examination options beyond the limitations of traditional tests [5].

Learner assessments are undoubtedly essential in the educational process. Examination scores inform the instructor on whether a student's progress is satisfactory or not. Online assessment systems provide instructors with many advantages, such as the creation of online examinations within a short time, administration of the examination through the computer, easy monitoring of answers during the examination, fast access to examination results without spending time on evaluation and correction, and easy calculation of the trends of the examination results. In addition, online examinations benefit learners by allowing them to not be physically present at a given location to take an examination, and the results can be made available to them immediately.

In online examinations, learners should register their names and passwords. The examination items will then be generated by the test bank according to the parameters set by the instructor. The items in online assessments are usually true/false and multiple choice questions, as well as involves reordering/rearrangement (matching, categorization, ranking, and others), completion, concept maps, and essays. Examination questions will appear on the screen, and then each learner will start answering them through his/her computer. At the end of the time limit, the examination will stop and the score will appear. The learner will then obtain the test results immediately.

3. ONLINE ASSESSMENT RELIABILITY

Security issues of e-learning systems have been discussed by many researchers [6]. Online examination is a challenge for e-learning security [7]. Currently, online assessments are mostly conducted at specific examination centers and require supervision mainly because, if administered in unsupervised locations, learners may acquire assistance from others to improve their examination results or have another individual take the examination for them. In such cases, instructors will become uncertain on who answered the examination questions, which conflicts the flexibility advantage of online education. Therefore, despite the expansion of the e-learning market locally and globally, a problem remains, especially with off-site examinations. Failure to verify the learners attending an examination is a major challenge in online learning environments.

Very little attention has been given in solving the problem of learners' unethical conduct. Moreover, many researchers criticized current e-learning systems for not focusing on the authentication of learners, particularly those who engage in online quizzes and examinations [8]. McGinity noted that biometrics has replaced the conventional password systems [9]. Another study highlighted the importance of detection mechanisms beyond the initial access to the e-learning system [10]. Therefore, a system must be developed to ensure that the person taking an examination is the student enrolled in the course. Yang and Verbauwhede suggested that biometrics systems provide better security than password systems [11]. Moreover, Hugl highlighted many technologies related to security that have not been used in e-learning [12]. These technologies include biometrics technologies that are increasingly becoming essential in a number of applications. Biometric authentication is the automatic recognition and identification of learners by using their physiological characteristics such as voice, hand geometry, fingerprints, and facial images. Generally, biometric authentication requires comparison of the stored data against the captured data.

No perfect biometric system that fits all needs has so far been created. All known systems have their advantages and disadvantages. A few studies had focused on improving e-learning security using biometric systems, but a limited number of them addressed the issue of continuous user authentication. In a recent study, Flior and Kowalski discussed a method for providing continuous biometric user authentication in online examinations via keystroke dynamics [13]. However, keystroke biometrics has its disadvantages, such as the major differences that can occur over time as a result of changes in typing pattern, tiredness of the hands after a period of typing, and improvement of skills.

In line with the previously mentioned concept, researchers are presently looking for the best biometric authentication method that will help validate the identification of the learner attending the examination and that will ensure that he/she is the same person as the one registered in the course without compromising his/her privacy. Face recognition systems are human-friendly because they require no contact and no additional hardware (given that most PCs and laptops come with a camera). More importantly, face recognition systems can be used for the continuous authentication of the learner during the entire examination period.

4. PROTOTYPE DEVELOPMENT: THE PROPOSED SOLUTION

Face recognition technologies operate by scanning the person's face and matching it with the stored image. The face recognition biometric system is a system that records distinguished facial features and stores the template in a server. In scanning the face, the camera identifies facial features and transmits the signal to the server where the scanned features are processed for matching

(See figure 1). Facial recognition identifies key features from the facial image. The system detects the face and captures image(s) of facial features that do not change over time, while avoiding those that change, such as facial expressions or hair. The first step in face recognition is the detection of the face in the image. Yang and his colleagues evaluated the main methods used for face detection, namely, knowledge-based method, appearance-based approach, feature invariant method, and template-matching method. The second step is the conduct of several approaches to model and recognize the facial image, such as direct correlation, elastic graph matching, neural networks, principal component analysis (PCA), and multiresolution analysis [14].

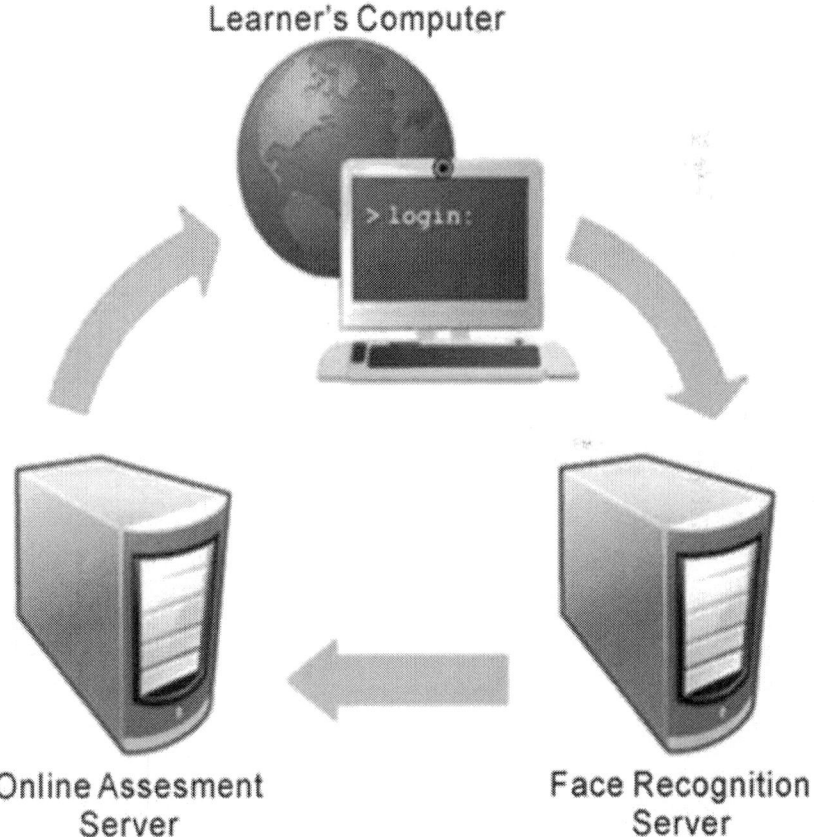

Figure 1. Proposed system solution.

The proposed examination system includes the development of a test bank on a specific field. The test bank contains a variety of question items (e.g., matching, ranking, essay, categorization, multiple-choice items, and true/false questions). These questions are classified into three difficulty levels: easy, moderate, and difficult. In designing the test, the instructor can specify the number of test questions, types of questions, and the difficulty associated to each question. The system will then automatically generate a random set of questions based on the criteria specified by the instructor. Therefore, each learner will receive a different set of questions but with the same difficulty level. The instructor is also allowed to design a new question and specify its difficulty. The newly formulated questions will be stored in the database and then added into the test bank. At the end of the examination time or after completing the test, the score and time spent by each learner during the examination will appear. The database will also save other data related to the registered users or learners, including profiles and images.

The face recognition system is integrated in the online assessment tool to identify and verify the learners allowed to access the exam and to continuously validate the learner's identity until the end of the examination. Specifically, during registration in the course (when used as part of an e-learning system) or for an examination, images of learners, in addition to other required data, are captured and stored in the database. Captured images are encrypted to protect learners' privacy. During examination time, the learner's identity is verified for attendance in the examination and is monitored by comparing the captured images with the one stored in the database. Figure 2 shows an overview of the proposed system architecture.

In addition, to address the issue of cheating in online examination systems, continuous checking is implemented, as shown in Figure 3. In the said figure, the learner is captured looking at the screen (left), reading (middle), and looking at an adjacent learner's PC (right). In the two-second video taken during the examination period, the images in the video are compared with each other to verify if the learner was looking somewhere else other than his/her screen. If, for all images within the two seconds, the learner was not looking at the screen and therefore was not focused in solving the examination questions, he/she will be warned by a change in background color, as shown inFigure 4. Failed authentication will also be made visible to the learners by the change in background color.

If the authentication failure continues for more than a few seconds, the system will stop and perform collaborative verification. In this stage, the system will ask the user to put his/her face in an appropriate position to capture a new image. If the error is repeatedly encountered, the examination will not be administered for suspected cheating.

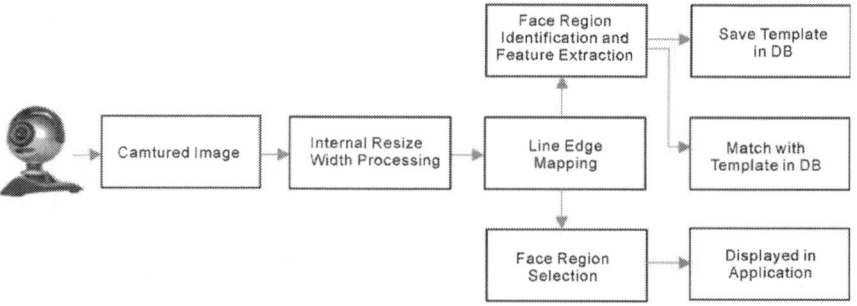

Figure 2. Face recognition system architecture.

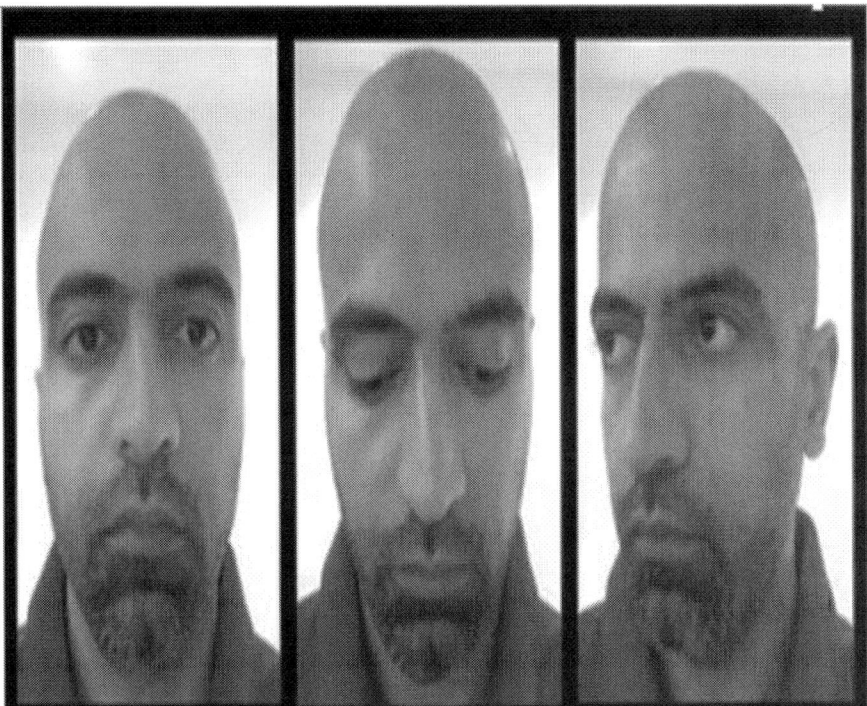

Figure 3. Images of learner at different positions.

In practice, for learners not to be disturbed and to minimize matching failures, a two-second video of the learner is recorded and the best image in terms of facial expressions, lighting, and resolution will be selected by an application previously installed on the client's side of the system. This image is sent to the server and will be compared with the stored image.

5. SURVEY RESULTS

Two surveys were conducted. The first survey targeted online instructors, and the second targeted potential users/students. The first investigation was conducted to identify the rate of image capture and the number of times the suspicious behavior is accepted without affecting learners' concentration during the examination. Moreover, e-learning instructors and experts (n = 8) were provided with the description of the proposed solution and were asked to rate the following statements using a five-point Likert scale (5—Strongly Agree to 1—Strongly Disagree): 1) the system will provide reliable results that reflect learners' achievements, 2) the system will effectively validate the learners' identity, 3) the probability of cheating during examinations will be minimized, 4) I will recommend this system to other instructors, and 5) the students will study harder. In addition, the instructors and experts were asked to specify the following information: 6) the capture rate of images for presence checks and 7) the acceptable number of times of suspicious behavior (warnings) during the examination. Figure 5 shows the results of the survey involving e-learning instructors and experts.

Table 1 presents the suggested rate of image capture and the suggested acceptable number of warnings.

Meanwhile, learners were engaged in an online quiz in a Blackboard environment and were provided with an explanation of the proposed system. The learners (n = 32) were asked to rate the following statements: 1) the probability of cheating during exams will be minimized, 2) the system's operation is easy to understand, 3) the system will benefit the student (e.g., encourage more focus in answering the examination), and 4) the system will encourage students to study harder. Figure 6shows the results of this survey involving learners.

6. CONCLUSIONS

Results show that almost all e-learning instructors agree with the given statements. However, some instructors expressed doubts regarding the effective authentication of learners and the reduced probability of cheating during examinations. Meanwhile, the students' survey showed a high percentage of neutral cases, mainly regarding the benefits that the system will provide to students.

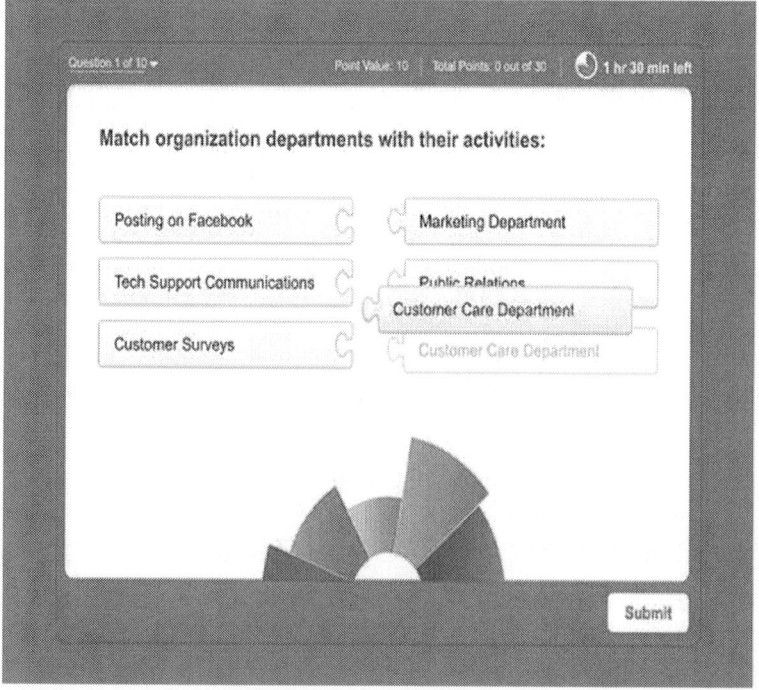

Figure 4. Red color as warning for learner with suspicious behavior.

Table 1. Suggested values from e-learning instructors/experts.

	Every 10 sec	Every 30 sec	Ever 1 min	Every 5 min	Every 10 min
The rate of images captured for presence checks	1	3	3	1	0
	1 time	2 times	3 times	4 times	5 or more
The number of times suspicious behavior is accepted during the exam	1	4	2	1	0

Two questions were asked to both e-learning instructors and students. These questions were whether the probability of cheating will decrease and whether the system will encourage students to study harder. Both parties believed that cheating will be reduced with the implementation of the facial recognition system and will motivate students to study harder. The facial recognition system should consider the image capture rate of 30 seconds for presence checks. This balanced value is appropriate to lessen the load on the system and to guarantee the level of reliability. Moreover, the learner with suspicious behavior should be warned twice.

Other applications of the facial recognition system include monitoring the attendance in virtual classrooms. In measuring the attendance rate, if the result of the student's verification is positive, the system will register an attendance mark for him/her. Otherwise, he/she will be marked absent.

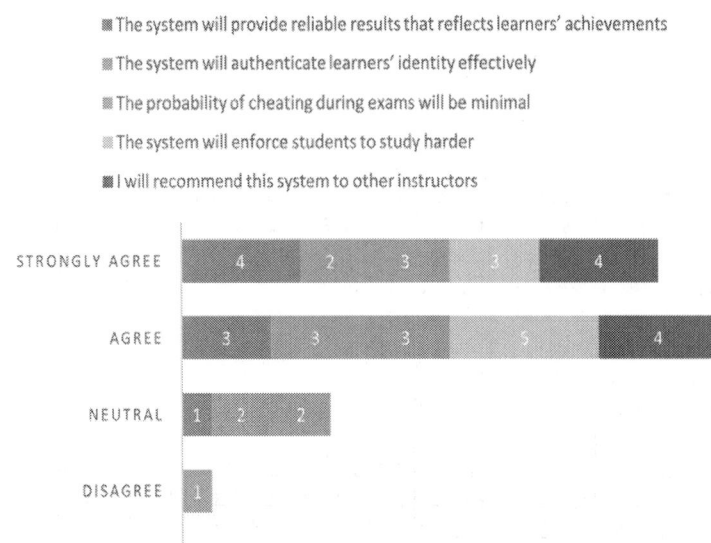

Figure 5. Results of e-learning instructors/experts survey.

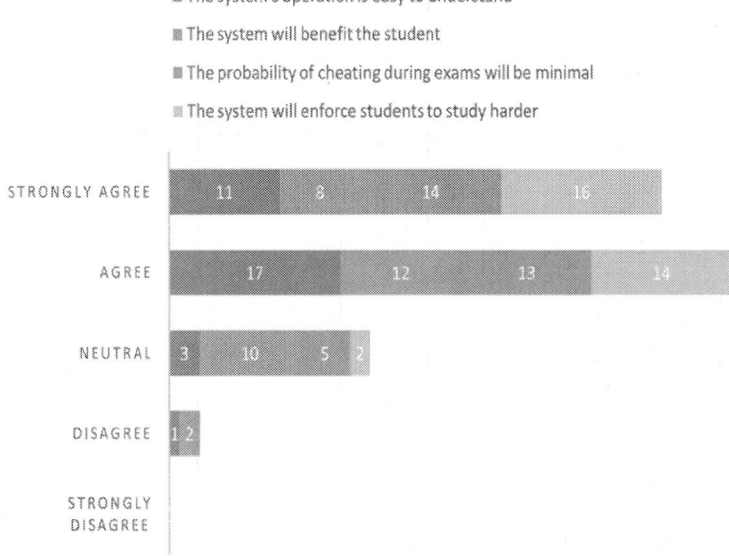

Figure 6. Results of e-learners survey.

REFERENCES

1. Assaf, W., Elia, G., Fayyoumi, A. and Taurino, C. (2007) Prospect of e-Learning: The Case of Jordan. e-Society 2007 —IADIS Multi Conference on Computer Science and Information Systems, Lisbon, 3-8 July 2007.
2. Ambient Insight (2010) The Worldwide Market for Self-Paced eLearning Products and Services: 2009-2014 Forecast and Analysis Report.
3. Rosenberg, M.J. (2001) E-Learning: Strategies for Delivering Knowledge in the Digital Age. McGraw-Hill, New York.
4. Hall, B. and Snider, A. (2000) Glossary: The Hottest Buzz Words in the Industry.
5. Scalise, K. and Gifford, B. (2006). Computer-Based Assessment in E-Learning: A Framework for Constructing "Intermediate Constraint" Questions and Tasks for Technology Platforms. Journal of Technology, Learning, and Assessment, 4.
6. Levy, Y. and Ramim, M. (2007) A Theoretical Approach for Biometrics Authentication of e-Exams. Nova Southeastern University, 93-101.
7. Huszti, A. and Petho, A. (2008) A Secure Electronic Exam System. Informatikafelsőoktatásban, 1-7.
8. Huang, W., Yen, D. C., Lin, Z.X. and Huang, J.H. (2004) How to Compete in a Global Education Market Effectively: A Conceptual Framework for Designing a Next Generation eEducation System. Journal of Global Information Management, 12, 84-107.
9. McGinity, M. (2005) Staying Connected: Let Your Fingers Do the Talking. Communications of the ACM, 48, 21-23.
10. Pillsbury, C. (2004) Reflections on Academic Misconduct: An Investigating Officer's Experiences and Ethics Supplements. Journal of American Academy of Business, 5, 446-454.
11. Yang, S. and Verbauwhede, I. (2003). A Secure Fingerprint Matching Technique. Proceedings of the 2003 ACM SIGMM workshop on Biometrics Methods and Applications, California, 89-94.
12. Hugl, U. (2005) Tech-Developments and Possible Influences on Learning Processes and Functioning in the Future. Journal of American Academy of Business, 6, 250-256.
13. Flior, E. and Kowalski, K. (2010) Continuous Biometric User Authentication in Online Examinations. Seventh International Conference on Information Technology IEEE Computer Society, Las Vegas, 12-14 April 2010, 488-492.
14. Yang, M., Kriegman, D. and Ahuja, N. (2002) Detecting Faces in Images: A Survey. IEEE Transactions on Pattern Analysis and Machine Intelligence, 24, 34-58.

CHAPTER 10

Performance Enhancement of Face Recognition in Smart TV Using Symmetrical Fuzzy-Based Quality Assessment

Yeong Gon Kim, Won Oh Lee, Ki Wan Kim, Hyung Gil Hong and Kang Ryoung Park

Division of Electronics and Electrical Engineering, Dongguk University, 26 Pil-dong 3-ga, Jung-gu, Seoul 100-715, Korea

ABSTRACT

With the rapid growth of smart TV, the necessity for recognizing a viewer has increased for various applications that deploy face recognition to provide intelligent services and high convenience to viewers. However, the viewers can have various postures, illumination, and expression variations on their faces while watching TV, and thereby, the performance of face recognition inevitably degrades. In order to handle these problems, video-based face recognition has been proposed, instead of a single image-based one. However, video-based processing of multiple images is prohibitive in smart TVs as the processing power is limited. Therefore, a quality measure-based (QM-based) image selection is required that considers both the processing speed and accuracy of face recognition. Therefore, we propose a performance enhancement method for face recognition through symmetrical fuzzy-based quality assessment. Our research is novel in the following three ways as compared to previous works. First, QMs are adaptively selected by comparing variance values obtained from candidate QMs within a video sequence, where the higher the variance value by a QM, the more meaningful is the QM in terms of a distinction between images. Therefore, we can adaptively select meaningful QMs that reflect the primary factors influencing the performance of face recognition. Second, a quality score of an image is calculated using a fuzzy method based on the inputs of the selected QMs, symmetrical membership functions, and rule table considering the characteristics of symmetry. A fuzzy-based combination method of image quality has the advantage of being less affected by the types of face databases because it does not perform an additional training procedure. Third, the accuracy of face recognition is enhanced by fusing the matching scores of the high-quality face images, which are selected based on the quality scores among successive face mages. Experimental results showed that the performance of face recognition using the proposed method was better than that of conventional methods in terms of accuracy.

Keywords: video-based face recognition in smart TV; quality measures; symmetrical fuzzy-based quality assessment; fusing matching scores

1. INTRODUCTION

With the rapid development of smart TVs, these are being used for various functionalities such as TV broadcasting, social network service (SNS), video on demand (VOD), television-commerce (T-commerce), and teleconferencing services. Recently, the applications of smart TV have been diversified for viewers' usability, and many commercialized smart TVs are equipped with a camera to provide intelligent services and high convenience to viewers [1,2,3]. In particular, user identification is required for some services such as T-commerce and T-banking in a smart TV environment. Considering the convenience of the viewers, biometrics can be used to provide user identification. As a typical example of biometric techniques, face recognition is a reasonable solution in a smart TV environment because it does not require the viewer to come in contact with the special sensor on the TV but can be executed only by the camera of the smart TV. Therefore, this can easily facilitate the use of face recognition in many applications with high convenience. However, the viewer can have various facial poses, illumination, and expression variations while watching TV, and the performance of face recognition is inevitably degraded due to these factors. There have been many studies focusing on these problems.

Two main approaches for face recognition that are robust to illumination changes have been proposed. One of them is to represent images with features that are insensitive to illumination changes [4,5]. The other approach is to construct a low-dimensional linear subspace for face images by considering different lighting conditions [6,7]. Pose variability is usually regarded as the most challenging problem in the field of face recognition. There are different algorithms for dealing with the pose variation problem [8]. In general, they can be categorized as follows: (1) the approach of extracting invariant features; (2) the multiview-based approach; and (3) the approach of using a three-dimensional (3D) range image. The approach of extracting invariant features uses some features in a face image that do not change under pose variation [9,10,11]. The multiview-based approach stores a variety of view images in the database to handle the pose variation problem or to synthesize new view images from a given image. Recognition is then carried out using both the given image and the synthesized images [12,13,14]. Face recognition from 3D range images is another approach being vividly studied by researchers. Because the 3D shape of a face is not affected by illumination and pose variations, the face recognition approach based on 3D shape has evolved as a promising solution for dealing with these variations.

The above approaches can handle some variations on face but still have drawbacks that restrict their application. That is, since some features invariant to one variation can be sensitive to other variations, it is difficult to extract features that are completely immune to all kinds of variation. Thus, it is unsafe to heavily rely on the selection of invariant features for a particular variation. These limitations of single image-based face recognition have motivated the development of video-based face recognition [15,16].

There are some major advantages of video-based face recognition. First, spatial and temporal information of faces in a video sequence can be used to improve the performance of single image-based face recognition. Second, recent psychophysical and neural researches have shown that dynamic information is very important for the human face recognition process [17]. Third, with additional information of various poses and face size, we can acquire more effective representations of faces such as a 3D face model [18] or super-resolution images [19], which can be used to improve the recognition performance. Fourth, video-based face recognition can adopt online learning techniques to update the model over time [20]. Even though there are obvious advantages of video-based recognition, there are some disadvantages such as the successive face images captured can have the factors of poor video quality, low image resolution, and pose, illumination, and expression variations. In spite of all these advantages and disadvantages, various kinds of approaches for video-based face recognition have been implemented.

A simple way to process the faces in a video sequence is to keep and use all the images in the sequence for face recognition. However, the use of all images of the video sequence can incur steep computational costs and does not guarantee optimal performance. Moreover, parts of these face images in the sequence are useless because of motion blur, non-frontal pose, and non-uniform illumination. Therefore, a method that chooses the best face images in terms of quality is required for a video sequence. This is based on face quality assessment (FQA), and the set of high-quality face images is denoted as Face Log [21]. Since the performance of face recognition is affected by multiple factors, the detection of one or two quality measures (QMs) is insufficient for FQA. An approach to simultaneously detect multiple QMs is the use of a fusion of QMs for FQA. Hsu *et al.* proposed a framework to fuse individual quality scores into an overall score, which is shown to be correlated to the genuine matching score (the matching score when an input face image is correctly matched with the enrolled one of the same person) of face recognition engine [22]. Many studies adopted a weighted quality fusion approach to combine QMs, and these weighted values for fusion were experimentally determined or obtained through training [23,24,25,26,27]. In addition, Anantharajah *et al.* used a neural network to fuse QM scores, where the neural network was trained to produce a high score for a high-quality face image [28]. On the other hand, Nasrollahi *et al.* proposed a method that uses a fuzzy combination of QMs instead of a linear combination [21]. A fuzzy inference engine was adopted for improving the performance without training, but a few fixed QMs were used in this method. In addition, they provided only the accuracy of quality assessment by their method, not showing the performance enhancement of face recognition.

All the previous studies used a few fixed QMs, and their measures are difficult to reflect primary factors that influence the performance of face recognition. For example, although some QMs that can assess facial pose and illumination variation are selected in a face recognition system, if other factors (such as blurring and expression variation) occur in the system, the performance of face recognition degrades.

Therefore, in this study, QMs are adaptively selected by comparing variance values obtained from candidate QMs within a video sequence, where the higher the variance value by a QM, the more meaningful is the QM in terms of a distinction between images. Therefore, the selected QMs can reflect primary factors that influence the performance, and the fusion of QMs is carried out by a symmetrical fuzzy system. Based on the selected high-quality images (Face Log), the performance of the proposed method is enhanced by fusing matching scores.

Table 1 shows a summary of comparisons between methods discussed in previous research and the proposed method.

The remainder of this paper is organized as follows: in Section 2, the proposed method is described. Experimental results and conclusions are presented in Section 3 and Section 4, respectively.

Table 1. Comparison of previous and proposed methods.

Category	Method			Strength	Weakness
Single image-based method	• Using the features of a single image for face recognition [4–14]			• Less time for image acquisition and processing	• Features invariant to one variation can be sensitive to other variations
	Not using image selection	• Using spatial and temporal information of faces in a video sequence [18–20]		• With redundant information, more sophisticated representations of faces can be reconstructed, which can be applied to improve recognition performance	• Using all images of the video sequence can incur steep computational costs and does not guarantee optimal performance • Parts of face images are useless because of motion blur, non-frontal pose, and non-uniform illumination
Video-based method	Selecting Face Log based on QMs	Using fixed QMs	Training for fusion [22,23,27,28]	• Optimal performance of face recognition can be obtained	• The performance of face recognition depends on the training set; therefore, it is difficult to select the training data with wide variations
			No training for fusion [21,24–26]	• Higher accuracy than the video-based method not using image selection and the single image-based method	• The performance improvement is limited because the fusion is experimentally or empirically determined with fixed QMs
	Adaptive selection of QMs and fusion of quality score without training (**proposed method**)			• QMs can be adaptively selected, and they can reflect primary factors that influence performance	• Additional processing time is required for the adaptive selection of QMs

2. PROPOSED METHOD

2.1. Overview of Proposed Method

Figure 1 shows the environment where the proposed face recognition system for smart TVs is applied. We used a universal serial bus (USB) camera of visible light (Webcam C600 by Logitech Corp. [29]) with an additional fixed (focal) zoom lens (1.75×), and the resolution of the captured image is 1600 × 1200 pixels with 3 (RGB) bytes per pixel. In our system, the camera has the interface of USB 2.0, and its maximum bandwidth is lower than 480 Mb/s. Therefore, the capturing speed of our camera is about 10 frames/s due to the limitation of the bandwidth of the USB interface. The Z distance between the user and the camera is 2–2.5 m, and the size of the TV screen is 60 inches.

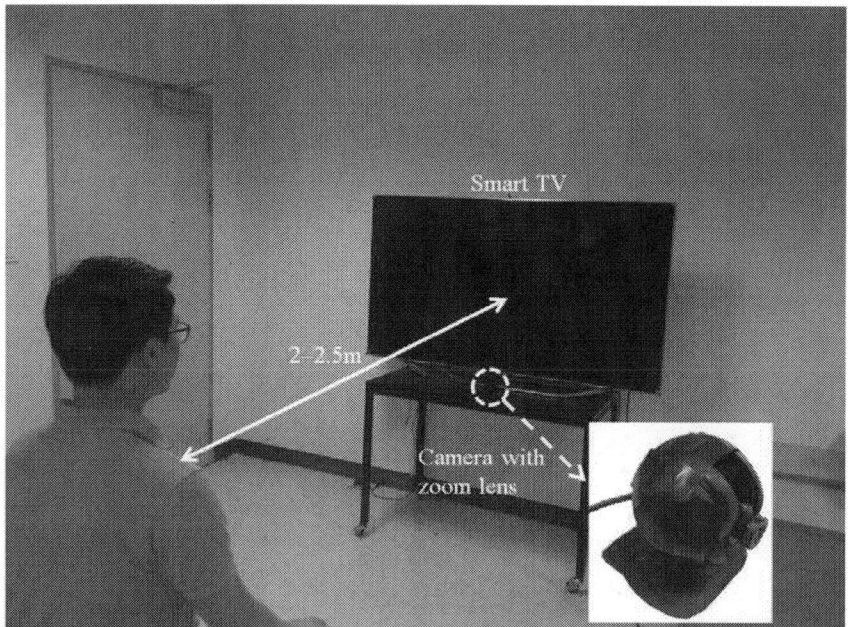

Figure 1. Environment in which the proposed face recognition system for smart TVs is used.

Figure 2 shows an overview of the proposed method. In the initial enrollment stage, we capture RGB color images of the user at a predetermined Z distance of 2 m (Step 1 in Figure 2) [30]. In Step 2, the user looks at five points (left-top, right-top, center, left-bottom, and right-bottom) on the TV screen by rotating his head. This step facilitates the acquisition of facial features (face texture, eyes, and nostrils) in Step 3, where we can obtain not only the information of facial features but also that of face histogram features according to the various poses of the head. In the recognition stage, RGB color images are captured by the camera (Step 4). In

Steps 5 and 6, the face regions of the captured images are detected by the adaptive boosting (Adaboost) algorithm [31], and then, the detected face region is tracked using the continuously adaptive mean shift (CamShift) algorithm [32]. In Step 7, the eye regions are detected by Adaboost and adaptive template matching (ATM) methods, and sub-block-based template matching is performed for nostril detection in Step 8. In addition, our method selects several face images based on the QMs and fuzzy-based quality assessment in Step 9. Finally, in Step 10, face recognition is carried out by fusing the matching scores of the selected images.

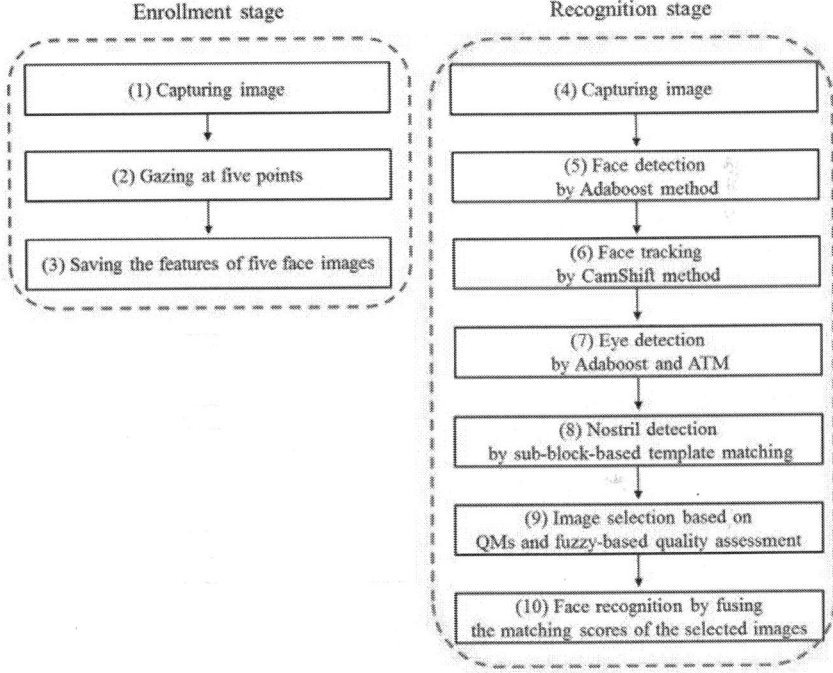

Figure 2. Overall procedure of the proposed method.

The third step addresses human detection in the combined difference image. After applying size filtering and a morphology operation based on the size of the candidate areas, noise is removed. The remaining areas are separated by a vertical and horizontal histogram of the detected regions using the intensity of the background. Detected regions that may have more than two human areas are merged. Therefore, further procedures are performed to separate the candidate regions and to remove noise regions. These are based on the information of the size and ratio (of the height to width) of the candidate regions considering the camera-viewing angle and perspective projection (see the details in Section 2.4). Finally, we obtain the correct human areas.

2.2. Face and Facial Feature Detection

In the proposed method, the five still images are acquired through the initial enrollment stage in order to enhance the accuracy of face recognition irrespective of the head poses. In the initial enrollment stage, the user is in front of the smart TV at a Z distance of 2 m and then, looks at five points on the TV with the head movements. Figure 3 shows the examples of face images in the enrollment stage. The face and eye regions of the captured image are detected using the Adaboost method [31]. In addition, the nostril area is detected using sub-block-based template matching [30]. Based on these facial features (the face, eye, and nostril), we not only save the information of these features to estimate the head pose of the image in the recognition stage but also enroll face histogram features according to the head pose to identify each user.

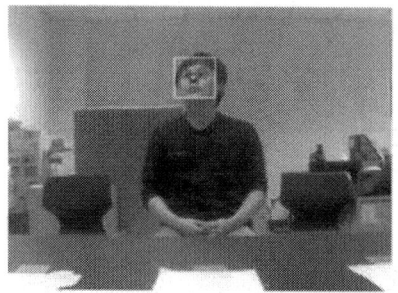 

Figure 3. Face and facial features obtained in the enrollment stage.

In the recognition stage, the information of successive images is used for face detection and tracking as shown in the Steps 5 and 6 of Figure 2. The face region detected by Adaboost is tracked using the CamShift method [30,32]. Even though the Adaboost method has a high detection rate, it is slightly time consuming. On the other hand, face tracking using the CamShift method has the advantages of higher processing speed and of being more robust to head pose variations.

The eyes are detected and tracked using Adaboost and ATM, respectively. Furthermore, the nostril area is detected using a nostril-detection mask based on sub-block-based template matching, and it is tracked using ATM [30]. With these facial features, we can obtain face images for assessing their quality. Figure 4 shows the examples of the detected and tracked face and facial features in the recognition stage.

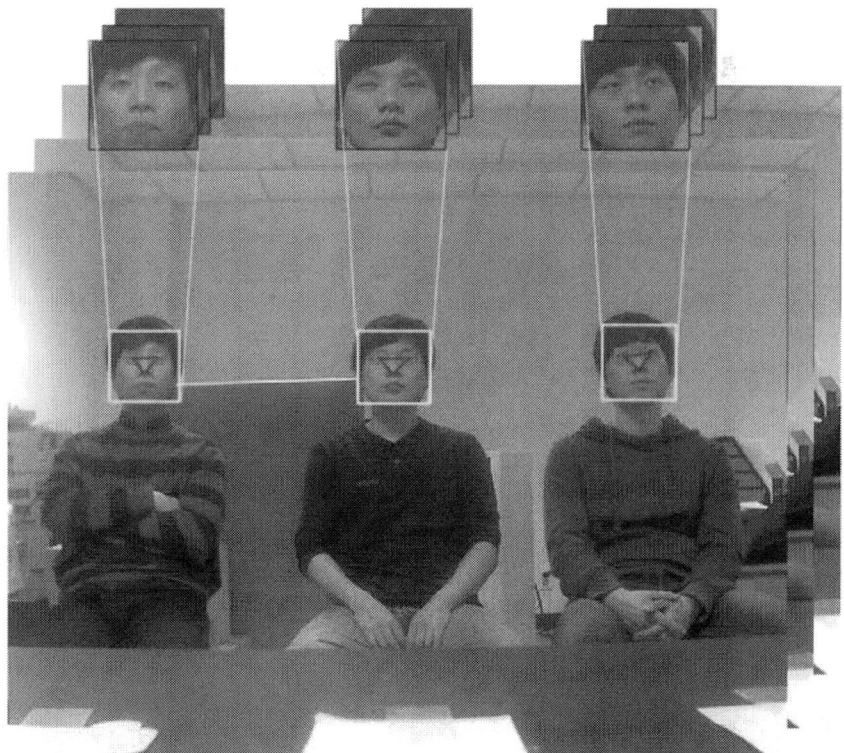

Figure 4. Face and facial features obtained in the recognition stage (Steps 5 through 8 of Figure 2).

2.3. Assessing the Quality of Face Images

In order to select high-quality face images, we need to determine a method for their assessment even though the quality of a face image is rather subjective. In this study, we assume that a high quality face image corresponds to the image's potential for correct identification. This section describes the methods of assessing the quality of face images. Six QMs are used to assess the quality of face images in our research. These QMs are as follows: head pose, illumination, sharpness, openness of the eyes, contrast, and image resolution. We outline these features in the following subsections:

2.3.1. Head Pose

One of the biggest challenges in face recognition is to identify individuals despite variations in the head pose. In general, the out-of-plane rotation (rotation in the horizontal or vertical direction) of face occurs while face recognition is carried out. Therefore, it is important that the QM for the head pose distinguishes between various rotations and assigns a higher score to the image closest to one of enrolled images in terms of the head pose. In this study, we estimate the head pose based on the detected eye and nostril points [30]. In order to estimate out-of-plane rotation, we utilize changes in distances between the two eyes, and those between the eyes and the nostril. Therefore, we calculate the rotation angles in the directions of X (horizontal) and Y (vertical) axes based on the distances between the detected eye and nostril points. As described in Steps 2 and 3 of Figure 2, facial feature points and face histogram features are obtained according to the head pose, and thereby, the head pose in an image in the recognition stage can be measured using the following equation:

$$P(t, a) = \sqrt{(xm_a - x_t)^2 + (ym_a - y_t)^2} \qquad (1)$$

where P(t,a) denotes the pose value; t, the face image number; and a, the pose number that represents one of the five head poses obtained in the enrollment stage. The calculated rotation angles in the directions of X and Y axes are xt and ytwith respect to the face image in the recognition stage. xma and yma are the average rotation angles of all the enrolled users according to head pose a. Therefore, we can obtain the final pose value F1 of the t-th image, which is the closest head pose value among the five pose values, as follows:

$$F_1 = \min_{1 \leq a \leq 5} P(t, a) \qquad (2)$$

2.3.2. Illumination Based on the Symmetrical Characteristics of Left and Right Face Regions

Variations caused by illumination changes are regarded as another significant challenge in face recognition. When a face is evenly lit up, shadows or a saturated area does not appear and uniform illumination exists in the entire face region. Based on this hypothesis, we divide the face region into left and right regions and then, calculate the difference between average values of each region by using the following equation:

$$I = \frac{1}{W \times H} \sum_{x=0}^{x=W-1} \sum_{y=0}^{y=H-1} |IMG(x,y)_t| \qquad (3)$$

$$F_2 = |I_L - I_R| \quad (4) \qquad\qquad (4)$$

where I denotes the average value of a face region. $IMG(x,y)$t represents the pixel value at the position (x,y) of the t-th image, and W and H denote the height and width of the image, respectively. F_2 represents the illumination value based on the difference between the average values of the left and right face regions. Based on the symmetrical characteristics of left and right face regions, F_2 inevitably becomes small in case of uniform illumination in the entire face region.

2.3.3. Sharpness

A user's head can usually move in front of the camera. Therefore, it is possible that the captured image is affected by motion blur, which decreases the quality of the image. Thus, defining a QM for sharpness is useful. Blurred images have less sharpness than well-focused images, and they are designed to yield a lower score for QM, as shown in the following equation [28]:

$$F_3 = \frac{\sum_{x=0}^{x=W-1} \sum_{y=0}^{y=H-1} |IMG(x,y)_t - LowPass(IMG(x,y)_t)|}{W \times H} \qquad (5)$$

where $IMG(x,y)_t$ denotes the pixel value at the position (x,y) of the t-th image and $LowPass(IMG(x, y)_t)$ represents the result of applying a low-pass filter to $IMG(x,y)_t$. By obtaining the difference between $IMG(x,y)_t$ and $LowPass(IMG(x, y)_t)$, F_3 can reflect the amount of mid- and high-frequency components of $IMG(x,y)_t$.

2.3.4. Openness of the Eyes

In this study, we check the openness of the eyes because information related to the eyes is important for face recognition. We use p-tile thresholding method to segment the detected eye image into the eye and background regions. After obtaining the eye region, we use component labeling to remove noise like hair or eyebrows. Finally, the obtained binarized image is projected onto the horizontal axis (x-axis), and the histogram of the black pixels is obtained. Figure 5a,b shows the examples of the binarized image and the corresponding histogram of open and closed eyes. As shown in Figure 5a, there are more black pixels in the mid area of the histogram than in the side areas in the case of an open eye (higher standard deviation of the number of black pixels). However, the numbers of black pixels are similar in both the mid and side areas in the case of a closed eye, as shown in Figure 5b (lower standard deviation of the number of black pixels).

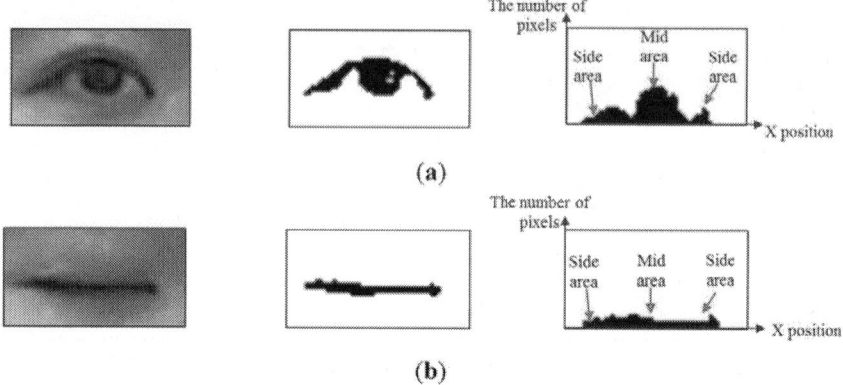

(a)

(b)

Figure 5. Examples of the binarized image and the corresponding histogram of open and closed eyes: (a) open eye; (b) closed eye.

Based on this characteristic, the standard deviation of the number of black pixels is calculated using the following equation:

$$F_4 = \frac{1}{n-1} \sum_{i=1}^{n} (x_i - \bar{x})^2 \tag{6}$$

where F_4 denotes the openness value calculated using the standard deviation of the number of black pixels projected onto the horizontal axis, and i and n represent the position of a projected pixel and the range of the histogram, respectively, as shown in Figure 5a,b. x_i indicates the number of black pixels at the i-th position, and $\bar{x}$ refers to the average of the number of black pixels. As shown in Figure 5a,b, we can estimate that F_4 of the open eye can exhibit a high value, whereas F_4 of the closed eye shows a low value.

2.3.5. Contrast

Contrast has a considerable influence on the quality of an image in terms of human visual perception as well as in an image analysis. A poorly illuminated environment affects the contrast and produces an unnatural image. In this study, the contrast value F5 is calculated using the methods discussed in [28,33] as follows:

$$F_5 = \frac{H_{q3} - H_{q1}}{I_r} \tag{7}$$

where H_{q3} and H_{q1} represent histogram bins at which the cumulative histogram has a value that is 75% and 25% of the maximum value, respectively. Ir denotes the intensity range of the image. In general, the histogram of a high-contrast image has a wider range of intensity than that of a low-contrast image. Therefore, F5 increases in the case of a high-contrast image.

2.3.6. Image Resolution

Image resolution is the easiest way to measure image quality. In general, high-resolution face images are preferred to low-resolution ones because such images can yield better recognition results. In order to reflect this characteristic in the quality assessment, we measure the inter-distance between two detected eyes as QM (F6) and assign the highest score to a high-resolution face image.

2.3.7. Normalization of Features for Quality Assessment

In order to compare a face image of a person with another face image of the same person in a video sequence, it is necessary to assign a quality score to each image. Because the ranges of the six QMs (F1–F6) are different from each other, we obtain normalized QMs in the range of 0–1 based on the minimum and maximum value of QM. As shown in Section 2.3.1, Section 2.3.2, Section 2.3.3, Section 2.3.4, Section 2.3.5 and Section 2.3.6, a good-quality image shows lower values of F1 and F2 and higher values of F3–F6. To make all the values of QMs consistent with each other, we ensure that the good-quality image has higher values of F1 and F2 throughout the normalization procedure.

2.4. Selecting High Quality Face Images Based on Fuzzy System Using Symmetrical Membership Function and Rule Tables Considering the Characteristics of Symmetry

After obtaining the six QMs (F1–F6), we adaptively select some of them as primary QMs (reflecting primary factors) that influence the performance of the recognition. This is because the factors degrading the performance of face recognition vary according to the video sequences. For example, in the first video sequence, the factors of illumination and pose variations are dominant, whereas those of variations in image resolution and sharpness become dominant in the second video sequence. Therefore, the use of a fixed number of QMs in all the video sequences cannot cope with these all these factors.

To overcome these problems, we propose a method in which four QMs are adaptively selected by comparing the variance values obtained from the six QMs within a video sequence, where the higher the variance value by a QM, the more meaningful is the QM in terms of a distinction between images. Therefore, these selected QMs can reflect the primary factors that influence the performance.

We use a fuzzy system to combine the values obtained by the four QMs and to obtain the final quality score of the image, as shown in Figure 6. Fuzzy systems are widely used in many applications such as edge detection [34], image segmentation [35,36] because of its aptitude to deal with nonlinearities and uncertainties. For example, Barghout [34] used the fuzzy system for edge detection and recognition. In another example, image segmentation problem was dealt with by a nested two-class fuzzy inference system [35]. In [36], he proposed the method of image segmentation based on iterative fuzzy-decision making.

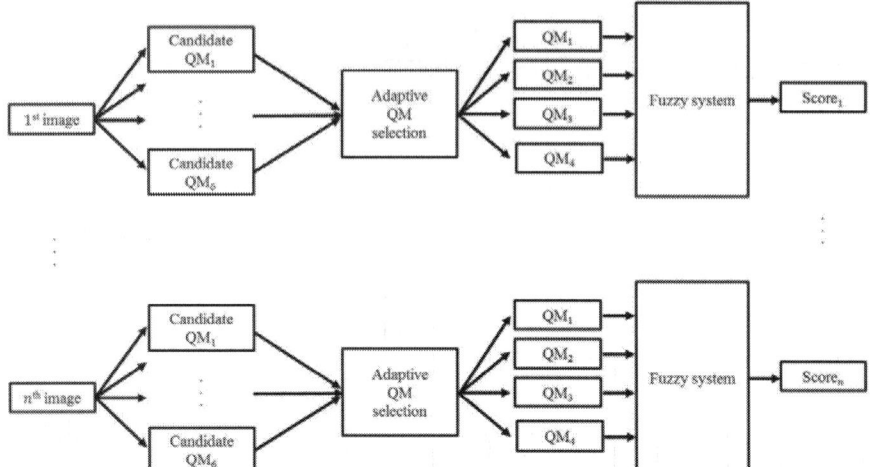

Figure 6. Procedure for calculating the final quality score of the input image using a fuzzy system.

In order to use the fuzzy system, we use Mamdani's method [37], which is utilized most commonly in many applications because of its simple structure of "min-max" operations. In addition, the universe of discourse is chosen to be the real number in the range of [0,1] for the input and output membership functions. The shape of membership function is often dependent on its applications, and can be determined through a heuristic process.

In this paper, we use symmetric triangular membership functions for the fuzzy system. There are two reasons for this. The first is that the triangular membership function is simple to implement and fast for computation [38,39]. Another reason is that it is difficult to adopt any particular shape of membership function based on a priori information about the distribution of input and output values. This is because the selected QMs for the fuzzy system are changeable according to the variance value obtained from each QM.

Therefore, the input membership function is designed according to the input value (QM_1–QM_4), as shown in Figure 7, and henceforth, we will call these QM_1–QM_4 values as factors 1–4.

The larger the values of factors 1–4, the better the quality of the image is. As shown in Figure 7, the input values of the fuzzy system are categorized into two types: low (L) and high (H). The output membership function of the fuzzy system is designed as shown in Figure 8. Here, the output value (Score in Figure 6) can be categorized as low (L), middle (M), and high (H). In general, the membership function is designed considering the distributions of input features and output values. Because we can assume that the distributions of L and H are similar to each other, we design that the membership functions of L and H become symmetrical based on 0.5. In addition, because we can assume the characteristics of left part of M distribution based on its mean (0.5) is similar to that of right part, the membership function of M is designed to become symmetrical based on its mean (0.5).

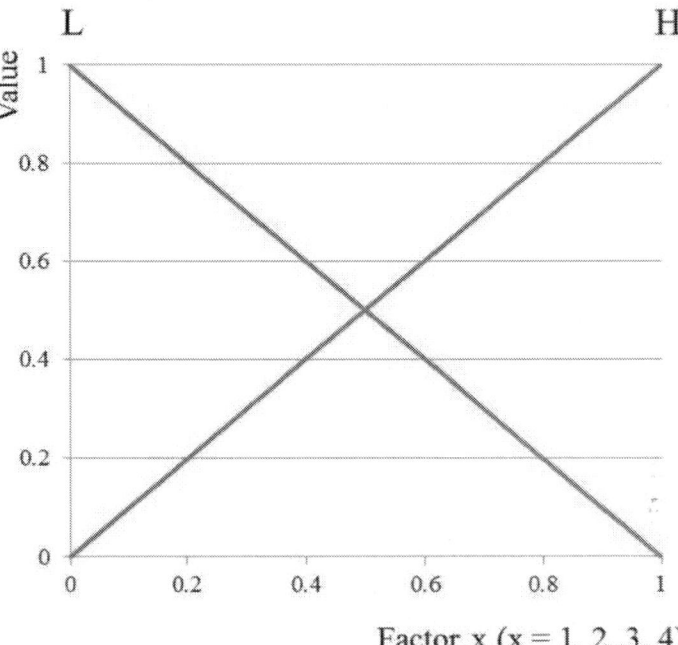

Figure 7. Symmetrical input membership function (SIMF).

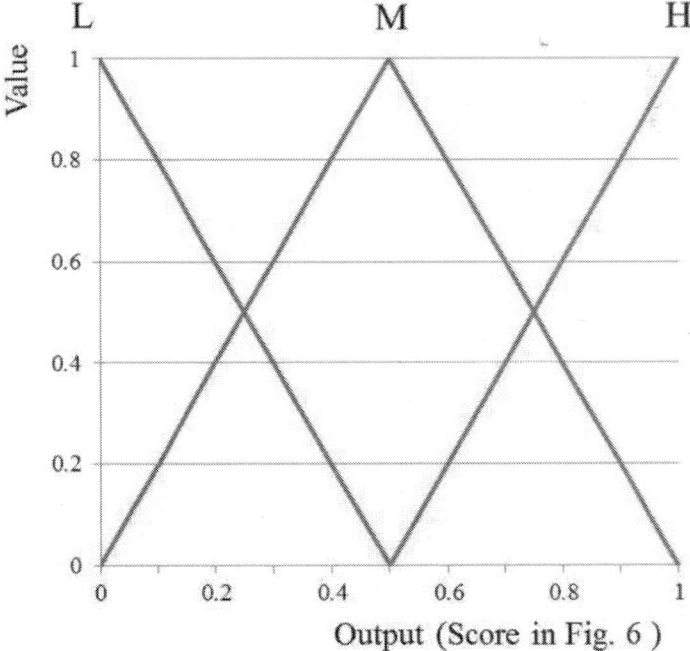

Figure 8. Symmetrical output membership function (SOMF).

As explained earlier, larger values of factors 1–4 represent a higher-quality image. With this hypothesis, we can define the relationships between the input (factors 1–4) and output values (scores) based on the fuzzy rules, as shown in Table 2. For example, if the values of all factors are low (L), image quality is considered to be low, and we assign the output value to be low (L). In addition, if the values of two factors are low (L) and those of the others are high (H), we assign the output value to be middle (M). If the values of all the factors are high (H), the image quality is considered to be high and we assign the output value to be high (H).

Because we consider that all the weights to factors 1–4 are similar, the fuzzy rule table of Table 2 is designed considering the characteristics of symmetry. That is, if output is L with the factors 1–4 of L, L, L, and L, respectively, the output becomes H with the factors 1–4 of H, H, H, and H, respectively. If the output is L with the factors 1–4 of L, H, L, and L, respectively, the output becomes H with the factors 1–4 of H, L, H, and H, respectively.

Table 2. Designed fuzzy rules of relationships between the input and output values considering the characteristics of symmetry.

Factor 1	Factor 2	Factor 3	Factor 4	Output
L	L	L	L	L
			H	L
		H	L	L
			H	M
	H	L	L	L
			H	M
		H	L	M
			H	H
H	L	L	L	L
			H	M
		H	L	M
			H	H
	H	L	L	M
			H	H
		H	L	H
			H	H

Based on the fuzzy membership functions and the fuzzy rules considering the characteristics of symmetry, we can obtain the output (score). Using the input membership function, one input value of a factor corresponds to two output values, as shown in Figure 9. Since there are four input values (factors

1–4), a total of eight output values can be obtained. For example, we can obtain a pair of two output values (0.2 (L) and 0.8 (H)) that are produced from factor 1 (0.8), as shown in Figure 9. Similarly, if we assume that the value of factors 2–4 is 0.8, same as that of factor 1, we can obtain the following four pairs of two output values: {(0.2 (L), 0.8 (H)), (0.2 (L), 0.8 (H)), (0.2 (L), 0.8 (H)), (0.2 (L), 0.8 (H))} from factors 1–4. From these four pairs, the following sixteen combination pairs of the output values are then obtained: {(0.2 (L), 0.2 (L), 0.2 (L), 0.2 (L)), (0.2 (L), 0.2 (L), 0.2 (L), 0.8 (H)), (0.2 (L), 0.2 (L), 0.8 (H), 0.2 (L)), ···, (0.8 (H), 0.8 (H), 0.8 (H), 0.8 (H))}. With one subset, one output value (0.2 or 0.8) and its symbol (L, M, or H) are determined based on the MIN or MAX method and the fuzzy rules presented in Table 2 [40]. MIN method selects one minimum output value among all values whereas MAX method selects one maximum output value among all values.

For example, we can determine the output value to be 0.8 based on the MAX method with one subset (0.2 (L), 0.2 (L), 0.2 (L), 0.8 (H)). Furthermore, its symbol can be determined as L using Table 2 (where factors 1–4 are L, L, L, and H, respectively, and the output is L, as shown in Table 2).

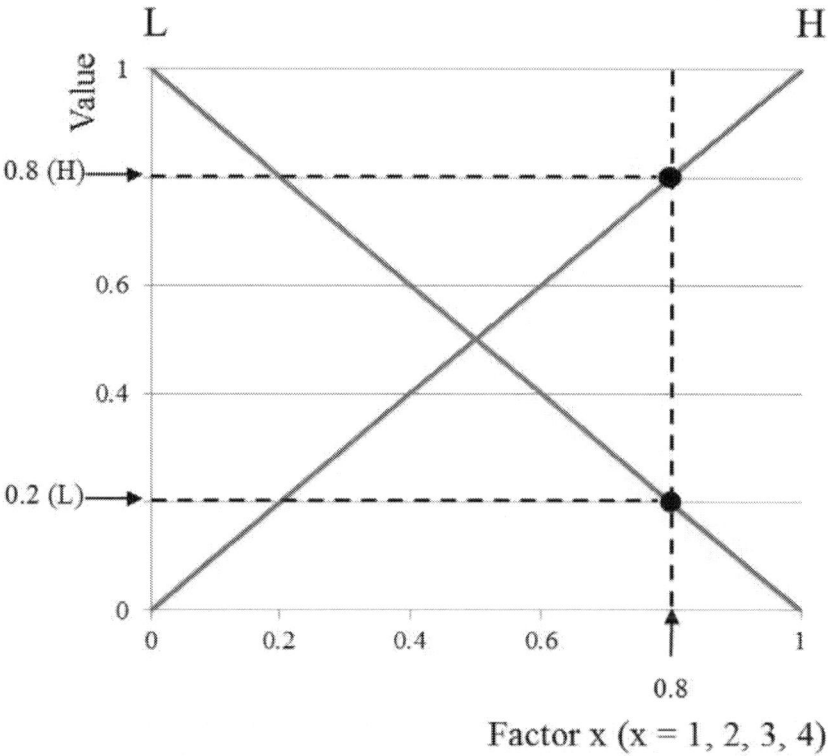

Figure 9. Obtaining two output values from one input factor using the input membership function.

Accordingly, we obtain 0.8 (L) from (0.2 (L), 0.2 (L), 0.2 (L), 0.8 (H)). We label the value (0.8 (L)) as the inference value (IV) in this paper. If the MIN method is applied, the obtained IV is 0.2 (L). Consequently, sixteen types of IVs are determined (from the sixteen combination pairs of the output values {(0.2 (L), 0.2 (L), 0.2 (L), 0.2 (L)), (0.2 (L), 0.2 (L), 0.2 (L), 0.8 (H)), (0.2 (L), 0.2 (L), 0.8 (H), 0.2 (L)), $\cdots$, (0.8 (H), 0.8 (H), 0.8 (H), 0.8 (H))} based on the MIN or MAX methods and the fuzzy rules presented in Table 2.

Based on these sixteen IVs, we can obtain the final score using a defuzzification step. Figure 10 shows an example of defuzzification on the basis of the membership function for the output value (score) and the IVs. With each IV, we can obtain either one or two outputs (scores), as shown in Figure 10. If the IV is 0.2 (L), its corresponding output is S_1. Thus, multiple outputs (S_1, S_2, $\cdots$, S_N) are obtained from the 16 IVs, and thereby, we can determine the final output score based on the defuzzification method [41]. Five defuzzification methods such as the first of maxima (FOM), last of maxima (LOM), middle of maxima (MOM), mean of maxima (MeOM), and center of gravity (COG) can be considered [41]. The FOM chooses the first output (S_2) among the outputs calculated using the maximum IV (0.8 (M)) as the output (score). The LOM chooses the last output (S_3) among the outputs calculated using the maximum IV (0.8 (M)) as the output (score). The MOM chooses the middle output (($S_2 + S_3$)/2) among the outputs using the maximum IV (0.8 (M)) as the output (score). Finally, the MeOM chooses the mean output (($S_2 + S_3$)/2) among the outputs by using the maximum IV (0.8 (M)) as the output (score) [41]. The output (score) of the COG is determined as S_5, as shown in Figure 10b, from the geometrical center (G in Figure 10b) of the union area of three regions (R_1, R_2, and R_3). In our research, we calculate the center based on the weighted average of all the regions defined by all the IVs.

Using the output scores ($Score_1$, $Score_2$, $\cdots$, $Score_n$ shown in Figure 6), we can construct a Face Log (face images of higher quality) for face recognition. Here, the score is stored for each face region of an observed person in a video sequence with n images, and we choose m face images by our method to build the Face Log in the order of the quality score. Figure 11 shows an example of a video sequence and the finally selected face images in the Face Log. The quality score of Figure 11b implies $Score_i$ of the i-th image of Figure 6.

2.5. Face Recognition

The selected m face images in the Face Log are used for face recognition. After obtaining the selected face image, the face region is redefined based on the positions of the detected and tracked eyes to normalize the size of the face region. This is because there can be variable sizes of the face regions depending on the Z distance between the camera and user's face. In addition, there can be illumination variations in the face region while a user is watching TV. In order to solve this problem, retinex filtering is performed for illumination normalization [42]. Using the face image after performing illumination normalization, features for face recognition are extracted.

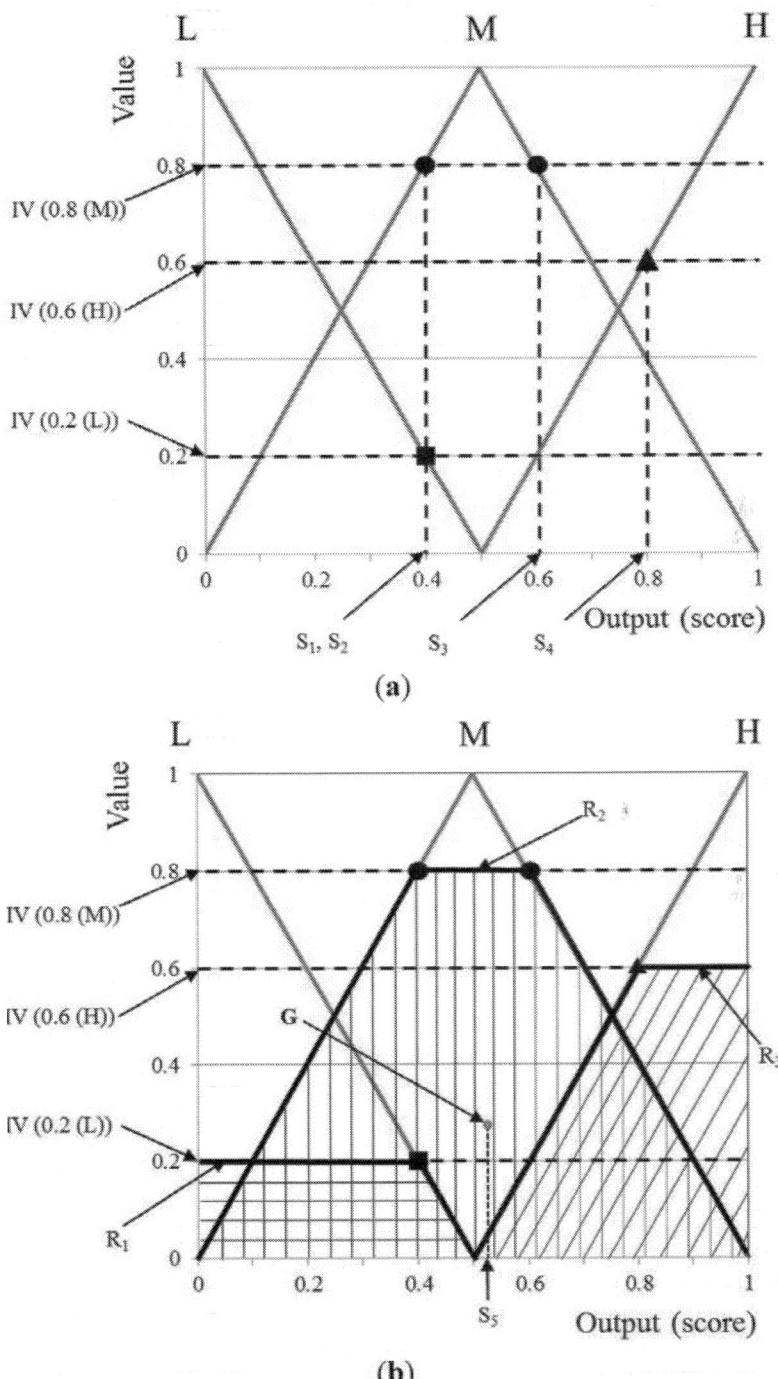

Figure 10. Obtaining the final output (score) by a defuzzification method: (**a**) first of maxima (FOM), last of maxima (LOM), middle of maxima (MOM), and mean of maxima (MeOM); (**b**) center of gravity (COG).

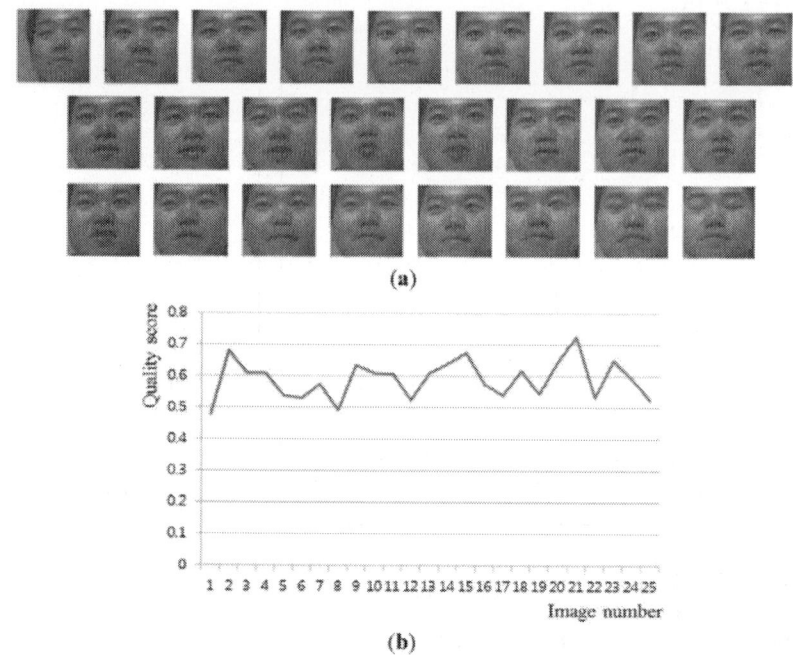

(a)

(b)

(c)

Figure 11. Example of a video sequence and the selected face images in Face Log: (**a**) all the n images in a video sequence; (**b**) quality score curve of the video sequence; (**c**) selected m images in Face Log.

Many techniques for extracting features for face recognition have been proposed in previous studies, such as principal component analysis (PCA) [43], linear discriminant analysis (LDA) [11], and local binary pattern (LBP) [44]. In our research, the normalized face image is used for extracting features based on multi-level binary pattern (MLBP) [45,46]. Figure 12illustrates the concept of MLBP feature extraction technique.

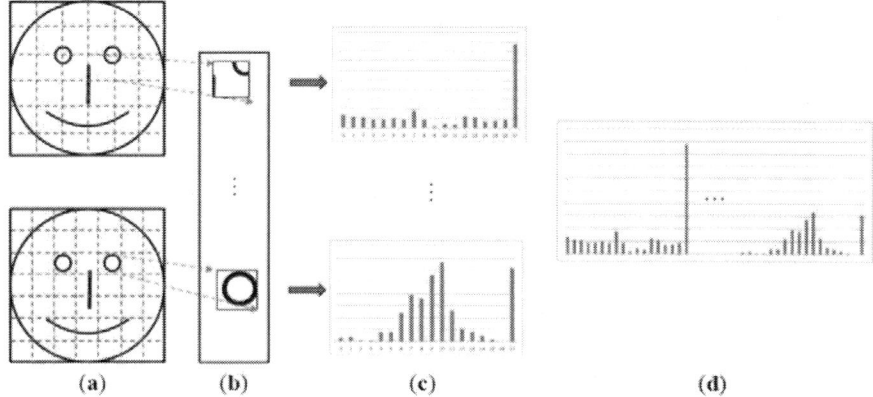

Figure 12. Extracting histogram features based on multi-level binary pattern (MLBP) at two levels: (**a**) a face image divided into sub-blocks; (**b**) sub-block regions; (**c**) histogram of (**b**) obtained by local binary pattern (LBP); (**d**) final histogram feature obtained by the histogram of (**c**).

The face region is divided into sub-blocks, and the LBP histograms from each block are obtained as shown in Figure 12c. The final histogram feature is concatenated from the blocks of all histograms in order to form the final feature vector for face recognition, as shown in Figure 12d.

In this paper, we use the chi-squared distance (matching score) to measure dissimilarities between the enrolled face histogram features and face histogram feature from the input image [40]. In order to deal with pose variations (horizontal and vertical rotation), the face histogram feature of the input image is matched with the five enrolled face histogram features (which are obtained in the enrollment stage of Figure 2), and then, the matching score is determined as the smallest value among the five.

Furthermore, we fuse the obtained matching scores of the selected face images in the Face Log as the final matching score. The weight values for the matching scores of a viewer are determined using Equation (8).

$$w_i = \frac{Score_i}{Score_1 + \cdots + Score_m} \tag{8}$$

where w_i denotes the i-th weight value for the matching score of an image in the Face Log, m represents the number of face images in Face Log, and $Score_i$ indicates the ith quality score obtained from Figure 6 within Face Log.

The final matching score of face recognition is calculated using the weight values of the selected face images, as shown in Equation (9). Using the matching scores ($MS_1, \cdots, MS_m$) obtained by MLBP and their weight values ($w_1, \cdots, w_m$) of face images in the Face Log, we obtain the fused matching score (FMS) by using Equation (9). Based on the smallest FMS, our system selects the genuine person in the enrolled data (the enrolled face image of the same person to the person of input image) corresponding to that of the input image.

$$FMS = w_1 \times MS_1 + \cdots + w_m \times MS_m \qquad (9)$$

3. EXPERIMENTAL RESULTS AND ANALYSES

The proposed method for face recognition was tested on a desktop computer with an Intel Core™ I7 3.5-GHz CPU and 8-GB RAM. The algorithm was developed using Microsoft Foundation Class (MFC)-based C++ programming and the Direct X9.0 software development kit (SDK).

Although there are a lot of face databases such as FEI [47], PAL [48], AR [49], JAFFE [50], YouTube Faces [51], Honda/UCSD video database [52], and IIT-NRC facial video database [53], most of them do not include all the factors such as the variations of head pose, illumination, sharpness, openness of eyes, contrast, and image resolutions in the video sequences of the database. Therefore, we constructed our own database (Database I), which includes all of these factors in the video sequences for the experiments.

When constructing our database, we defined 20 groups based on 20 people who participated in the test. The three people in each group carried out three trials using the proposed system by varying the Z distance (2, 2.5 m) and the sitting positions (left, middle, and right). Participants randomly and naturally looked at any point on the TV screen with their eyes blinking as if they were watching TV. During this period, successive images were acquired. The database contained a total of 31,234 images for measuring the performance of the proposed method. In addition, an additional five images per person in each group were obtained at the distance Z of 2 m at the enrollment stage. Figure 13 shows the examples of images captured for the experiments.

This database contains the variations in head pose, illumination, sharpness, and openness of the eyes, as shown in Figure 14. It is observed that the head pose is the most influential quality measure to construct the Face Log, whereas image resolution has no influence as a QM because participants have few Z-distance variations. In Figure 14, the bar in the first (row) denotes the number of selected QMs as the first order. Similarly, the bar in the fourth (row) denotes the number of selected QMs as the fourth order. The vertical axis of Figure 14 shows the number of selected QMs.

In the first experiment, we measured the accuracy of the face recognition method based on genuine acceptance rate (GAR) where we set the number of enrolled persons to three. As explained in Section 2.4, either the MIN or MAX method can be chosen for obtaining the IV. Further, the final output score can be obtained by using one of the five defuzzification methods (FOM, LOM, MOM, MeOM, or COG). Therefore, we compared the accuracies of face recognition using the MIN or MAX methods and compared them according to the defuzzification method, as shown in Table 3. Here, the number of face images in a video sequence is 10; we also measured the

Figure 13. Captured images for the experiments.

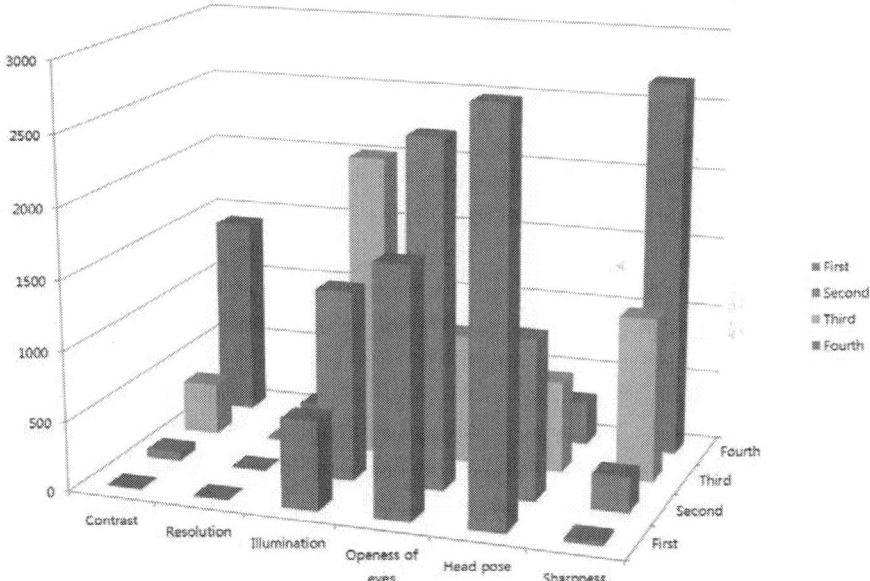

Figure 14. Results of adaptively selected quality measures.

accuracies of face recognition by changing the number of selected images in the Face Log. In Table 3, no fusion means all the matching scores (MS_1, ..., MS_m) of Equation (9) are used for calculating the GAR not using the fusion method of Equation (8). Fusion of Table 3 means our method, and therefore, the FMS of Equation (9) is used for calculating GAR.

Experimental results showed that the methods based on MIN and COG generally show higher face recognition accuracies than the other methods. The best accuracy (92.94%) was obtained with the Fuzzy MIN rule and COG in the case of fusing five selected images.

Table 3. Comparison of face recognition accuracies using the MIN or MAX methods according to the defuzzification method (unit: %).

Method		Number of Selected m Images								
		1	2		3		4		5	
		No Fusion	No Fusion	Fusion	No Fusion	Fusion	No Fusion	Fusion	No Fusion	Fusion
Fuzzy MIN rule	FOM	92.05	89.74	91.95	91.35	92.3	91.36	92.42	91.32	92.44
	LOM	91.61	91.05	91.78	91.3	91.95	91.28	92.17	91.33	92.23
	MOM	92.29	91.35	91.88	91.43	92.05	91.45	92.38	91.45	92.45
	MeOM	92.23	91.54	92.29	91.53	92.63	91.48	92.92	91.41	92.82
	COG	92.27	91.61	92.42	91.63	92.73	91.63	92.89	91.58	**92.94**
Fuzzy MAX rule	FOM	92.12	91.49	91.74	91.49	91.98	91.46	92.1	91.45	92.16
	LOM	92.31	91.38	91.9	91.41	92.18	91.36	92.2	91.36	92.46
	MOM	91.77	91.31	92.01	91.53	92.43	91.46	92.4	91.43	92.42
	MeOM	92.43	91.47	92.09	91.52	92.68	91.51	92.72	91.47	92.72
	COG	92.37	91.4	92.33	91.58	92.77	91.42	92.72	91.54	92.86

In the next experiment, we performed the additional experiments for measuring the accuracies of face recognition in terms of GAR according to the number of QMs as shown in Table 4. Since we use the fuzzy system to combine the values obtained by the selected QMs and to obtain the final quality score of the image, the number of QMs should be at least two. Therefore, we compared the accuracies of face recognition according to the number of QMs and defuzzification method. Experimental results showed that the greater the number of QMs, the higher the accuracy of face recognition becomes.

In the third experiment, in order to show the accuracy changes of face recognition according to the membership function, the accuracy of symmetric input and output membership functions (SIMF and SOMF (Figure 7 and Figure 8)) is compared with asymmetric ones (Figure 15). Here, asymmetric input membership functions (AsIMF 1 and 2) are defined based on each cutoff point at input value, respectively, as shown in Figure 15a,b. Similarly, asymmetric output membership functions (AsOMF 1 and 2) are defined based on each cutoff point at output value, respectively, as shown in Figure 15c,d.

Table 4. Comparison of the accuracies of the proposed method according to the number of quality measures (QMs) (unit: %).

Method		Number of QMs		
		2	3	4
Fuzzy MIN rule	FOM	92.14	92.22	92.44
	LOM	92.05	92.17	92.23
	MOM	91.95	92.10	92.45
	MeOM	92.10	92.25	92.82
	COG	92.58	92.84	**92.94**
Fuzzy MAX rule	FOM	91.40	92.02	92.16
	LOM	92.14	92.35	92.46
	MOM	92.20	92.63	92.42
	MeOM	92.31	92.74	92.72
	COG	92.16	92.86	92.86

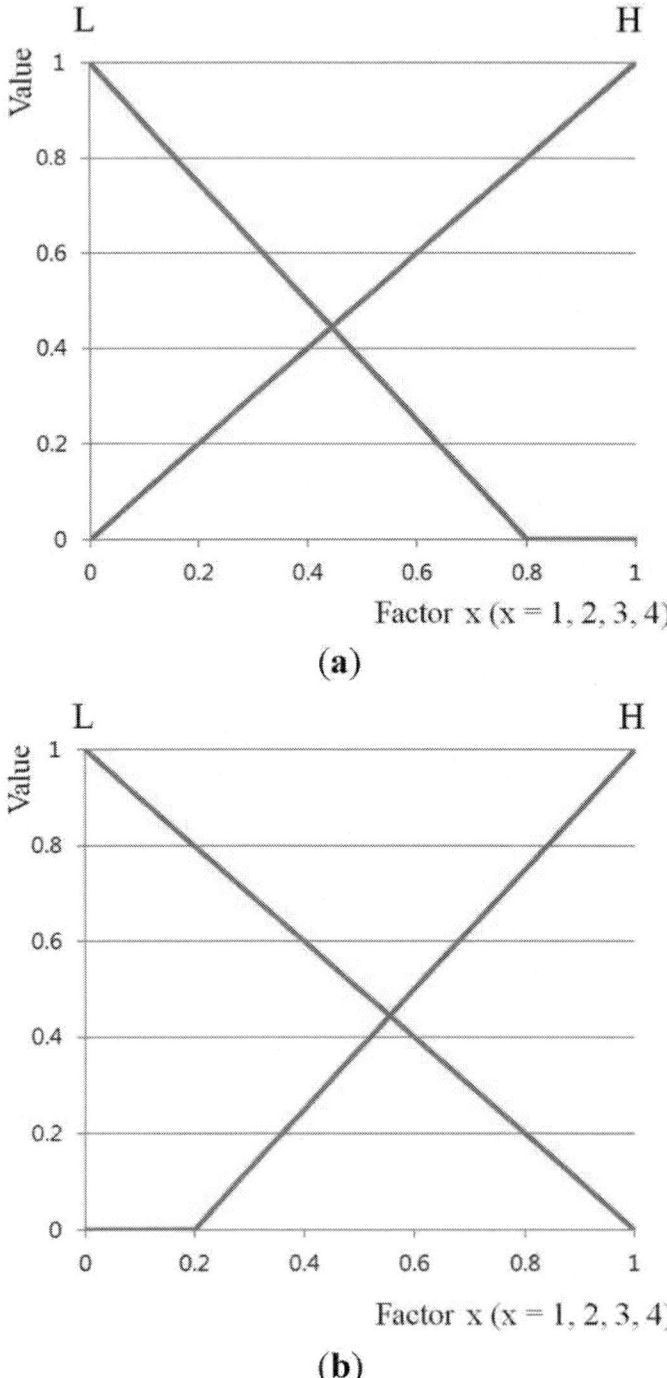

(a)

(b)

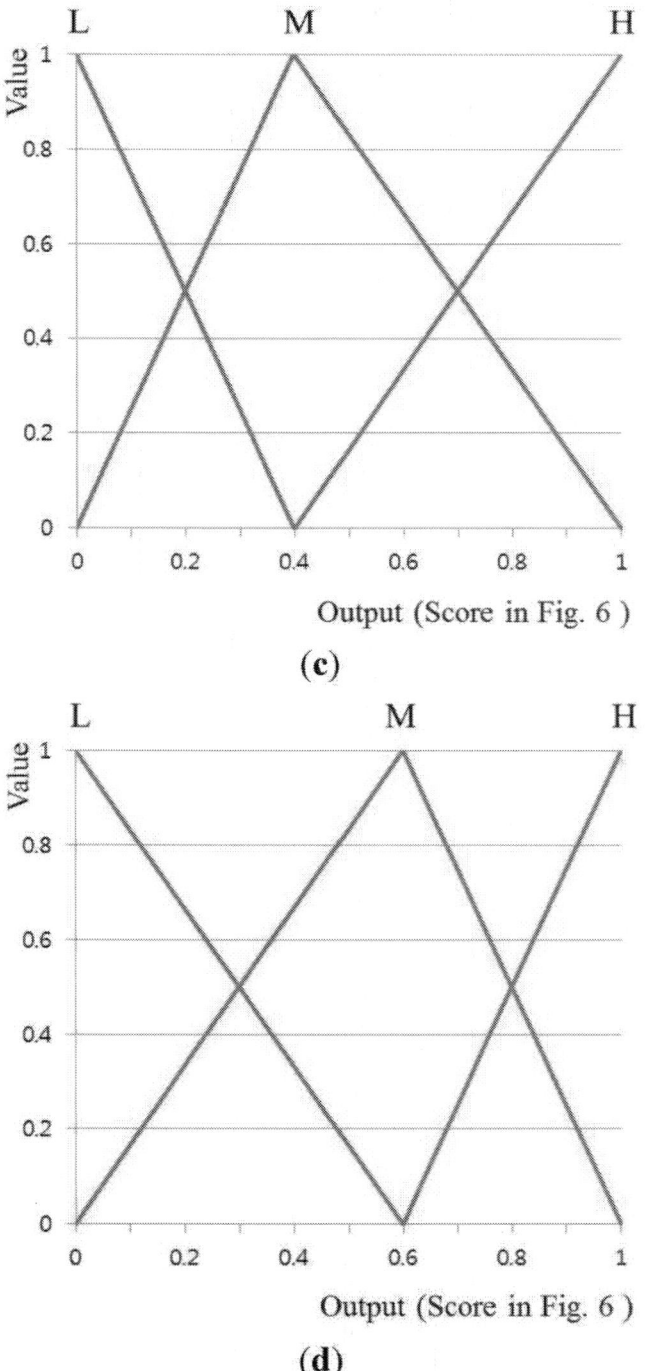

Figure 15. Asymmetrical triangular membership functions: (**a**) asymmetric input membership function (AsIMF) 1; (**b**) AsIMF 2; (**c**) AsOMF 1; (**d**) AsOMF 2.

Table 5 shows the accuracies of face recognition according to the membership function and defuzzification method. It is observed that the accuracy of symmetric membership functions is slightly higher but very similar to that of the others (based on asymmetric membership function). On the basis of these results, it can be shown that the symmetric triangular shape for the membership function is more considerable than others in terms of not only accuracy but also implementation (explained in Section 2.4).

In the next experiment, as shown in Table 6, we measured the accuracies of face recognition according to the number of face images in a video sequence in terms of GAR. In Table 6, fusion means our method, and therefore, the *FMS* of Equation (9) is used for calculating GAR. It is observed that the accuracy of face recognition improves with an increase in the number of face images.

Table 5. Comparison of the accuracies of the proposed method according to the shape of the membership function (unit: %).

Method		SIMF and SOMF (Figures 7 and 8)	Membership Function			
			AsIMF 1 and SOMF (Figures 15a and 8)	AsIMF 2 and SOMF (Figures 15b and 8)	SIMF and AsOMF 1 (Figures 7 and 15c)	SIMF and AsOMF 2 (Figures 7 and 15d)
Fuzzy MIN rule	FOM	92.44	92.55	92.31	92.27	92.44
	LOM	92.23	92.53	92.56	92.56	92.74
	MOM	92.45	92.39	92.23	92.42	92.59
	MeOM	92.82	92.47	92.44	92.49	92.67
	COG	**92.94**	92.71	92.51	92.60	92.88
Fuzzy MAX rule	FOM	92.16	92.07	92.33	92.17	92.37
	LOM	92.46	92.68	92.44	92.59	92.87
	MOM	92.42	92.65	92.56	92.44	92.71
	MeOM	92.72	92.58	92.47	92.59	92.60
	COG	92.86	92.75	92.45	92.45	92.84

Table 6. Comparison of the accuracies of the proposed method according to the number of face images in a video sequence (unit: %).

Number of Selected *m* Images		Number (*n*) of Face Images in a Video Sequence			
		10	15	20	25
Fusion	2	92.42	93.12	93.63	93.92
	3	92.73	93.59	93.89	94.33
	4	92.89	93.57	94.06	94.59
	5	92.94	93.72	94.18	**94.71**

In addition, the accuracy of the proposed method was compared with a fixed quality measure-based approach [21]. Using the results from Figure 14, we selected four influencing QMs (head pose, illumination, sharpness, and openness of eyes) for a fixed quality measure-based approach. In this experiment, n and m are set to 25 and 5, respectively. As described in Section 2.4, since the fixed quality measure-based approach cannot assess other quality measures that affect the accuracy of face recognition, its accuracy is lower than that of the proposed method. Furthermore, we compared the accuracy of a previous method that uses all images in a video sequence [40], and the accuracy of fixed quality measure-based approach [21] to the accuracy of our method, the results of which are shown in Table 7. Experimental results showed that the proposed method has a higher face recognition accuracy than the other methods.

Table 7. Comparison of the accuracy of the proposed method with that of the other methods (unit: %).

Method	Using all Images in a Video Sequence [46]	Fixed Quality Measures-Based Approach [21]	Adaptive Quality Measure-Based Approach (Proposed Method)
Accuracy	89.74	93.09	**94.71**

In Figure 16 and Figure 17, we show the face images of correctly recognized results. As shown in Figure 16a,b, although there exist the images where hand occludes a part of face, face area is not correctly detected, or eyes are closed, our method can exclude these bad-quality images. On the basis of these results, we can confirm that good-quality images are correctly selected by our method and that they are correctly matched with the enrolled face image of the same person.

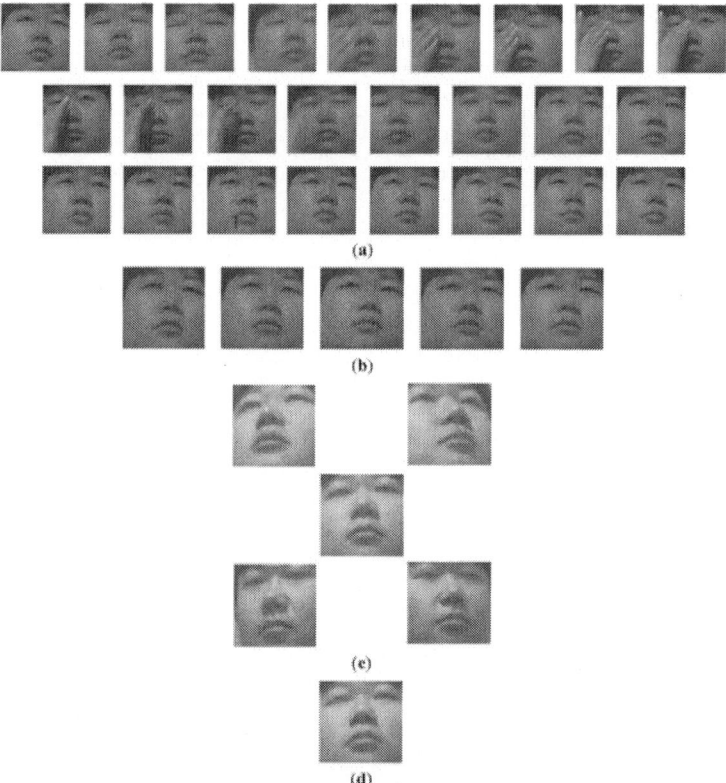

Figure 16. Example of correctly recognized results: (**a**) a sequences of 25 images; (**b**) selected five images; (**c**) enrolled face images (gazing at five positions of TV); (**d**) correctly matched face image among face images of enrolled persons.

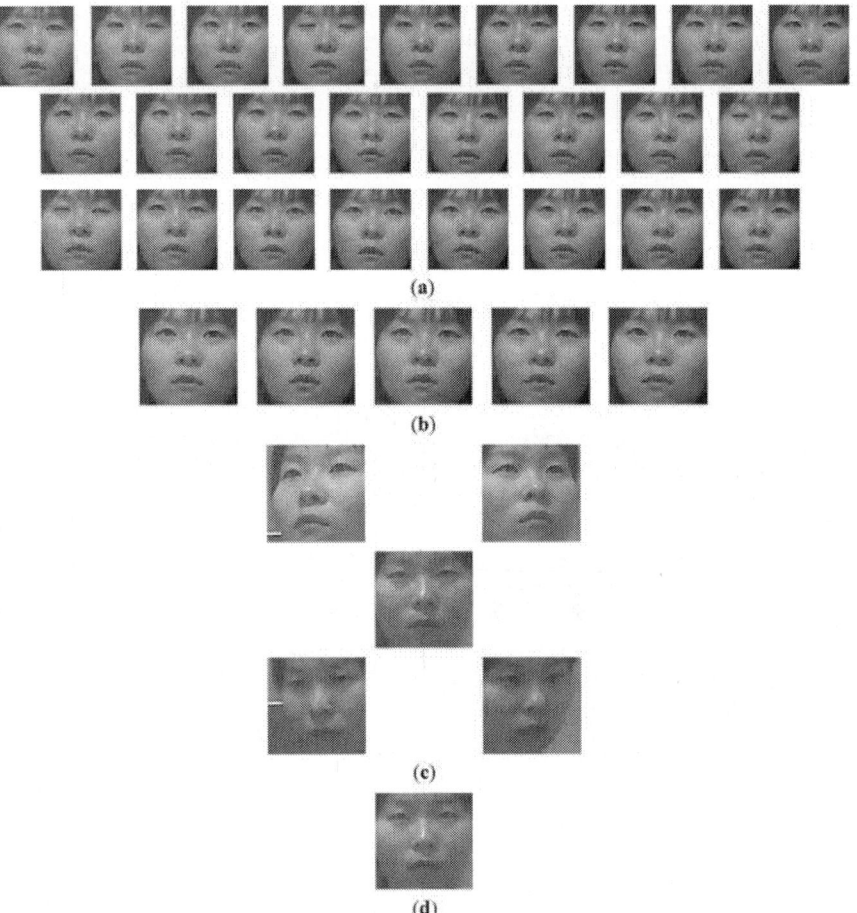

Figure 17. Example of correctly recognized results: (**a**) a sequences of 25 images; (**b**) selected five images; (**c**) enrolled face images (gazing at five positions of TV); (**d**) correctly matched face image among face images of enrolled persons.

In Figure 18 and Figure 19, we show the face images of incorrectly recognized results. As shown in these Figures, although there exist the images where eyes are closed, our method can exclude these bad-quality images, and the face images of good quality are correctly selected. However, they are incorrectly matched with other person's enrolled face image. This is because there exists a size difference between the face areas of enrolled image and input one due to the incorrect detection of face and eye region (Steps 6 and 7 of Figure 2), which leads to misrecognition. These can be solved by increasing the accuracy of re-definition of face area based on more accurate detection algorithm. Because our research is not focused on accurate detection, but focused on correctly selecting the face images of good quality, this research of accurate detection would be studied as a future work.

We used the images which were acquired while each user naturally watches TV as shown in Figure 3 for our experiments. In case of the severe rotation of head as shown in Figure 20, the face and facial features are difficult to be detected. In addition, because these cases do not happen when a user looks at a TV normally, we did not use these images for our experiments.

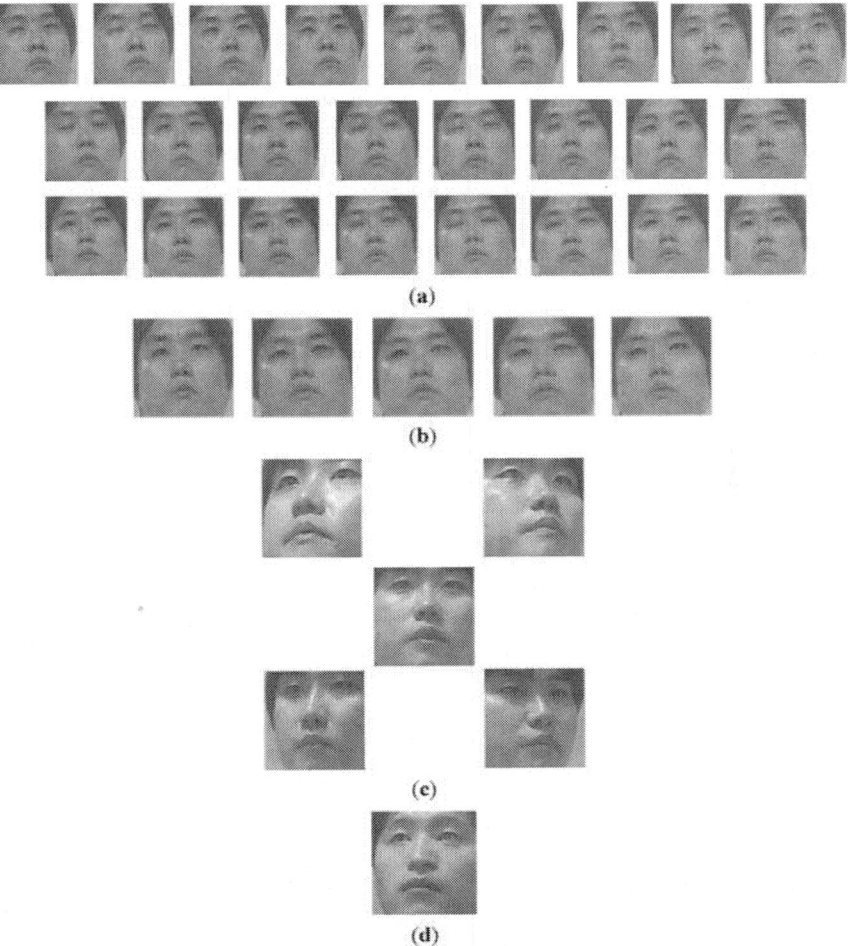

Figure 18. Example of incorrectly recognized results: (**a**) a sequences of 25 images; (**b**) selected five images; (**c**) enrolled face images (gazing at five positions of TV); (**d**) incorrectly matched face image among face images of enrolled persons.

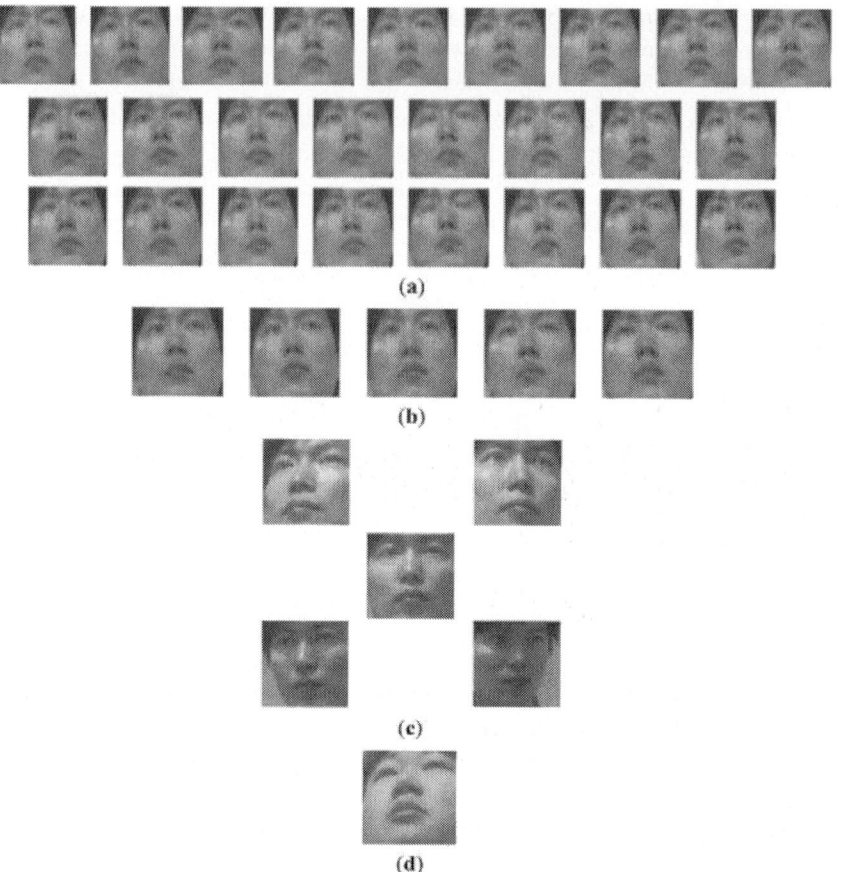

Figure 19. Example of incorrectly recognized results: (**a**) a sequences of 25 images; (**b**) selected five images; (**c**) enrolled face images (gazing at five positions of TV); (**d**) incorrectly matched face image among face images of enrolled persons.

When a user looks at a TV normally, the primary factors for determining degree of head pose change are the size of the TV and viewing distance between the user and TV. The relationship between the TV size and optimal viewing distance is already defined in [56]. That is, the larger the TV size is, the farther the optimal viewing distance should be. Moreover, the smaller the TV size is, the nearer the optimal viewing distance should be.

Based on this [56], we show the three cases, (a) when users are watching TV of 50 inches at the viewing distance of about 2 m; (b) when users are watching TV of 60 inches at the viewing distance of about 2.4 m; and (c) when users are watching TV of 70 inches at the viewing distance of about 2.8 m as shown in Figure 21a–c, respectively. As shown in the upper images ofFigure 21a–c, the degrees of head pose change are almost similar all in these cases when each user gazes at the same (lower-left) position on the TV

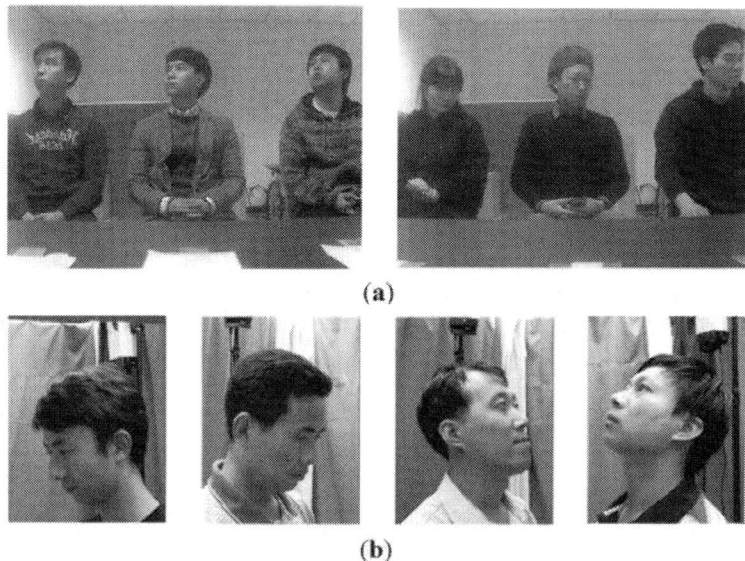

Figure 20. Examples of severe rotation of head: (**a**) Images captured by our system; (**b**) images from CAS-PEAL-R1 database [54,55].

although the image resolution of each user decreases according to the increase of the viewing distance. As shown in the lower images of Figure 21a–c, the correct regions of face and facial features are detected by our method, and our method of measuring the quality of face image is also working successfully, which shows that the performance of our method is not affected by using the smaller or larger sized TV if considering the optimal viewing distance.

The degree of head pose change is different according to the Z distance even though the user looks at the same location on the TV screen of same size as shown in Figure 22a–c. When a user is at a nearer distance from TV (in the case of Figure 22a), the degree of head pose change becomes larger than that of the case of Figure 22c. On the other hand, when the user is at a farther distance (in the case of Figure 22c), the degree of head pose change becomes smaller than that of the case of Figure 22a although the image resolution of the user of Figure 22c decreases.

As shown in the right-lower images of Figure 22a–c, the correct regions of face and facial features are detected by our method, and our method of measuring the quality of face image is also working successfully, which shows that the performance of our method is not affected by the farther or nearer distance between the user and TV.

In addition, we performed the experiments with additional open database. By using open database, the CAS-PEAL-R1 database [54,55], we could increase the number of subjects and show the fidelity of the proposed method irrespective of the kind of database. The CAS-PEAL-R1 database contains 30,863 images of 1040 subjects (595 males and 445 females). In the CAS-PEAL-R1 database, most images (21 poses × 1040 individuals) with pose variations were acquired according to the different camera position.

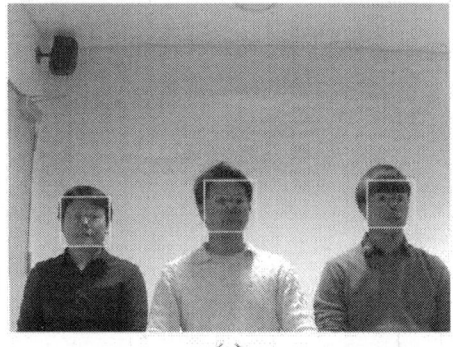

(a)

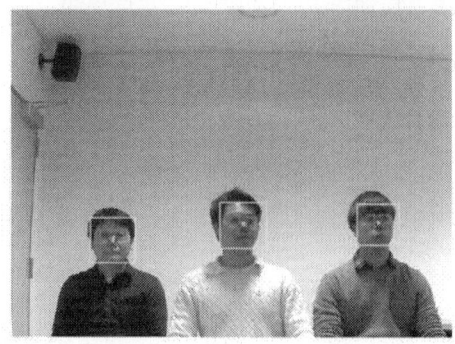

(b)

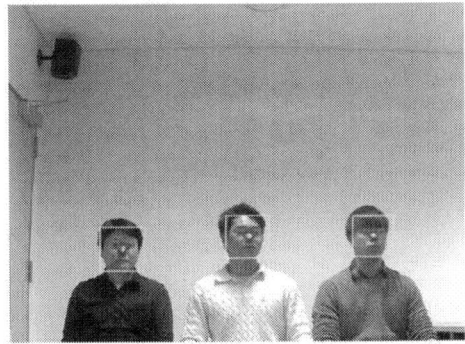

(c)

Figure 21. Examples of head pose change according to the size of the TV with the optimal viewing distance. In (a)–(c), the upper and lower figures represent the original images and result ones including the detected regions of face and facial features, respectively: (a) when users are watching TV of 50 inches at the viewing distance of about 2 m; (b) when users are watching TV of 60 inches at the viewing distance of about 2.4 m; (c) when users are watching TV of 70 inches at the viewing distance of about 2.8 m.

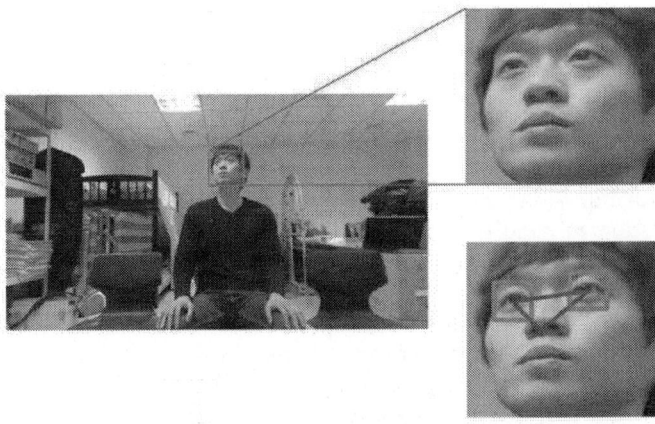

(a)

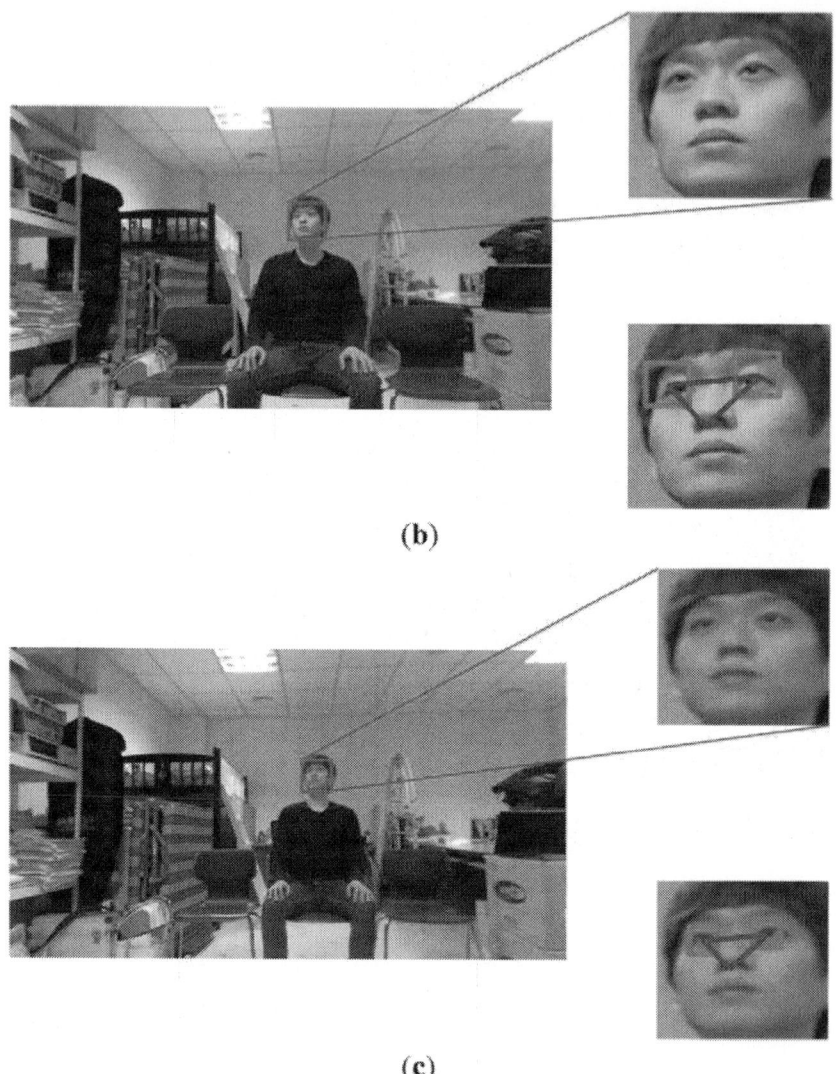

Figure 22. Examples of head pose change according to the Z distance with the TV of 60 inches: (**a**) 1.5 m; (**b**) 2 m; (**c**) 2.5 m.

However, we do not cover high degree of rotation for our experiment because severe rotation of head does not happen when a user looks at a TV normally. Therefore, we used only nine images for each subject under different poses as shown inFigure 23a. In addition, each subject has the images under at least six illumination changes, those with six different expressions, and those with different image resolutions within the face region as shown in Figure 23b–d, respectively.

Consequently, a total of 2410 images (Database II) from 100 subjects were used for our experiments because the remainder images from 940 subjects contain few of variations. The image resolution is 360 × 480 pixels.

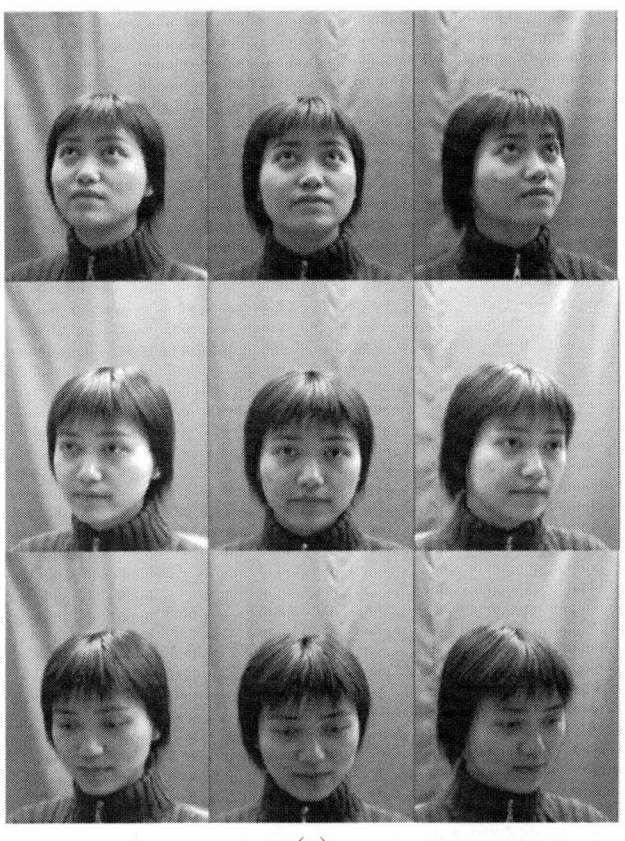

(a)

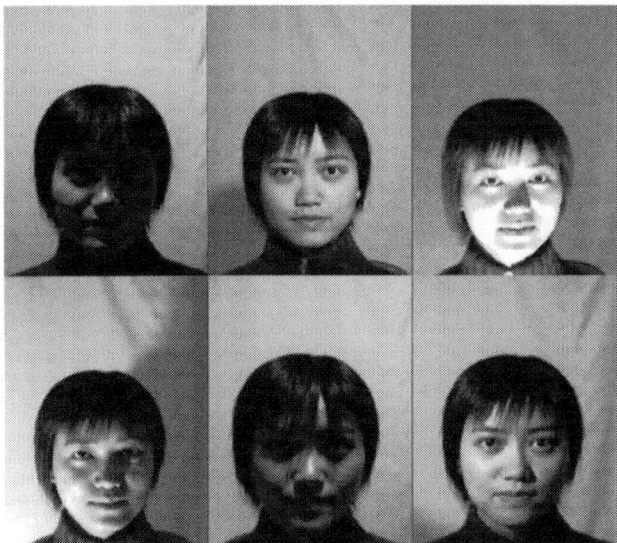

(b)

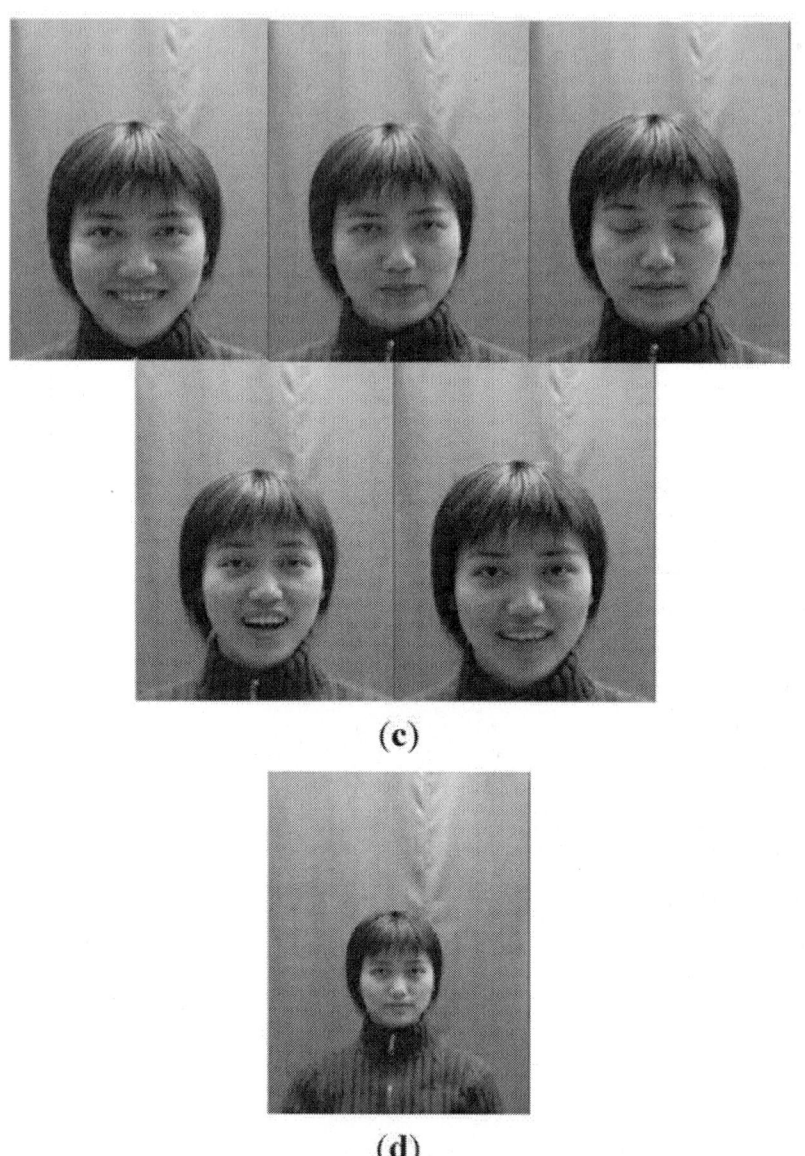

Figure 23. Examples of one subject with different variations in Database II: (**a**) pose variation; (**b**) illumination variation; (**c**) expression variation; (**d**) image resolution variation.

Experimental results showed that the Database II includes variations in head pose, contrast, and illumination as shown inFigure 24. It is observed that the illumination is the most influential QM to construct the Face Log, whereas image resolution and openness of eyes have no influence as a QM. That is because the number of images where the Z distance variations of users occur is small among the whole images of Database II. In addition, the number of images where participants close their eyes is small among the whole Database II.

As the next experiment, we measured the accuracy of the face recognition method based on GAR where we set the number of enrolled persons to three. Using the results from Figure 24, the accuracy of the proposed method was compared with a fixed quality measure-based approach [21]. Here, we selected three influencing QMs (illumination, contrast, and head pose) for a fixed quality measure-based approach. In addition, we compared the accuracy of a previous method that uses all still images [46], and the accuracy of fixed quality measure-based approach [21] to the accuracy of our method as shown in Table 8. Experimental results showed that the accuracy of the proposed method is higher than those of other methods, and its accuracy is similar to that of Table 7 using our own database. From that, we can confirm that the accuracy of our method is less affected by the kinds of database and the number of participants in database.

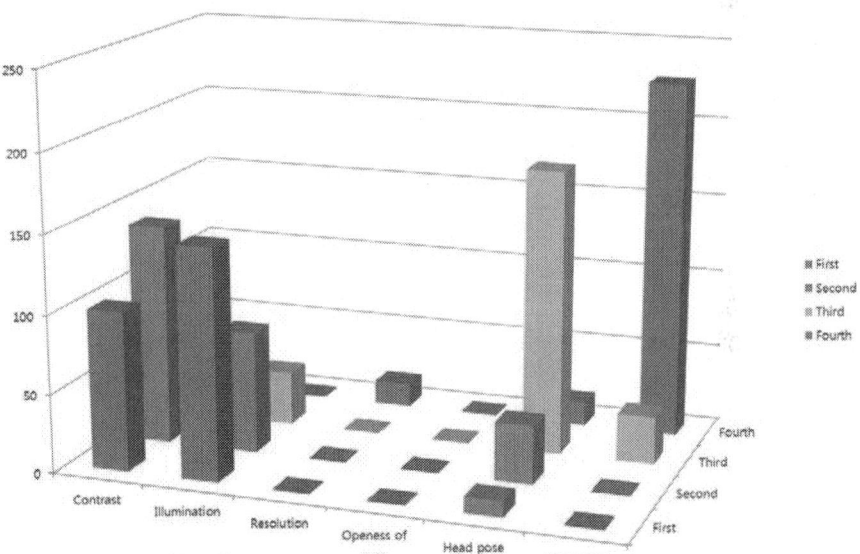

Figure 24. Results of adaptively selected QMs with Database II.

Table 8. Comparison of the accuracy of the proposed method with that of the other methods with Database II (unit: %).

Method	Using all Images in a Video Sequence [46]	Fixed Quality Measures-Based Approach [21]	Adaptive Quality Measure-Based Approach (Proposed Method)
Accuracy	79.2	95.72	97.67

Because we consider the TV watching environment in doors, the case that user's face is so dark that the facial features cannot be discriminative does not occur in our experiments. However, we consider the cases that shadows exist on both sides of face or in the entire face to a degree as shown in Figure 25b,d. Each number below Figure 25a–d represents the QM value (F2 of

Equation (4)) of illumination. As shown in Figure 25, the F2 of the case without shadows (Figure 25(a)) is higher than those of the case of shadows on both sides of face (Figure 25b) or in the entire face (Figure 25d). In addition, the F2 ofFigure 25a is higher than that of the case of shadows in the right side of face (Figure 25c). The reason why the F2 ofFigure 25a is higher than those of Figure 25b,d is that it is difficult that the shadows are uniform on the both sides of the face due to the 3 dimensional shape of face even if shadows on both sides of face or in the entire face. Therefore, the F2 (based on the difference between the average values of left and right sides of the face) becomes larger than that without shadow ofFigure 25a.

From these results, we can find that our QM of illumination (F2 of Equation (4)) can produce the correct quality value with the face images including the shadows on both sides of face or in the entire face to a degree.

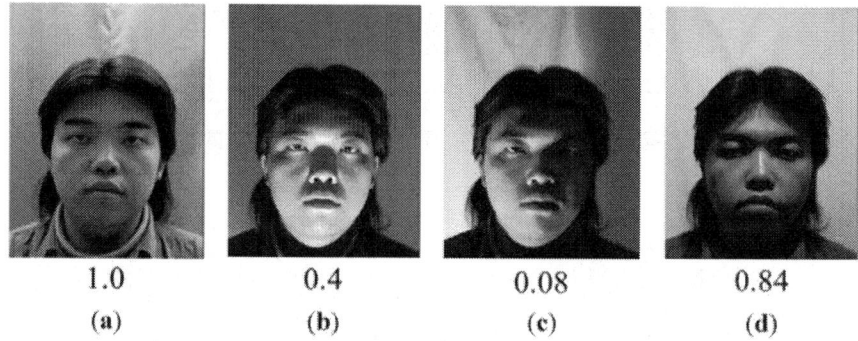

| 1.0 | 0.4 | 0.08 | 0.84 |
| (a) | (b) | (c) | (d) |

Figure 25. Examples of the F2 value according to the illumination condition with Database II: **(a)** uniform illumination; **(b)** shadows on both sides of face; **(c)** shadows on right side of the face; **(d)** shadows in the entire face.

Generally, an image of higher resolution is regarded as containing more information than that of lower resolution. Likewise, the face images of high-resolution are preferred to those of low-resolution in terms of recognition. This is because the face images of high-resolution can yield better recognition results. As shown in Figure 22 and Figure 26, the nearer the Z distance between a user and TV is, the larger the size of face box becomes, which increases the inter-distance between two detected eyes and the image resolution (width and height) of face region. Therefore, the inter-distance between two detected eyes has almost proportional relationship with the image resolution. Based on these characteristics, we measure the distance between two detected eyes as QM (F6) (explained in Section 2.3.6) and assign the higher score to the face image of higher resolution.

The accurate detection of eye regions has influence on the performance of three QMs (head pose (F1 explained in Section 2.3.1), openness of the eyes (F4 explained in Section 2.3.4), and image resolution (F6 explained in Section 2.3.6)). In addition, the performance of QM (F1) is also affected by the accuracy of nose detection.

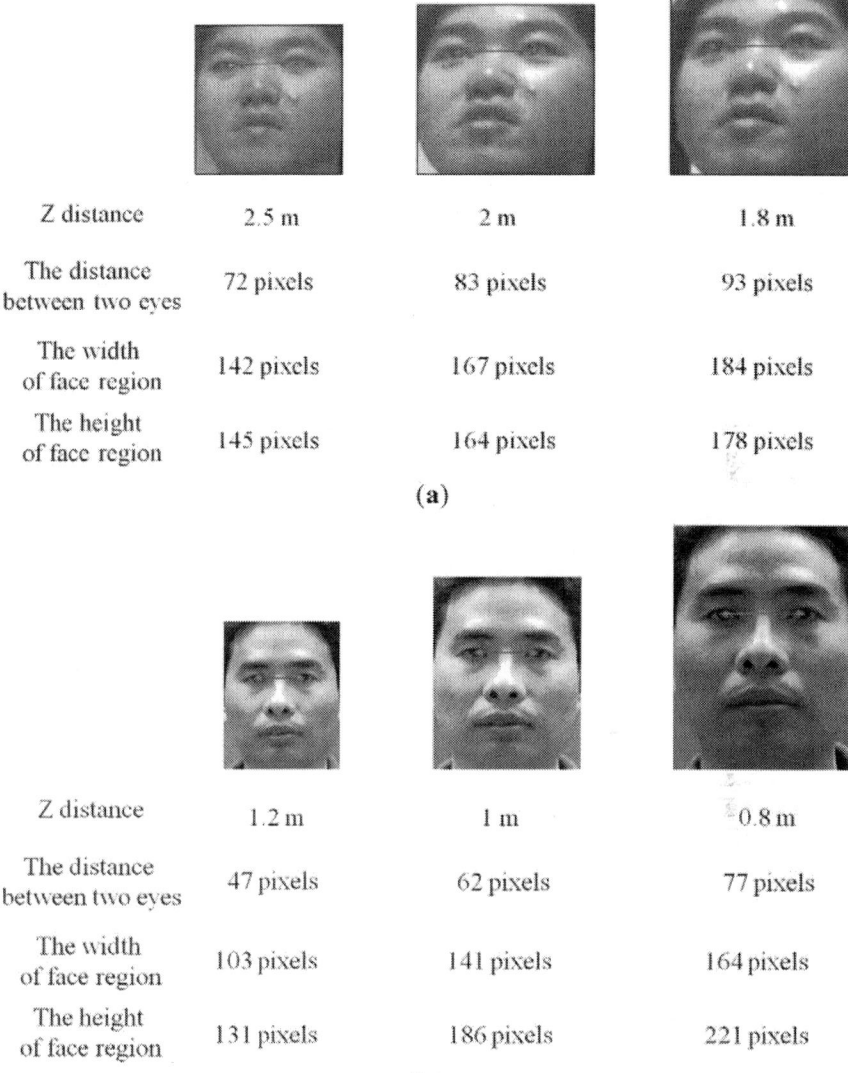

Figure 26. Relationship between the inter-distance between two eyes and image resolution of face region: (**a**) examples from our own database; (**b**) examples from database II.

As the next experiments, we measured the accuracies of face, eye and nostril detection. For this, we manually depicted bounding boxes on the face (eye or nostril) areas in the images as ground truth regions. The detection results were evaluated using the Pascal overlap criterion [57]. In order to judge true/false positives, we measured the overlap of the detected and ground truth boxes. If the area of overlap (AO) between the detected box

(BD) and ground truth box (BGT) of Figure 27exceeds 0.5 using Equation (10), we count the result as a correct detection.

$$A_O = \frac{\text{area}(B_D \cap B_{GT})}{\text{area}(B_D \cup B_{GT})} \tag{10}$$

where $B_D \cap B_{GT}$ denotes the intersection of the detected and ground truth boxes and $B_D \cup B_{GT}$ is their union [57].

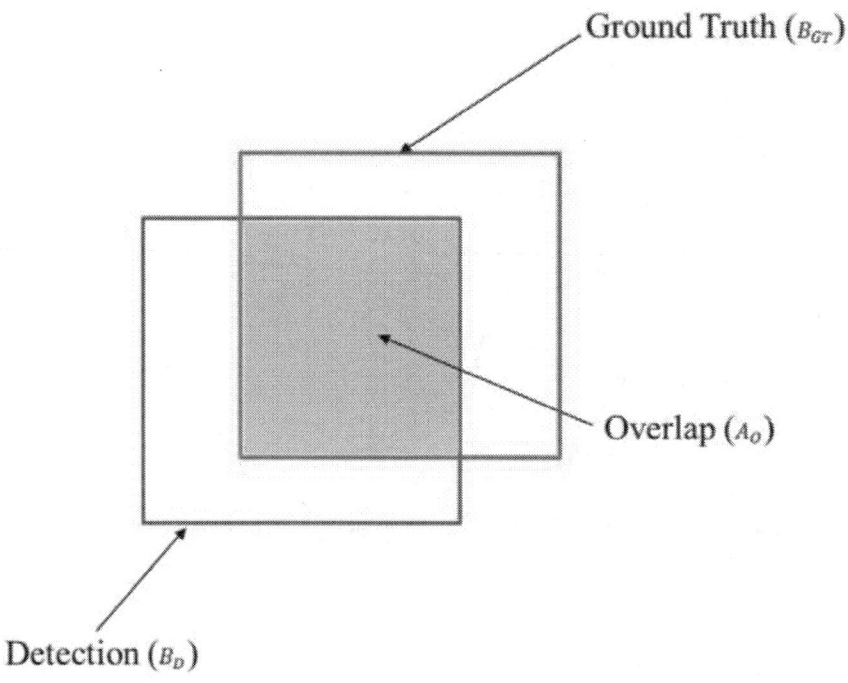

Figure 27. Overlapped area of ground truth and detected boxes.

Based on the Equation (10), we can count the number of true positive and false positive detection cases. Here, true positive means that the faces (eyes or nostrils) are correctly detected as faces (eyes or nostrils), and false positive is the case where non-faces (non-eyes or non-nostrils) are incorrectly detected as faces (eyes or nostrils). Consequently, the accuracies of face, eye and nostril detection are measured using the following Equations (11) and (12) [58],

$$Recall = \frac{N_{tp}}{M} \tag{11}$$

$$Precision = \frac{N_{tp}}{N_{tp} + N_{fp}} \quad (12)$$

where M is the number of faces (eyes or nostrils), N_{tp} is the number of true positives, and N_{fp} is the number of false positives. As shown in Equations (11) and (12), the maximum and minimum values of the recall and precision are 1 and 0, respectively. The higher values of the recall and precision represent a higher accuracy of the detection of face (eye or nostril).

Experimental results showed that the recall and precision of face detection were 99.5% and 100%, respectively. In addition, with the images where face regions were successfully detected, the recall and precision of eye detection were 99.73% and 100%, respectively, and those of nostril detection were 99.67% and 100%, respectively.

In general, the optimization of the quality metrics can be performed by principal component analysis (PCA), linear discriminant analysis (LDA), or neural network, *etc.* However, most of these methods require the additional training procedure with training data, which makes the performance of system be affected by the kinds of training data. In order to solve this problem, we obtain the optimal weight (quality score) of the face image by using the schemes of adaptive QM selection and fuzzy system as shown in Figure 6. Because these schemes do not require the additional (time-consuming) training procedure, the performance of our method is less affected by the kinds of database, which was experimentally proved with two databases (our own database and Database II) as shown in Table 5 and Table 6.

As shown in Figure 28, we show the correlation of the matching score (similarity) by face recognition and quality score. In both the matching score and quality score, the higher values represent the higher matching similarity and better quality, respectively. With the 25 face images of Figure 28a, we show the graph of correlation between the quality scores (which are obtained by our method of Figure 6) and the matching scores (by our face recognition method of multi-level binary pattern (MLBP) of Figure 12) as shown in Figure 28b. The values of the quality score and matching score are respectively normalized so as to be represented in the range from 0 to 1. Figure 28b shows that these two scores are much correlated. In addition, we calculate the correlation value between the quality scores and matching scores. The calculated correlation value is about 0.87. The correlation value ranges from −1 to 1. 1 and −1 mean the positive and negative correlation cases, respectively. 0 represents the uncorrelated case [59,60]. Based on the correlation value, we can find that the quality scores are much correlated to the matching scores by face recognition, and the higher quality score corresponds to the good face recognition. The reason why there are no error bars in Figure 28b is that this figure shows one example where the correlation coefficient between quality scores and matching scores is obtained from the data of a single human individual.

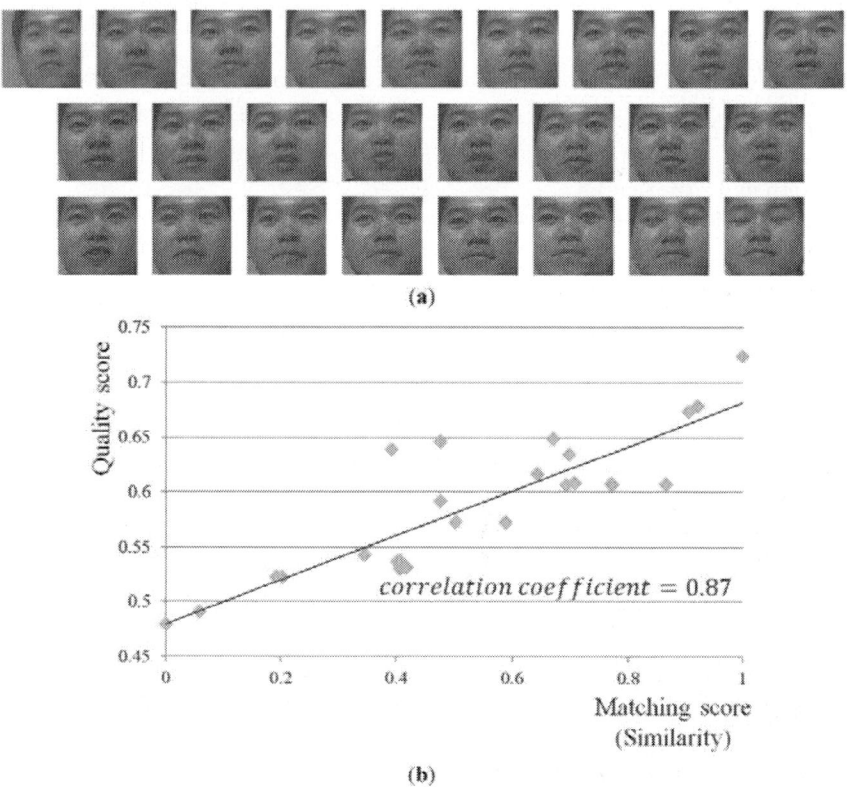

(a)

(b)

Figure 28. Example of a video sequence and correlation between the matching score by face recognition and quality score: (**a**) all the 25 images in a video sequence; (**b**) the graph of correlation between the quality scores and matching scores.

Although the weighted quality fusion approach can be considered as an alternative, this approach usually requires the weight values for fusion. The weight values can be determined by the experience of developer, but it cannot guarantee the optimal weight values to be determined irrespective of the variety of input data. To solve this problem, the weight values should be obtained through the time-consuming training procedure, which makes the performance of system be affected by the kinds of training data. However, our fuzzy-based fusion approach of image quality has the advantage of being less affected by the types of face databases because it does not perform an additional training procedure, which was experimentally proved with two databases (our own database and Database II) as shown in Table 7 and Table 8. This is the reason why we choose fuzzy approach over weighted quality metrics.

In general, the quality metrics which should have higher weights can be changed according to the kinds of database. As shown in Figure 14, the quality metrics of head pose (F1 of Equation (2)) and openness of eye (F4 of Equation (6)) are more dominant than others because the variations of head pose and eye openness/closure are frequent in our database. However, Figure 24 shows that the quality metrics of illumination (F2 of Equation (4)) and contrast (F5 of Equation (7)) are more dominant than others because the variations of illumination and contrast are frequent in Database II. Therefore, the quality metrics which should have higher weights can be changed according to the kinds of database, and our method (based on adaptive QM selection and fuzzy-based fusion as shown in Figure 6) can select the optimal quality metrics and quality score considering the variety of database. In addition, this was experimentally proved with two databases (our own database and Database II) as shown in Table 7 and Table 8.

4. CONCLUSIONS

In this paper, we proposed a new performance enhancement method of face recognition by the adaptive selection of face images through symmetrical fuzzy-based quality assessment. To select high-quality face images from a video sequence, we measured the qualities of face images on the basis of four QMs that were adaptively selected by comparing the variations of each of the six QMs. These QMs were combined using a fuzzy system (based on symmetrical membership function and rule table considering the characteristics of symmetry) into one quality score for the face image. After obtaining the quality score of the face image, high-quality images from the Face Log were selected in the order of the quality score of each image. The performance of face recognition was enhanced by fusing the matching scores of the high-quality face images.

Experimental results show that the proposed method outperforms the other methods in terms of accuracy. Misrecognition errors are caused by a size difference between the face areas of enrolled image and input one due to the incorrect detection of face and eye region. These can be solved by increasing the accuracy of re-definition of face area based on more accurate detection algorithm. Because our research is not focused on accurate detection, but focused on correctly selecting the face images of good quality, research in accurate detection could be studied as a future work.

ACKNOWLEDGMENTS

This work was supported by the SW R&D program of MSIP/IITP (10047146, Real-time Crime Prediction and Prevention System based on Entropy-Filtering Predictive Analytics of Integrated Information such as Crime-Inducing Environment, Behavior Pattern, and Psychological Information).

AUTHOR CONTRIBUTIONS

Yeong Gon Kim and Kang Ryoung Park designed the overall system of proposed face recognition. In addition, they wrote and revised the paper. Won Oh Lee, Ki Wan Kim, and Hyung Gil Hong implemented the methods of face and eye detection. Also, they helped the data collection and experiments.

REFERENCES

1. Lin, K.-H.; Shiue, D.-H.; Chiu, Y.-S.; Tsai, W.-H.; Jang, F.-J.; Chen, J.-S. Design and Implementation of Face Recognition-aided IPTV Adaptive Group Recommendation System Based on NLMS Algorithm. In Proceedings of International Symposium on Communications and Information Technologies, Gold Coast, Australia, 2–5 October 2012; pp. 626–631.
2. Lee, S.-H.; Sohn, M.-K.; Kim, D.-J.; Kim, B.; Kim, H. Smart TV Interaction System Using Face and Hand Gesture Recognition. In Proceedings of International Conference on Consumer Electronics, Las Vegas, NV, USA, 11–14 January 2013; pp. 173–174.
3. Chae, Y.N.; Lee, S.; Han, B.O.; Yang, H.S. Vision-based Sleep Mode Detection for a Smart TV. In Proceedings of International Conference on Consumer Electronics, Las Vegas, NV, USA, 11–14 January 2013; pp. 123–124.
4. Gao, Y.; Leung, M.K.H. Face Recognition Using Line Edge Map. *IEEE Trans. Pattern Anal. Mach. Intell.* **2002**, *24*, 764–779.
5. Yilmaz, A.; Gökmen, M. Eigenhill *vs.* Eigenface and Eigenedge. *Pattern Recognit.* **2001**, *34*, 181–184.
6. Basri, R.; Jacobs, D.W. Lambertian Reflectance and Linear Subspaces. *IEEE Trans. Pattern Anal. Mach. Intell.* **2003**, *25*, 218–233.
7. Georghiades, A.S.; Belhumeur, P.N.; Kriegman, D.J. From Few to Many: Illumination Cone Models for Face Recognition under Variable Lighting and Pose. *IEEE Trans. Pattern Anal. Mach. Intell.* **2001**, *23*, 643–660.
8. Du, S.; Ward, R. Face Recognition under Pose Variations. *J. Frankl. Inst. Eng. Appl. Math.* **2006**, *343*, 596–613.
9. Wiskott, L.; Fellous, J.-M.; Krüger, N.; Malsburg, C.V.D. Face Recognition by Elastic Bunch Graph Matching. *IEEE Trans. Pattern Anal. Mach. Intell.* **1997**, *19*, 775–779.
10. Cootes, T.F.; Edwards, G.J.; Taylor, C.J. Active Appearance Models. *IEEE Trans. Pattern Anal. Mach. Intell.* **2001**, *23*, 681–685.

11. Belhumeur, P.N.; Hespanha, J.P.; Kriegman, D.J. Eigenfaces *vs.* Fisherfaces: Recognition Using Class Specific Linear Projection. *IEEE Trans. Pattern Anal. Mach. Intell.* **1997**, *19*, 711–720.

12. Blanz, V.; Grother, P.; Phillips, P.J.; Vetter, T. Face Recognition Based on Frontal Views Generated from Non-Frontal Images. In Proceedings of International Conference on Computer Vision and Pattern Recognition, San Diego, CA, USA, 20–25 June 2005; pp. 454–461.

13. Blanz, V.; Vetter, T. Face Recognition Based on Fitting a 3D Morphable Model. *IEEE Trans. Pattern Anal. Mach. Intell.* **2003**, *25*, 1063–1074.

14. Vetter, T. Learning Novel Views to a Single Face Image. In Proceedings of International Conference on Automatic Face and Gesture Recognition, Killington, VT, USA, 14–16 October 1996; pp. 22–27.

15. Wang, H.; Wang, Y.; Cao, Y. Video-based Face Recognition: A Survey. *World Acad. Sci. Eng. Technol.* **2009**, *60*, 293–302.

16. Barr, J.R.; Bowyer, K.W.; Flynn, P.J.; Biswas, S. Face Recognition from Video: A Review. *J. Pattern Recognit. Artif. Intell.* **2012**, *26*, 1266002:1–1266002:53.

17. O'Toole, A.J.; Roark, D.A.; Abdi, H. Recognizing Moving Faces: A Psychological and Neural Synthesis. *Trends Cognit. Sci.* **2002**, *6*, 261–266.

18. Chowdhury, A.R.; Chellappa, R.; Krishnamurthy, S.; Vo, T. 3D Face Reconstruction from Video Using A Generic Model. In Proceedings of International Conference on Multimedia and Expo, Lausanne, Switzerland, 26–29 August 2002; pp. 449–452.

19. Baker, S.; Kanade, T. Limits on Super-Resolution and How to Break Them. *IEEE Trans. Pattern Anal. Mach. Intell.* **2002**, *24*, 1167–1183.

20. Liu, X.; Chen, T.; Thornton, S.M. Eigenspace Updating for Non-stationary Process and Its Application to Face Recognition. *Pattern Recognit.* **2003**, *36*, 1945–1959.

21. Nasrollahi, K.; Moeslund, T.B. Complete Face Logs for Video Sequences Using Face Quality Measures. *IET Signal Process.* **2009**, *3*, 289–300.

22. Hsu, R.-L.V.; Shah, J.; Martin, B. Quality Assessment of Facial Images. In Proceedings of Biometrics Consortium Conference, Baltimore, MD, USA, 19–21 September 2006; pp. 1–6.

23. Fourney, A.; Laganière, R. Constructing Face Image Logs that are Both Complete and Concise. In Proceedings of Canadian Conference on Computer and Robot Vision, Montreal, QC, Canada, 28–30 May 2007; pp. 488–494.

24. Nasrollahi, K.; Moeslund, T.B. Hybrid Super Resolution Using Refined Face Logs. In Proceedings of International Conference on Image Processing Theory, Tools and Applications, Paris, France, 7–10 July 2010; pp. 435–440.

25. Bagdanov, A.D.; Bimbo, A.D.; Dini, F.; Lisanti, G.; Masi, I. Posterity Logging of Face Imagery for Video Surveillance. *IEEE Multimedia* **2012**, *19*, 48–59.

26. Bharadwaj, S.; Bhatt, H.; Vatsa, M.; Singh, R.; Noore, A. Quality Assessment based Denoising to Improve Face Recognition Performance. In Proceedings of IEEE Computer Society Conference on Computer Vision and Pattern Recognition Workshops, Colorado Springs, CO, USA, 20–25 June 2011; pp. 140–145.

27. Wei, Z.; Li, X.; Zhuo, L. An Automatic Face Log Collection Method for Video Sequence. In Proceedings of International Conference on Internet Multimedia Computing and Service, Huangshan, China, 16–18 August 2013; pp. 376–379.

28. Anantharajah, K.; Denman, S.; Sridharan, S.; Fookes, C.; Tjondronegoro, D. Quality Based Frame Selection for Video Face Recognition. In Proceedings of International Conference on Signal Processing and Communication Systems, Gold Coast, Australia, 12–14 December 2012; pp. 1–5.

29. Webcam C600. Available online: https://support.logitech.com/en_us/product/5869 (accessed on 13 August 2015).

30. Lee, W.O.; Kim, Y.G.; Shin, K.Y.; Nguyen, D.T.; Kim, K.W.; Park, K.R.; Oh, C.I. New Method for Face Gaze Detection in Smart Television. *Opt. Eng.* **2014**, *53*, 053104:1–053104:12.

31. Viola, P.; Jones, M.J. Robust Real-Time Face Detection. *Int. J. Comput. Vis.* **2004**, *57*, 137–154.

32. Boyle, M. *The Effects of Capture Conditions on the CAMSHIFT Face Tracker*; Technical Report 2001-691-14; Department of Computer Science, University of Calgary: Calgary, Canada, 2001.

33. Tripathi, A.K.; Mukhopadhyay, S.; Dhara, A.K. Performance Metrics for Image Contrast. In Proceedings of International Conference on Image Information Processing, Himachal Pradesh, India, 3–5 November 2011; pp. 1–4.

34. Barghout, L. System and Method for Edge Detection in Image Processing and Recognition. U.S. Patent PCT/US2006/039812, 20 April 2006.

35. Grycuk, R.; Gabryel, M.; Korytkowski, M.; Scherer, R.; Voloshynovskiy, S. From Single Image to List of Objects Based on Edge and Blob Detection. In *Artificial Intelligence and Soft Computing*; Springer International Publishing: Cham, Switzerland, 2014; Volume 8468, pp. 605–615.

36. Barghout, L. Spatial-Taxon Information Granules as Used in Iterative Fuzzy-Decision-Making for Image Segmentation. In *Granular Computing and Decision-Making: Interactive and Iterative Approaches*;

Springer International Publishing: Cham, Switzerland, 2015; Volume 10, pp. 285–318.

37. Mamdani, E.H.; Assilian, S. An experiment in linguistic synthesis with a fuzzy logic controller. *Int. J. Man Mach.* **1975**,*7*, 1–13.

38. Hurwitz, D.S.; Wang, H.; Knodler, M.A., Jr.; Ni, D.; Moore, D. Fuzzy sets to describe driver behavior in the dilemma zone of high-speed signalized intersections. *Transport. Res. Part F Traffic Psychol. Behav.* **2012**, *15*, 132–143.

39. Wu, J.; Wu, Y. Detecting Moving Objects Based on Fuzzy Non-Symmetric Membership Function in Color Space. *Int. J. Eng. Ind.* **2011**, *2*, 62–69.

40. Klir, G.J.; Yuan, B. *Fuzzy Sets and Fuzzy Logic-Theory and Applications*; Prentice-Hall: Upper Saddle River, NJ, USA, 1995.

41. Leekwijck, W.V.; Kerre, E.E. Defuzzification: Criteria and Classification. *Fuzzy Sets Syst.* **1999**, *108*, 159–178.

42. Hines, G.; Rahman, Z.-U.; Jobson, D.; Woodell, G. Single-Scale Retinex Using Digital Signal Processors. In Proceedings of Global Signal Processing Conference, Santa Clara, CA, USA, 27–30 September 2004.

43. Turk, M.; Pentland, A. Eigenfaces for Recognition. *J. Cogn. Neurosci.* **1991**, *3*, 71–86.

44. Ahonen, T.; Hadid, A.; Pietikäinen, M. Face Description with Local Binary Patterns: Application to Face Recognition.*IEEE Trans. Pattern Anal. Mach. Intell.* **2006**, *28*, 2037–2041.

45. Nguyen, D.T.; Cho, S.R.; Shin, K.Y.; Bang, J.W.; Park, K.R. Comparative Study of Human Age Estimation with or Without Pre-classification of Gender and Facial Expression. *Sci. World J.* **2014**, *2014*, 1–15.

46. Lee, W.O.; Kim, Y.G.; Hong, H.G.; Park, K.R. Face Recognition System for Set-Top Box-Based Intelligent TV. *Sensors***2014**, *14*, 21726–21749.

47. FEI Face Database. Available online: http://fei.edu.br/~cet/facedatabase.html (accessed on 13 August 2015).

48. Minear, M.; Park, D.C. A lifespan database of adult facial stimuli. *Behav. Res. Methods* **2004**, *36*, 630–633.

49. AR Face Database. Available online: http://www2.ece.ohio-state.edu/~aleix/ARdatabase.html (accessed on 13 August 2015).

50. The Japanese Female Facial Expression (JAFFE) Database. Available online: http://www.kasrl.org/jaffe.html (accessed on 13 August 2015).

51. Wolf, L.; Hassner, T.; Maoz, I. Face Recognition in Unconstrained Videos with Matched Background Similarity. In Proceedings of IEEE Conference on Computer Vision and Pattern Recognition, Providence, RI, USA, 20–25 June 2011; pp. 529–534.

CHAPTER 11

Multi-Layer Sparse Representation for Weighted LBP-Patches Based Facial Expression Recognition

Qi Jia, Xinkai Gao, He Guo, Zhongxuan Luo and Yi Wang

School of Software, Dalian University of Technology, Dalian 116621, China

ABSTRACT

In this paper, a novel facial expression recognition method based on sparse representation is proposed. Most contemporary facial expression recognition systems suffer from limited ability to handle image nuisances such as low resolution and noise. Especially for low intensity expression, most of the existing training methods have quite low recognition rates. Motivated by sparse representation, the problem can be solved by finding sparse coefficients of the test image by the whole training set. Deriving an effective facial representation from original face images is a vital step for successful facial expression recognition. We evaluate facial representation based on weighted local binary patterns, and Fisher separation criterion is used to calculate the weighs of patches. A multi-layer sparse representation framework is proposed for multi-intensity facial expression recognition, especially for low-intensity expressions and noisy expressions in reality, which is a critical problem but seldom addressed in the existing works. To this end, several experiments based on low-resolution and multi-intensity expressions are carried out. Promising results on publicly available databases demonstrate the potential of the proposed approach.

Keywords: facial expression recognition; local binary patterns; weighted patches; sparse representation; multi-layer model

1. INTRODUCTION

Expression is a basic way to express human's emotion, and it is also an effective non-verbal communication method. Therefore, automatic facial expression recognition (AFER) has become more and more important and plays an important role in computer vision. Facial expression recognition has wide application prospects in areas such as human-computer interface (HCI), image retrieval and psychological research.

According to different facial expression features, existing facial expression analysis approaches can be categorized into static images-based methods and dynamic image sequences-based methods. Methods based on dynamic image sequences are often used to enhance recognition performance by use of longitudinal changes information in facial image sequences [1,2]. However, they require more facial images and more computation time to analyze the facial information, which are very challenging in many situations. Thus wide attention has been paid to static images-based methods [3], and our work is also based on this.

Static images-based methods mainly consist of two stages: feature extraction and classifier design. Deriving an effective facial representation from original face images is a vital step for successful facial expression recognition. Many efforts have been devoted to two classical methods: geometric-based methods and appearance-based methods [4,5,6,7]. Geometric-based methods analyze the relationship of facial components and extract facial features with shape and locations [8]. However, in reality it is difficult to detect and track the feature points of facial expressions accurately and fast. As a contrast, the features of appearance-based methods are extracted from the whole face, such as Gabor-wavelet feature, local binary pattern feature. Gabor-wavelet feature [9] is widely used to describe facial expression with a bank of Gabor filters. Zhang *et al.* [10] utilized the features of facial elements and muscle movements by extracting patch-based 3D Gabor features. However, it is time-consuming to calculate a mass of different Gabor coefficients. In [11] the input images are first subjected to local, multi-scale Gabor-filter operations, and then the resulting Gabor decompositions are encoded using radial grids, imitating the topographical map-structure of the human visual cortex. There are also some similar approaches that divide a face image into certain small blocks, and apply some feature extraction algorithms, such as local binary pattern (LBP) analysis and scale invariant feature transformation (SIFT) [12,13,14] in order to get a local texture description. The local binary pattern (LBP) feature [9,12] improves expression recognition performance and efficiency because of its tolerance of illumination changes and compactness. However, an important fact that has often been neglected is that some patches such as the eyes, mouth, and nose provide more important information in facial expression recognition [15]. In our work, an improved LBP descriptor with spatial feature information is used. The Fisher separation criterion is employed to automatically calculate corresponding weights for all the facial patches without supervision.

On the other hand, in the stage of classifier design, many classical methods have been commonly adopted, such as linear discriminant analysis (LDA), linear programming (LP) and support vector machines (SVM), *etc.* LDA is a subspace learning method which can represent facial deformations in lower feature space. LP recognizes facial expressions by simultaneous feature selection and classifier training. SVM is a popular and powerful classifier technique for many classification problems. SVM maps the feature vectors into higher dimensional feature space by appropriate kernels, and finds the maximal margin to separate different facial expressions. Shan et al. [16] compared such several classifier methods and concluded that SVM achieved better performance and was more robust for low resolution expression images. In [17], SVM is combined with a local Steerable Pyramid Transform (SPT) feature for facial expression verification. A newly developed method shows good discrimination ability by integrating curvelet transform and online sequential extreme learning machine with radial basis function (RBF) hidden nodes [18]. Sparse representation-based classification (SRC), which is derived from the compressive sensing (or compressive sampling—CS) theory [19,20], has been successfully used for face recognition [21]. Other research based on SRC have also shown promising results in the field of facial expression recognition during the past few years, and achieved better performances than many classical algorithms [22,23,24].

Although numerous methods have been proposed over the years, facial expression recognition still remains a significant challenge to recognize different human expression styles as well as their large inter-personal variations. In order to improve recognition rates in reality, many factors that complicate facial expression analysis should be taken into account, such as low-resolution, low-intensity and noisy facial expressions. Many efforts have been paid to solve low-resolution conditions. Tian [4] performed experiments on low resolution with geometric features and Gabor features, and a three-layer neural network was used as classifier. Then, Shan et al. [16] improved the recognition rate with LBP feature, instead of geometric and Gabor features, which are difficult to detect and track facial components in low-resolution images, and LBP features are more robust over a range of low resolutions. Moreover, Shan's work also showed that compared to a neural network, SVM could provide better performance with the same expression features. On the other hand, most works choose extreme facial images with high intensity from the whole sequences to construct the experimental database. However, there are usually many different expression intensities that human beings can express in reality. Some low intensity expressions are very similar, such as fear and sadness, which always have low recognition rates in most existing methods. For this problem, Ekman [25] developed the Facial Action Coding System (FACS), which is used to identify facial expressions called micro-expressions. It usually occurs in high-state situations which are very brief in duration, lasting only 1/25 to 1/15 of a second. This system measures the relaxation or contraction of each individual muscle and assigns a facial action unit. Thus, FACS must train and analyze a mass of facial expression images with supervision, and it is very difficult to perceive the action units of low-intensity expressions. Most of the systems developed for AFER cater only to controlled laboratory circumstances in which much of the expression performing is rehearsed and performed by trained actors. What is not currently known is how well these systems adapt to suboptimal circumstances in which there is image noise and occlusion in the images. Some works solve the problem by various filter methods, which is a usual way to handle noisy images [26]. Ouyang et al. [27] used Histograms of Oriented Gradient (HOG) descriptors and LBP conjunction with SRC separately to get two judgment vectors, and managed to fuse them to achieve a better performance. In this paper, we focus on the ability of the classifiers to handle noise.

In our former work, we formulated the static facial expression recognition problem by a sparse representation (SR) method, which casts the recognition problem as finding a sparse approximation of the test image in terms of the whole training set. SR can simultaneously reduce feature dimensions and classify expressions without training the classifier preliminarily [28]. In this paper, we utilize sparse representation for facial expression recognition, in which the training facial expression images are used as the dictionary to code an input testing facial image as a sparse linear combination of them via l1-norm minimization. Intuitively, the desired representation for the testing facial expression is sparse, which should only be represented in terms of training instances of the same expression.

In this paper, we focus on the performance of SR on low resolution or noisy facial images. Moreover, considering the situation of multi-intensity based on the static facial expression, the implement of a multi-layer sparse representation (MLSR) model is elaborated, which is used to improve the recognition performance of multi-intensity facial expressions, especially for low intensity expression. More importantly, MLSR can be used in many similar classification problems. Experiments results show the robustness of our methods on low-resolution, multi-intensity and noisy facial expressions. Further, we also discuss the across-dataset which is a common problem in recognition tasks.

The paper is organized as follows: Local Binary Patterns are introduced in Section 2. Section 3 proposes the sparse representation method for facial expression recognition. Multi-layer sparse representation method is presented in Section 4. Experiments on facial expression recognition are listed in Section 5. Section 6 concludes this paper.

2. EXPRESSION FEATURE EXTRACTION

2.1. Local Binary Patterns (LBP)

In this part, facial image is divided into several patches based on LBP features. LBP is a novel low-cost image descriptor for texture classification which can present salient micro-patterns of facial images. The original LBP operator was introduced by Ojala *et al.* [29], in which the image pixel is labeled as a binary class by setting a threshold for the difference between the center pixel and its neighbors as shown in Equation (1):

$$LBP = \sum_{j=0}^{7} T(g_j - g_c) \cdot 2^j, T(x) = \begin{cases} 1, x \geq 0 \\ 0, x < 0 \end{cases} \tag{1}$$

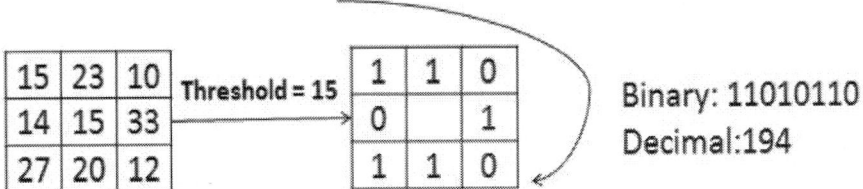

Figure 1. The basic LBP operator.

The concatenation of the neighboring labels is then used as a unique descriptor for each pattern, and its decimal number is the LBP feature of the center point. Figure 1 gives a simple example. In our experiments, the size of each patch is 20 × 20 pixels. The limitation of the basic LBP operator is that the 3 × 3 patch is too small to capture dominant features with large scale structures. Hence the operator is extended with different sizes [30]. Using circular neighborhoods and interpolating the pixel values allow any radius and number of pixels in the patch. Further extension of LBP is to use uniform patterns [31]. LBP is called uniform if it contains at most two bitwise transitions from 0 to 1 or *vice versa* when the binary string is considered circular. For example, 00110000 and 11100001 are uniform patterns. It is observed that uniform patterns account for nearly 90% of all patterns in the (8,1) patch and for about 70% in the (16,2) patch in texture images.

Here we adopt a LBP operator $LBP_{P,R}^{u2}$, where (P,R) denotes P sampling points on a circle of radius of R. The subscript represents using the operator in a (P,R) neighborhood; the superscript $u2$ indicates using only uniform patterns and labeling all remaining patterns with a single label.

2.2. Weighted Patches

The histogram contains information about the distribution of the local micro-patterns, such as edges, spots and flat areas over the whole image. Thus a histogram of a labeled image $p(x,y)$ can be defined as follows:

$$H_i = \sum_{x,y} S(p(x,y)=i), \ i=0,...,n-1, \ (x,y) \in R_j \qquad (2)$$

where n is the number of different labels produced by the LBP operator. $S(T) = 1$ when T is true and $S(T) = 0$ when T is false.

Further, for efficient facial representation, feature extracted should also retain spatial information. Hence the facial image is divided into m small patches $R_0, R_1,..., R_m$ and a spatially enhanced histogram is defined as:

$$H_i = \sum_{x,y} S(p(x,y)=i), \ i=0,...,n-1, \ (x,y) \in R_j \qquad (3)$$

Hence for each patch, $LBP_{8,2}^{u2}$ operator will produce a 59 dimension descriptor. As shown in Figure 2a, a histogram with 59 labels is adopted to accumulate the value on each dimension for all the patches. For example, for a facial image with 120 × 160 pixels, there are 48 (6 × 8) patches (20 × 20 pixels for each patch). Then, the feature dimension of each facial image is 2832 (59 × 48). To exclude redundancy data, the feature data is projected into a PCA subspace. In the dimension reduction step 98% information is kept according to the reconstruction error.

Moreover, it is expected that some face patches (e.g., eyes, mouth, and nose) provide more important information than others in facial expression recognition. Therefore, different facial patches should have different weights. Shan et al. [12] designed the weights empirically and gave higher weights for interesting facial patches manually, which is not automatic and reasonable. In order to learn suitable weights for each patch automatically, an approach based on fisher separation criterion [32] is used to calculate weighs as follows.

We set C to be the number of facial expression classes. Let the similarities of different samples of the same facial expression compose the intra-expression similarity class, and samples from different facial expressions compose the extra-expression similarity class. For patch R_j of the image, the mean value $M_I(j)$ and the variance $S_I^2(j)$ of the intra-expression similarity class can be computed as follows:

$$M_I(j) = \frac{1}{C}\sum_{m=1}^{C}\frac{2}{N_m(N_m-1)}\sum_{k=2}^{N_m}\sum_{n=1}^{k-1}\Psi(H_j^{(m,n)},H_j^{(m,k)}) \quad (4)$$

$$S_I^2(j) = \sum_{m=1}^{C}\sum_{k=2}^{N_m}\sum_{n=1}^{k-1}(\Psi(H_j^{(m,n)},H_j^{(m,k)})-M_I(j))^2 \quad (5)$$

Similarly, the mean value $M_E(j)$ and the variance $S_E^2(j)$ of the extra-expression similarity class can be calculated as follows:

$$M_E(j) = \frac{2}{C(C-1)}\sum_{m=1}^{C-1}\sum_{n=m+1}^{C}\frac{1}{N_mN_n}\sum_{k=1}^{N_m}\sum_{l=1}^{N_n}\Psi(H_j^{(m,k)},H_j^{(n,l)}) \quad (6)$$

$$S_E^2(j) = \sum_{m=1}^{C-1}\sum_{n=m+1}^{C}\sum_{k=1}^{N_m}\sum_{l=1}^{N_n}(\Psi(H_j^{(m,k)},H_j^{(n,l)})-M_E(j))^2 \quad (7)$$

where $H_j^{(m,n)}$ denotes the histogram extracted from R_j region of the nth sample in m^{th} class. N_m is the sample number in m^{th} class in the dataset. Here, we use the histogram intersection $\Psi(H^1, H^2)$ as the similarity

$$\Psi(H^1, H^2) = \sum_{n=1}^{L} \min(h_n^1, h_n^2),$$

measurement of two histograms. Where H^1 and H^2 are two histograms, and L is the number of bins in the histogram.

Finally, the weight $W(j)$ for R_j is computed as Equation (8), and the weighted feature extracted from R_j region of the nth sample in the mth class is $HW_j^{(m,n)}$. Figure 2b shows the final expression feature with weighted patches. The patches are expressed in gray level from 0 to 255. The darker the patch is, the larger the weight is. It is obvious that the weight is set reasonably:

$$W(j) = \frac{(M_I(j) - M_E(j))^2}{S_I^2(j) + S_E^2(j)}, HW_j^{(m,n)} = W(j) \times H_j^{(m,n)} \qquad (8)$$

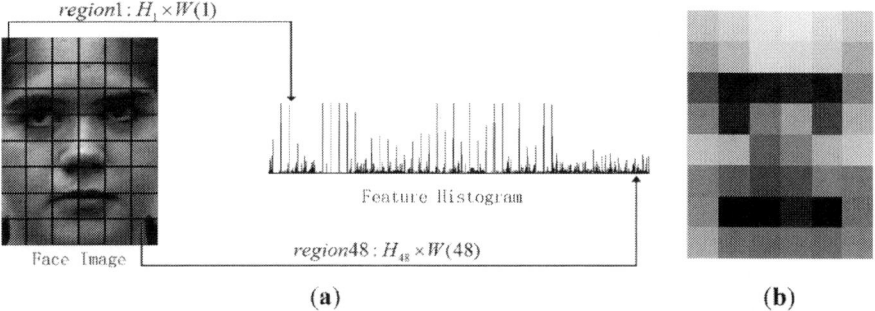

Figure 2. LBP feature extracting with weighted patches. (**a**) Histogram of LBP features; (**b**) Weighted patches.

3. FACIAL EXPRESSION RECOGNITION BASED ON SPARSE REPRESENTATION

The SR based classification (SRC) for facial expression is conducted by evaluating which class of training samples can achieve the minimum reconstruction error with the input testing image by the sparse coding coefficients. Suppose we have C expression classes, and let $A = [A_1, A_2,..., A_c]$ be the concatenation of the n training samples from all the C classes, where $n = n_1 + n_2 + ... + n_c$, then the linear representation of y_0 can be written in terms of all training samples as $y_0 = A\alpha$.

However, usually, the feature vector of an image is very high-dimensional and the number of training samples is limited, so A is over-determined. The least-squares solution can exhibit severe bias for "the curse of dimensionality" caused by over-determined. Luckily, the CS theory has shown that a sparse signal can be recovered from a small number of its linear measurements with high probability [20]. According to CS, a sparse signal $x \in R^n$ should be recovered from the following linear random projections: $y = \Phi x$, where $y \in R^m$ is the measurement vector, $\Phi \in R^{m \times n}(m \ll n)$ is a random projection matrix. For facial expression features, the projection from the image space to the feature space can be represented as a measure matrix Φ. Therefore, the measure vector is $y = \Phi y0 = \Phi A \alpha$.

Considering the existence of error, the formulation can be rewritten as $y = \Phi A \alpha + e_0 = \overline{A} \alpha + e_0$, where e0 is a noise term with bounded energy $\|e_0\|_2 < \varepsilon$ and $\overline{A} = \Phi A$. Then our goal is to solve the convex optimization problem: compute min $\|\alpha\|_1$ subject to $\|y - \overline{A} \alpha\|_2 \le \varepsilon$. The sparse representation based facial expression recognition algorithm is summarized in Algorithm 1. The result is shown in Figure 3. As we can see, there are six basic emotions (anger, disgust, fear, happy, sad, surprise), and the test image shows obvious difference to the other expressions, especially in Figure 3d,f. Meanwhile, the other four expressions are a little bit similar in some instance, as shown in Figure 3a–c,e. In most existing methods, the result will get worse when the test image is low-intensity facial expression. However, the SR based classification method shows stable recognition rate, which will be demonstrated in the next Section and evaluated in the experiments.

Algorithm 1. Facial Expression Recognition based on Sparse Representation

Input: a matrix of training images $A \in R^{d \times n}$ for C expression classes, a linear feature transform $\Phi \in R^{m \times d}$, a test image $y_0 \in R^n$, and an error tolerance ε.

Output: identity $(y) = \text{argmin } r_i(y)$.

1. Compute $y = \Phi y_0$ and $\overline{A} = \Phi A$, and normalize y and columns of $\overline{A}$ to unit length.

2. Solve the following convex optimization problem

$$\min \|\alpha\|_1 \text{ subject to } \|y - \overline{A} \alpha\|_2 \le \varepsilon$$

3. Compute the residuals:

$$r_i(y) = \|y - \overline{A} \delta_i(\alpha)\|_2, \text{ for } i = 1, 2, ..., C$$

where $\delta_i(\cdot): R^n \to R^n$ is the characteristic function which selects the coefficients associated with the ith class.

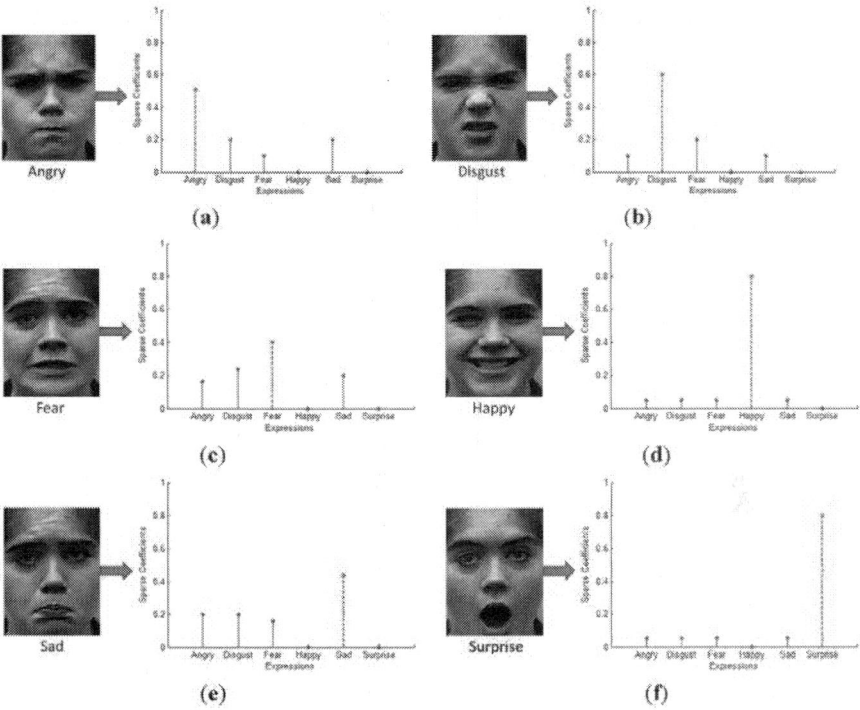

Figure 3. Classification result for different testing expressions (**a**) Angry; (**b**) Disgust; (**c**) Fear; (**d**) Happy; (**e**) Sad; (**f**) Surprise.

4. MULTI-INTENSITY FACIAL EXPRESSION RECOGNITION BASED ON MULTI-LAYER SPARSE REPRESENTATION

Most existing research in facial expression analysis mainly focuses on recognizing extreme facial expressions. However, in real-world applications, such as human-computer interaction and data-driven animation, low-intensity or subtle facial expressions are more universal than high-intensity or exaggerated expressions. Some low-intensity facial expressions are very similar, such as anger and disgust, which always have low recognition rate in most existing approaches. The essential reason is that these methods neglect the correlations among different intensities. As shown in Figure 4, Figure 4a,c are "anger", while Figure 4b,d are "disgust". In the high intensity version, Figure 4a,b have obvious differences at the nose. However, in the low intensity form, Figure 4c,d are nearly the same. In a traditional classifier, Figure 4b,d are trained in the same class, in which the details of low intensity expression may be weakened by high intensity images.

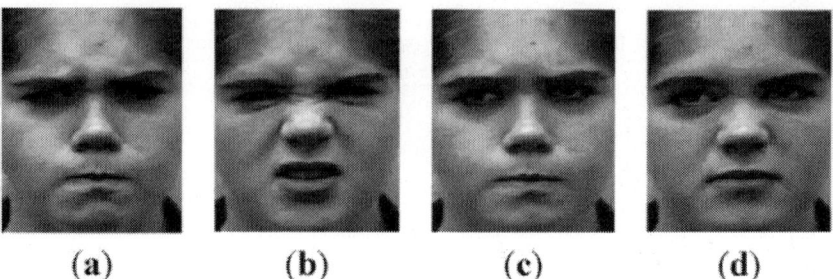

(a)   **(b)**   **(c)**   **(d)**

Figure 4. Anger and Disgust in different intensity (**a**) Anger in high intensity; (**b**) Disgust in high intensity; (**c**) Anger in low intensity; (**d**) Disgust in low intensity.

Driven by this purpose, a multi-layer sparse representation (MLSR) method is proposed with more detailed categories to improve the recognition performance of multi-intensity facial expressions. Our notion is motivated by the latest work of Liu *et al.* [30], which casts sparse representation as nonnegative curds and whey (NNCW). The MLSR model consists three layers: the first layer is used to calculate sparse correlations among each expression intensity of each expression class; the second layer codes the testing expression with all intensity groups of each expression class; the third layer is set to represent the testing facial expression among different expressions. Therefore, the MLSR technique can present the correlations within a same intensity and the disparities between different intensities, which are beneficial to the situation of multi-intensity facial expressions. As shown in Figure 5, on the first layer of MLSR, each expression is divided into D different intensity. All the images with the same expression and the same intensity are trained together, which means images with the same expression and different intensity are trained separately. Therefore, an easily confused low intensity expression has stronger correlations with the corresponding intensity class.

Considering the correlations are also strong in the same expression of different intensity. On the second layer, we combine different intensity of each expression separately to strengthen the correlations. Finally, the last layer gives the classification results.

The entire MLSR method is summarized in Algorithm 2. We assume that there are C facial expression classes and each expression class has D types of intensity (Note each facial expression has the same D here). The images set of the jth type of intensity of the ith-class expression is defined as $A_{i,j}$, and $N_{i,j}$ is the image number in $A_{i,j}$.

On the first layer, linear regression models of $B(B = C \times D)$ are considered by treating y_0 as the response, and each image from $A_{i,j}$ has one basis: $y_0 = A_{i,j}\alpha_{i,j}$. That is, the regression problem of the jth type of intensity of the ith-class expression is $y = \Phi_{i,j}y_0 + e_0^{i,j} = \Phi_{i,j}A_{i,j}\alpha_{i,j} + e_0^{i,j} = \overline{A}_{i,j}\alpha_{i,j} + e_0^{i,j}$, where $i = 1,...,C$, $j = 1,...,D$, $\alpha_{i,j} \in R^{N_{i,j} \times 1}$ denoted in Section 3. The result leads to B sparse representations of y, which can be express as $Z_{i,j} = A_{i,j}\hat{\alpha}_{i,j}$ and $\hat{\alpha}_{i,j}$ is reconstruction coefficients learned from samples within the jth type of intensity of the ith-class expression.

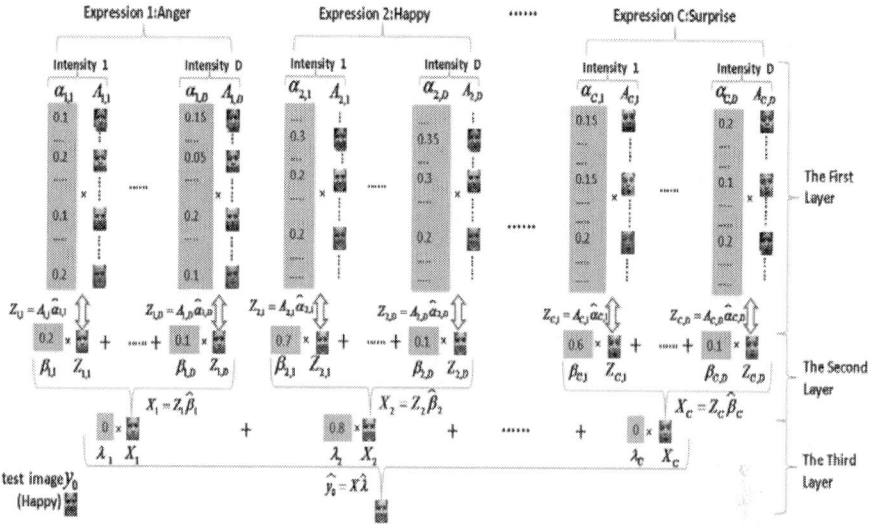

Figure 5. Multi-layer sparse representation model.

On the second layer, C linear regression models are built. The representation of y_0 is refined by the representations of y_0 obtained in the first stage: $y_0 = Z_i\beta_i, Z_i = [Z_{i,1},...,Z_{i,D}], \beta_i \in R^{D\times 1}$. That is, the ith regression problem is based on: $y = \Phi_i y_0 + e_0^i = \Phi_i Z_i \beta_i + e_0^i = \bar{Z}_i \beta_i + e_0^i$. The result of the second layer leads to C sparse representations of y, which can be expressed as $X_i = Z_i\hat{\beta}_i$. $\hat{\beta}_i$ is reconstruction coefficients learned from samples within the first layer's representations for the ith-class expression.

Algorithm 2. Multi-intensity Facial Expression Recognition based on Multi-Layer Sparse Representation

Input: training images $[A_{1,1},...,A_{1,D},...,A_{C,1},...,A_{C,D}]$; measure matrixes of each layer $[\Phi_{1,1},...,\Phi_{1,D},...,\Phi_{C,1},...,\Phi_{C,D}], [\Phi_1,...,\Phi_C], \Phi$;

A test image $y_0 \in R^n$, and error tolerance ε.

Output: identity $(y) = \arg\min r_i(y)$.

1. The First Layer: Solve the following convex optimization problem in each intensity of each expression class.

$\min \|\alpha_{i,j}\|_1$ subject to $\|y - \bar{A}_{i,j}\alpha_{i,j}\|_2 \leq \varepsilon$ for $i = 1,...,C, j = 1,...,D$

2. The Second Layer: Solve the the testing facial expression and all intensity groups in the same expression. following convex optimization problem between

$\min \|\beta_i\|_1$ subject to $\|y - \bar{Z}_i\beta_i\|_2 \leq \varepsilon$ for $i = 1,...,C$

3. The Third Layer: Solve the following convex optimization problem of the testing facial expression among different expressions

$\min \|\lambda\|_1$ subject to $\|y - \bar{X}\lambda\|_2 \leq \varepsilon$

4.Compute the residuals:

$r_i(y) = \|y - \hat{\lambda}_i \bar{X}_i\|_2$ for $i = 1,...,C$

Finally, the third layer further refines the representation of y_0 by using the representation of y_0 obtained in the second stage: $y_0 = X\lambda$, $X = [X_1,\dots, X_C]$, $\lambda = [\lambda_1,\dots, \lambda_C] \in R^{C\times 1}$.

As we can see, the final regression problem is:

$$y = \Phi y_0 + e_0 = \Phi X\lambda + e_0 = \bar{X}\lambda + e_0 \qquad (9)$$

Then, the Dynamic Group Sparsity (DGS) [33] algorithm is used to solve the convex optimization problem, which is more accurate and faster than conventional lasso methods.

In order to explain the roles of each layer in detail, a low intensity expression "Fear" is selected as shown in Figure 6a. In the SR classifier, it is recognized as "Anger" as shown in Figure 6b. However, it exhibits obvious discrimination with other expressions with MLSR, as shown in Figure 6c.

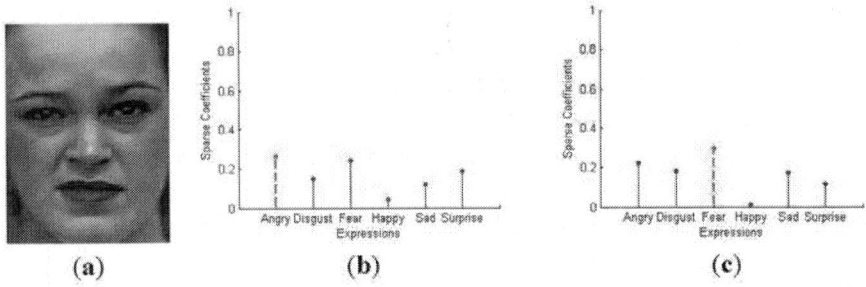

Figure 6. Recognition result of "Fear" with SR and MLSR. (**a**) "Fear" in low intensity; (**b**) Recognition result with SR; (**c**) Recognition result with MLSR.

5. EXPERIMENTS

5.1. Data Set

The Cohn-Kanade Dataset (CK) [34], Extended Cohn-Kanade Dataset (CK+) [35] and the JAFFE Dataset [36] are used in our experiments, which are the most comprehensive datasets for facial expression recognition. The CK+ Dataset's participants were instructed by an experimenter to perform a series of 23 facial displays. Each display began in a neutral face. Image sequences were digitized into either 640 × 490 or 640 × 480 pixel arrays with 8-bits gray-scale or 24-bit color values. On the other hand, the JAFFE Dataset contains 213 images of seven facial expressions (six basic facial expressions and one neutral expression) posed by 10 Japanese female models. Ten expressers posed three or four examples of each expression.

First of all, face images are normalized to a fixed distance between the centers of two eyes. The fixed distance is 60 pixels. It is observed that the width of a face is roughly two times of the distance, and the height is roughly three times. Hence, facial images of 120 × 160 pixels are cropped from original frames based on the two eyes location. Some parameters can be optimized for the LBP feature selection. The first one is the LBP operator, and another is the number of patches divided, which has been illuminated in Section 2.

5.2. Sparse Representation for 7-Class Facial Expressions Recognition

In our experiments, 309 image sequences are selected from the CK+ Database. The only selection criterion is that a sequence can be labeled as one of the six basic emotions (angry, disgust, fear, happy, sad, surprise). The database is presented the inventory of the six basic expressions as Table 1. For each sequence, the neutral face and three peak frames are used for prototypic expression recognition, resulting in 1236 images (135 Angry, 177 Disgust, 75 Fear, 207 Happy, 84 Sad, 249 Surprise and 309 Neutral). The sequences come from 106 subjects. To evaluate the generalization performance to novel subjects, a 10-fold cross validation testing scheme is adopted as [12]. However, our results have no comparability to Shan's work because of different database, experimental setups, pre-processing procedures, weighted patches, and classifier. But we can also draw many conclusions from the following experiments.

First, the performance of weighted patches which is introduced in Section 2 is analyzed. As shown in Table 2, the recognition results based on SR method with weighted patches is 87%, which performs better than the method without weights of 85.1%. On the other hand, to verify the effectiveness of sparse representation for facial expression recognition, SVM is used as the classifier for comparison, which has been a popular technique for facial expression recognition. Here we use the SVM implementation with linear kernel and RBF kernel in the public available machine learning library SPIDER [37], and both methods are carried out with weighted patches proposed in our paper. Moreover, in order to highlight the classification performance of our methods (SR and MLSR), the following experiments about SVM are all performed with weighted patches which is to keep the same input for comparison. In sparse representation's implementation, DGS algorithm [33] is used to solve the convex optimization problem. The comparison in Table 2 illustrates that sparse representation method outperforms SVM under the same experimental conditions.

Table 1. Number of each basic expression's sequences.

Emotion	Numbers
Angry	45
Disgust	59
Fear	25
Happy	69
Sadness	28
Surprise	83
Total	309

Table 2. Comparisons between SR and SVM.

Methods	Recognition Results
Sparse Representation (patches without weights)	85.1%
Sparse Representation (patches with weights)	87%
SVM with linear kernel (patches with weights)	85.4%
SVM with RBF kernel (patches with weights)	86.5%

5.3. Data Set Performance Over Different Resolutions

In many facial expression recognition applications, the input facial images are often low-resolution. We further evaluated the SR-based algorithm over a range of image resolutions, investigating its performance against low-resolution images. Similarly, SVM (linear kernel) is used for comparison. The lower resolution images are down-sampled from the original images. To get enough features from the lower resolution images, lower resolution images are resized to the size of the original images (120×160) by simple interpolation methods and then extract expression feature as above.

The recognition rates of 6-class expressions (without neutral expression) are shown in Figure 7. It is observed that the SR method is more effective for the great mass of resolutions than the SVM method. It indicates that sparse representation is robust for low-resolution expression images as SVM method. Besides, the recognition rate for the original image is 92.6%, which is higher than the result in Table 2, for the former data set is 6-class and the latter is 7-class.

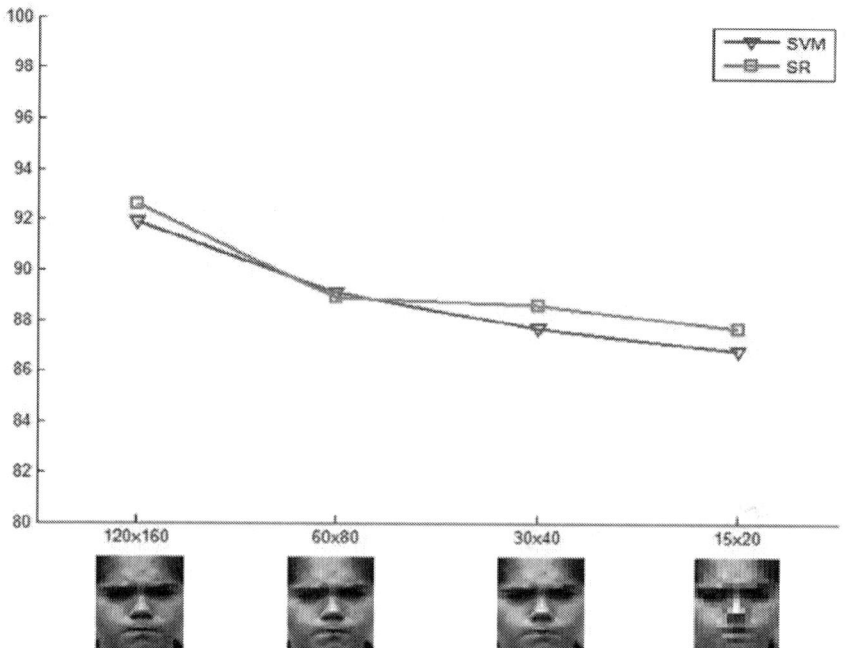

Figure 7. Comparisons on image resolutions between SR and SVM.

5.4. Mlsr For Multi-Intensity Facial Expression Recognition

5.4.1. Recognition Results for Different Intensity

In this experiment, three frames are chosen from each expression sequence, as shown in Table 1. The chosen three frames have different expression intensity: one is high intensity which is the last frame in the sequence, one is moderate intensity which is selected in the middle of the sequence, and one is low intensity which is in the front half of the sequence. Therefore, our experiment consists of 6-class expressions and each expression includes three types of expression intensity. In Figure 8, there is an example of multi-intensity expressions. Half of all the sequences are used for training, and the rest half are used to test. Therefore, the testing is also person-independent. To verify the performance of MLSR, SR and SVM are used for comparing, which are presented in Figure 9 and Table 3.

As we can see in Figure 8, the testing data are divided into three parts (high intensity, moderate intensity, low intensity), and the recognition rate is calculated separately. For high intensity and moderate intensity, MLSR do not get better result than SR, but MLSR presents better accuracy for low-intensity expressions. However, in Figure 9c, the fear and sad expression show no differences like the other expressions for MLSR and SR. This is because the low-intensity sequences of fear and sad expressions are much more alike and difficult to distinguish, even for human beings. It is predictable that MLSR would achieve some improvements for high intensity and moderate intensity, if the database is more plentiful. Table 3 shows the

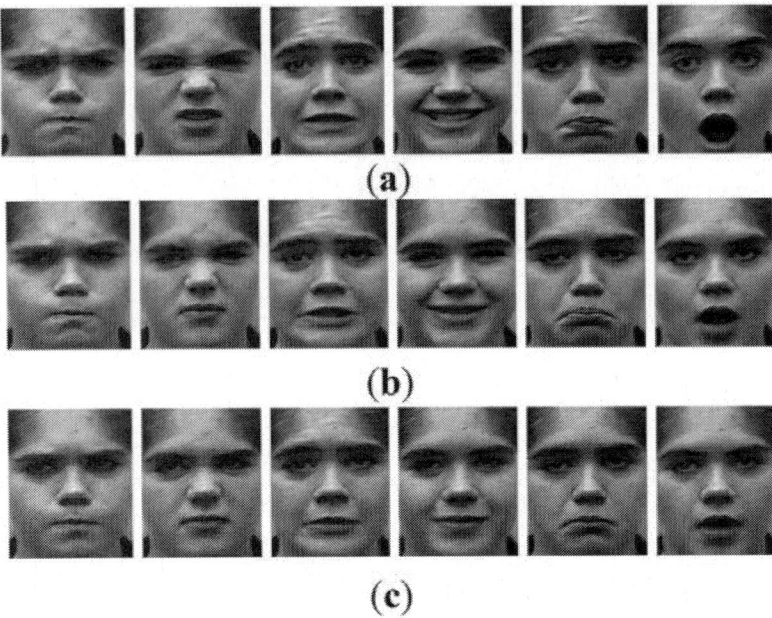

Figure 8. Multi-intensity example of six expressions: Angry, Disgust, Fear, Happy, Sad, Surprise. (**a**) High intensity; (**b**) Moderate intensity; (**c**) Low intensity.

results of multi-intensity expression recognition between MLSR and SVM. Likewise, MLSR and SVM are similar for high intensity and moderate intensity. However, for low intensity, MLSR is better than SVM.

5.4.2. Comparison with State-of-The-Art Performance

In order to follow the same protocol, The Cohn-Kanade (CK) Dataset [34] is introduced to make a comparison between MLSR and state-of-the-art approaches. And the person-independent strategy is used for cross-validation.

The CK database contains facial expression images of 97 university students (35% males and 65% females. 15% African–American and 3% Asian or Latino). Each person has a set of image sequences from neutral to certain facial displays coded by action units. For our study, we first select those sequences (from 92 persons) that can be labeled using one of the six basic expressions. For every person, we select three images for each expression the same as in the CK+ database (one is high intensity which is the last frame in the sequence, one is moderate intensity which is selected in the middle of the sequence, and one is low intensity which is in the front half of the sequence). Therefore, we select 296 sequences and there are 296 × 3 = 888 images in total for our experiments. All the images are separated into 10 sets, and all images of one subject are included in the same set. To evaluate the generalization performance to novel subjects, a 10-fold cross validation testing scheme is adopted.

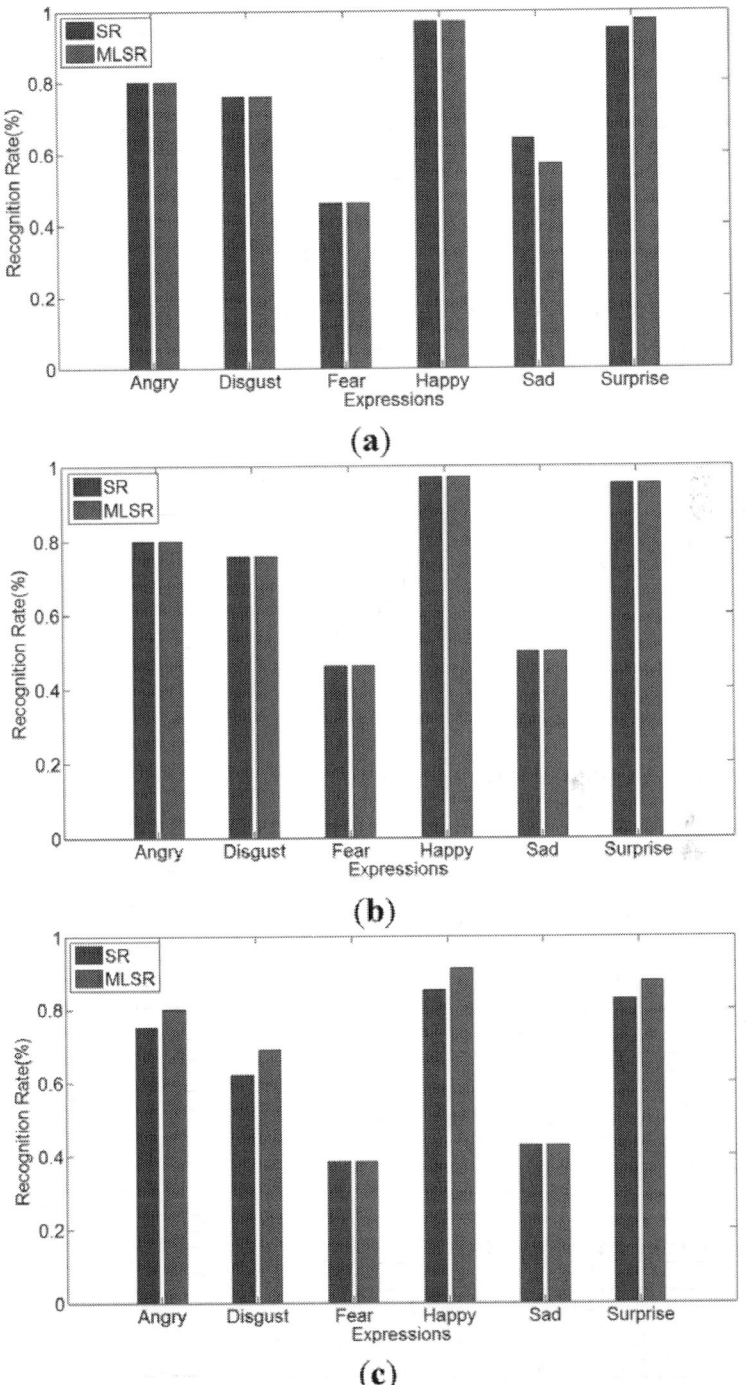

Figure 9. The comparison between MLSR and SR about multi-intensity recognition. (**a**) High intensity; (**b**) Moderate intensity; (**c**) Low intensity.

Table 3. Comparisons on intensities between MLSR and SVM.

Methods	High Intensity	Moderate Intensity	Low Intensity
MLSR	83.56%	82.26%	76.35%
SVM	84.07%	81.43%	71.96%

Note that the results of different algorithms may not be directly comparable because of differences in experimental setups, the number of subjects and so on, but they can still indicate the discriminative performance of each approach.

Table 4. Comparison with state-of-the-art performance of the proposed algorithm on the CK database.

Methods	Subjects	Measure	Recognition Rate
[6]	97	5-fold	90.90%
[7]	90	-	93.66%
[10]	92	leave-one-subject-out	94.48%
[13]	97	10-fold	96.26%
[14]	90	leave-one-subject-out	96.33%
[12]	96	10-fold	88.4% (92.1%)
[11]	94	10-fold	91.51%
[18]	92	leave-one-subject-out	95.17%
MLSR	92	10-fold	92.3%

Table 4 shows a similar comparison with respect to the CK database. The first five methods belong to a dynamic approach which based on video sequences, and the last four approaches, including ours, belong to a static approach which is normally applied to static images. The dynamic approach uses the relations between the images in the videos, which needs more information. Most methods give better results than the static one, and [14] got the best recognition accuracy of 96.33%. However, in real-world applications, such as human-computer interaction and data-driven animation, static images are more universal than videos for expression recognition. Among the static approaches, our recognition accuracy is 92.3% which is just lower than [18]. In [18] a curvelet transform is used to get the features of the expression, and it is more sensitive to noise. However, our approach can remain robust over noise images, which is shown in the next Section.

5.5. Performance Over Noise

In order to illustrate the robustness of SR over noise, Gaussian noise is imposed on the database with the standard deviation varying from 5 to 20. Some examples are shown in Figure 10. Figure 10a is the expression of anger corrupted by the noise whose standard deviations are 5, 10, 15, 20, respectively. Figure 10b,c are expressions of disgust and sadness, accordingly. The expressions are blurred when the noise is severe.

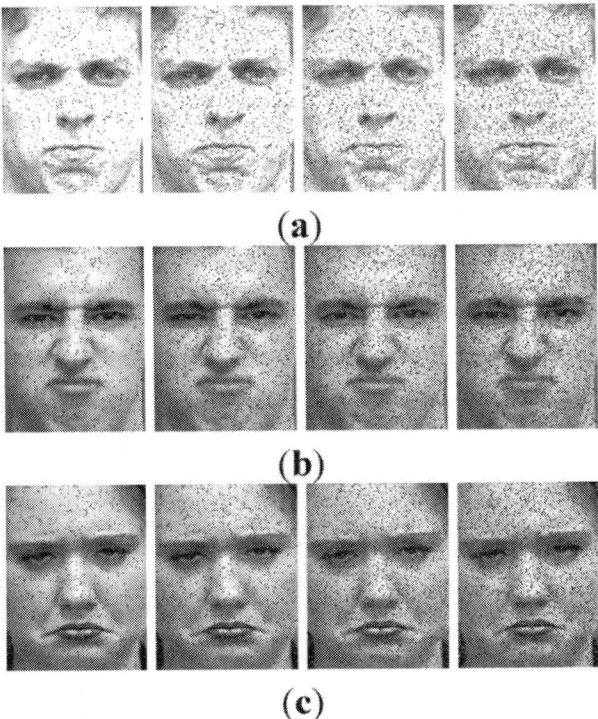

Figure 10. Facial expression with noise. (**a**) Anger; (**b**) Disgust; (**c**) Sadness.

The experiments are performed on 7-class facial expressions, and the recognition rates are shown in Table 5. The recognition results of each expression class are illustrated in each column. For each expression class, the recognition rates drop gradually as the noise are heaver. There is a sensible difference between different expression classes. For some distinguishable expression such as disgust, happy and surprise, the recognition rates are as high as 84%–95%. Even when the noise is severe at 20%, the recognition rates are around 84%–91%. However, the recognition rates of some undistinguishable expressions are lower relatively, which is around 50%–77%. The average recognition rates over each noise degree are shown in the last column. The recognition rate of each expression class multiplied by the percent of its number over the whole database gives the results, which are around 77%–84%. The results show that SR can remain robust against noise in different degrees.

Table 5. Recognition rate of SR over noisy images.

Expression \ Noise	Angry	Disgust	Fear	Happy	Neutral	Sad	Surprise	Average
5	77.048	92.325	68.055	94.662	78.125	56.945	92.870	83.712
10	71.238	89.231	67.499	93.015	77.438	53.055	91.144	81.559
15	67.370	85.845	61.945	92.301	75.915	50.056	89.147	79.205
20	61.946	83.952	56.111	91.190	73.364	49.168	87.897	76.856

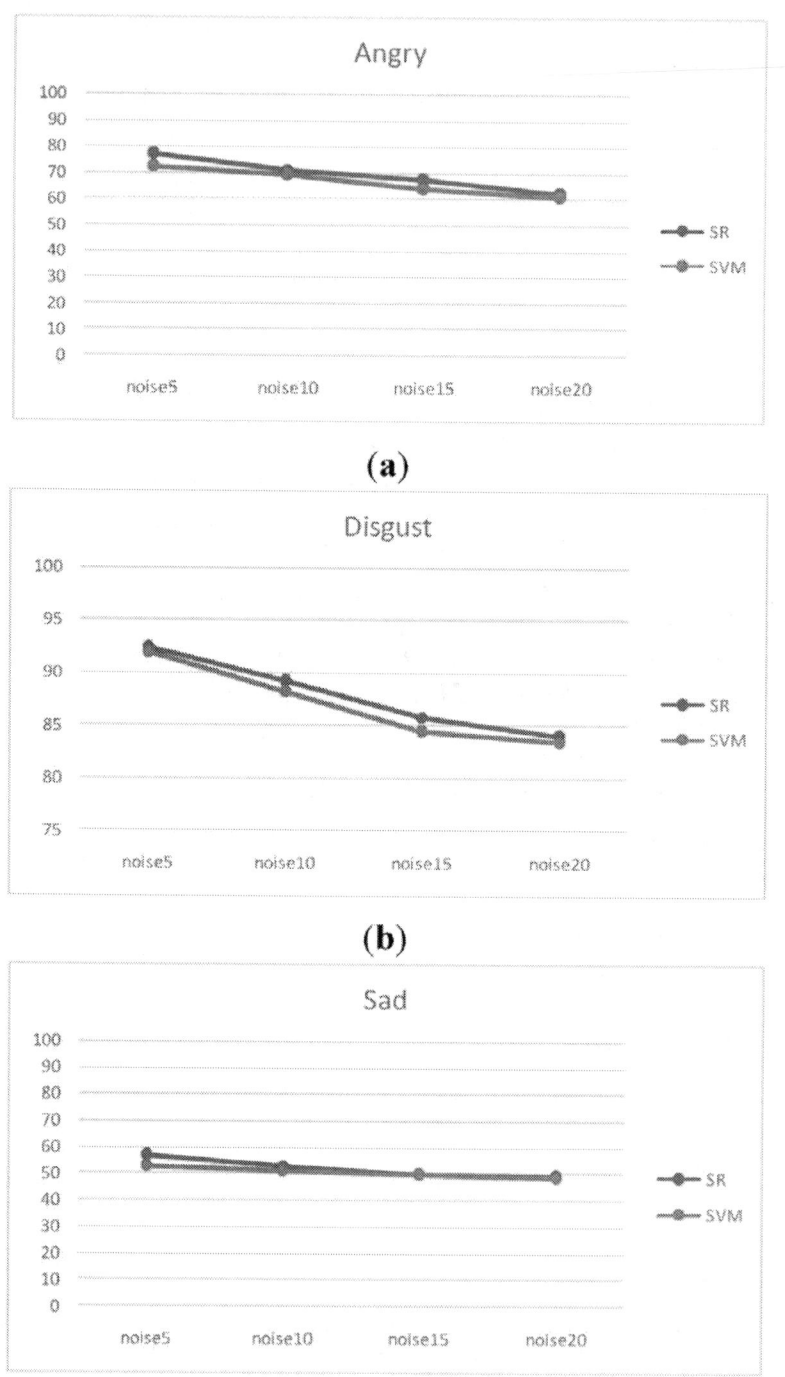

Figure 11. The comparison between SR and SVM against noise. (**a**) Angry;
(**b**) Disgust; (**c**) Sad.

We also make a comparison with SVM. The performance is nearly equivalent to SR. However, for some expressions such as angry, disgust and sad, SR can achieve better performance than SVM. The result is shown in Figure 11. The recognition rate of both methods drops gradually as the noise increases. SR steadily performs better than SVM when the noise power increases.

5.6. CROSS-DATABASE EXPERIMENT

In general, most researchers do experiments using a single dataset for training and testing. That is, part of the dataset is employed for training the classifier, and the rest is used for testing the classifier. However, in the real world, we can't ensure that the test dataset comes from the same dataset as the training dataset. Therefore, to verify a classifier's robustness between different datasets, across-dataset experiments become very essential. Shan *et al.* [12] performed across-dataset experiments which trained on the CK database and due to the different controlled environments, such as atmospheres, image equipment and illumination. Therefore, the current expression classifier which is trained on a controlled environment's database can only work well for a testing database with the same controlled environment. In this paper, we choose the Extended Cohn-Kanade Dataset as training database, and the JAFFE Dataset as testing database. Likewise, we take SR and SVM for comparisons, as shown in Table 6. Both SR and SVM methods achieve very low recognitions which are even below 50%.

Table 6. Comparisons on across-database between SR and SVM.

Methods	SR	SVM
Train: CK+ Test: JAFFE	40.5%	39.4%

Further, the lack of standard evaluation criteria of basic expressions is also a crucial reason. Different expression databases have different evaluation criteria. They are consistent in general, but there are variances which influence the recognition. Figure 12 shows an example from the CK+ and JAFFE Datasets, in which the same angry expressions come from different databases. The left one is from the CK+ database, and the right one is from the JAFFE database. There are obvious differences between the two images, such as the eyebrows, mouth, and cheek. Therefore, the problem of how to establish a standard evaluation criterion for basic expressions remains to be resolved.

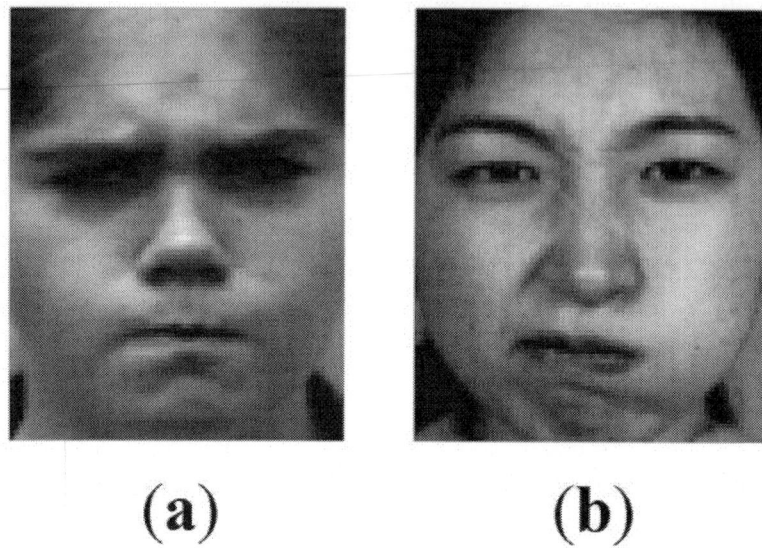

(a)   **(b)**

Figure 12. The same expression from different datasets. (**a**) The CK+ dataset; (**b**) The JAFFE dataset.

6. CONCLUSIONS

In this paper, a novel facial expression recognition based on sparse representation is proposed. First of all, the Fisher separation criterion is introduced to calculate the weight of LBP patches, which enhances the effect of important regions, such as the mouth and eyes. The most important contribution of this paper is the introduction of sparse representation as a facial expression recognition method. We have compared the proposed method with SVM, and low-resolution and noisy facial images are also taken into account. Furthermore, considering multi-intensity facial expression in the real-world, a multi-layer sparse representation (MLSR) model is constructed to improve the recognition rate. Promising results on standard database demonstrate the potential of the proposed approach. To this end, across-database experiments represent a future challenge in facial expression recognition.

ACKNOWLEDGMENTS

We would like to thank Jeffery Cohn for granting access to the Extended Cohn-Kanade Dataset, and Michael J. Lyons for granting access to the JAFFE Dataset. This work was supported by National Natural Science Foundation of China (61402072, 61402077, 61432003, 61328206, 11171052).

AUTHOR CONTRIBUTIONS

Qi Jia conceived and designed the experiments and wrote part of the paper. Xinkai Gao performed the experiments and also wrote part of the paper. He Guo supervised the work and contributed materials. Zhongxuan Luo and Yi Wang analyzed the data.

REFERENCES

1. Fang, H.; Mac Parthalain, N.; Aubrey, A.J.; Tam, G.K.L.; Borgo, R.; Rosin, P.L.; Grant, P.W.; Marshall, D.; Chen, M. Facial expression recognition in dynamic sequences: An integrated approach. *Pattern Recogn.* **2014**, *32*, 740–748.
2. Zeng, Z.; Pantic, M.; Roisman, G.; Huang, T. A survey of affect recognition methods: Audio, visual, and spontaneous expressions. *IEEE Trans. Pattern Anal. Mach. Intell.* **2009**, *31*, 39–58.
3. Sorci, M.; Antonini, G.; Cruz, J.; Robin, T.; Bierlaire, M.; Thiran, J.-P. Modelling human perception of static facial expressions. *Image Vis. Comput.* **2010**, *28*, 790–806.
4. Tian, Y. Evaluation of face resolution for expression analysis. In Proceedings of the 2004 IEEE Computer Society Conference on Computer Vision and Pattern Recognition Workshops (CVPRW'04), Washington, DC, USA, 27 May–2 June 2004; pp. 82–88.
5. Littlewort, G.; Bartlett, M.; Fasel, I.; Susskind, J.; Movellan, J. Dynamics of facial expression extracted automatically from video. *Image Vis. Comput.* **2006**, *24*, 615–625.
6. Yeasin, M.; Bullot, B.; Sharma, R. From facial expression to level of interest: A spatio-temporal approach. In Proceedings of the 2004 IEEE Computer Society Conference on Computer Vision and Pattern Recognition (CVPR'04), Washington, DC, USA, 27 June–2 July 2004; pp. 922–927.
7. Aleksic, P.S.; Katsaggelos, A.K. Automatic facial expression recognition using facial animation parameters and multi-stream HMMS. *IEEE Trans. Inf. Foren. Secur.* **2006**, *1*, 3–11.
8. Tian, Y.L.; Kanade, T.; Cohn, J. Recognizing action units for facial expression analysis. *IEEE Trans. Pattern Anal. Mach. Intell.* **2011**, *23*, 97–115.
9. Abdulrahman, M.; Gwadabe, T.R.; Abdu, F.J.; Eleyan, A. Gabor wavelet transform based facial expression recognition using PCA and LBP. In Proceedings of Signal Processing and Communications Applications Conference (SIU), Trabzon, Turkey, 23–25 April 2014; pp. 2265–2268.

10. Zhang, L.; Tjondronegoro, D. Facial expression recognition using facial movement features. *IEEE Trans. Affect. Comput.* **2011**, *2*, 219–229.

11. Gu, W.; Xiang, C.; Venkatesh, Y.V.; Huang, D.; Lin, H. Facial expression recognition using radial encoding of local Gabor features and classifier synthesis. *Pattern Recogn.* **2012**, *45*, 80–91.

12. Shan, C.; Gong, S.; McOwan, P.W. Robust facial expression recognition using local binary patterns. In Proceedings of the 2005 IEEE International Conference on Image Processing (ICIP'05), Genova, Italy, 11–14 September 2005; pp. 370–373.

13. Zhao, G.Y.; Pietikainen, M. Dynamic texture recognition using local binary patterns with an application to facial expressions. *IEEE Trans. Pattern Anal. Mach. Intell.* **2007**, *29*, 915–928.

14. Li, Z.S.; Imai, J.; Kaneko, M. Facial expression recognition using facial-component-based bag of words and PHOG descriptors. *Inf. Med. Technol.* **2010**, *5*, 1003–1009.

15. Eisenbarth, H.; Alpers, G.W. Happy Mouth and Sad Eyes: Scanning Emotional Facial Expressions. *Emotion* **2011**, *11*, 860–865.

16. Shan, C.; Gong, S.; McOwan, P.W. Facial expression recognition based on local binary pattern: A comprehensive study. *Image Vis. Comput.* **2009**, *27*, 803–816.

17. Uçar, A. Facial expression recognition based on significant face components using steerable pyramid transform. In Proceedings of the International Conference on Image Processing, Computer Vision and Pattern Recognition (IPCV'13), Las Vegas, NV, USA, 22–25 July 2013; Volume 2, pp. 687–692.

18. Uçar, A.; Demir, Y.; Güzeliş, C. A new facial expression recognition based on curvelet transform and online sequential extreme learning machine initialized with spherical clustering. *Neural Comput. Appl.* **2014**, 1–12.

19. Donoho, D.L. Compressed sensing. *IEEE Trans. Inf. Theor.* **2006**, *52*, 1289–1306.

20. Baraniuk, R.G. Compressive sensing. *IEEE Signal Process. Mag.* **2007**, *24*, 118–121.

21. Wright, J.; Yang, Y.; Ganesh, A.; Sastry, S.S.; Ma, Y. Robust face recognition via sparse representation. *IEEE Trans. Pattern Anal. Mach. Intell.* **2009**, *31*, 210–227.

22. Zhang, S.; Zhao, X.; Lei, B. Robust Facial Expression Recognition via Compressive Sensing. *Sensors* **2012**, *12*, 3747–3761.

23. Huang, M.; Wang, Z.; Ying, Z. New Method for Facial Expression Recognition Based on Sparse Representation Plus LBP. In Proceedings of the 3rd International Congress on Image and Signal Processing (CISP), Yantai, China, 16–18 October 2010; pp. 1750–1754.

24. Liu, P.; Han, S.; Tong, Y. Improving Facial Expression Analysis Using Histograms of Log-Transformed Nonnegative Sparse Representation with a Spatial Pyramid Structure. In Proceedings of the IEEE International Conference and Workshops on Automatic Face and Gesture Recognition (FG), Shanghai, China, 22–26 April 2013; pp. 1–7.

25. Ekman, P. Darwin, deception, and facial expression. *Ann. N. Y. Acad. Sci.* **2003**, *1000*, 205–221.

26. Ou, J.; Bai, X.; Pei, Y.; Ma, L.; Liu, W. Automatic Facial Expression Recognition Using Gabor Filter and Expression Analysis. In Proceedings of the 2010 Second International Conference on Computer Modeling and Simulation, Sanya, China, 22–24 January 2010; pp. 215–218.

27. Ouyang, Y.; Sang, N. A Facial Expression Recognition Method by Fusing Multiple Sparse Representation Based Classifiers. In Proceedings of the 10th International Symposium on Neural Networks, Dalian, China, 4–6 July 2013; pp. 479–488.

28. Jia, Q.; Liu, Y.; Guo, H.; Luo, Z.X.; Wang, Y.X. Sparse Representation Approach for Local Feature based Expression Recognition. In Proceedings of the International Conference on Multimedia Technology, Hangzhou China, 26–28 July 2011; pp. 4788–4792.

29. Ojala, T.; Pietikäinen, M.; Harwood, D. A comparative study of texture measures with classification based on featured distribution. *Pattern Recogn.* **1996**, *29*, 51–59.

30. Liu, Y.; Wu, F.; Zhang, Z.; Zhuang, Y. Sparse representation using nonnegative curds and whey. In Proceedings of the IEEE Conference on Computer Vision and Pattern Recognition (CVPR), San Francisco, CA, USA, 13–18 June 2010; pp. 3578–3585.

31. Ojala, T.; Pietikäinen, M.; Menp, T. Multiresolution grayscale and rotation invariant texture classification with local binary patterns. *IEEE Trans. Pattern Anal. Mach. Intell.* **2002**, *24*, 971–987.

32. Duda, R.; Hart, P.; Stork, D. *Pattern Classification*, 2nd ed.; Wiley Interscience: New York, NY, USA, 2000.

33. Huang, J.; Huang, X.; Metaxas, D. Learning with dynamic group sparsity. In Proceedings of International Conference on Computer Vision, Kyoto, Japan, 29 September–2 October 2009; pp. 64–71.

34. Kanade, T.; Cohn, J.F.; Tian, Y. Comprehensive database for facial expression analysis. In Proceedings of the Fourth IEEE International Conference on Automatic Face and Gesture Recognition, Grenoble, France, 28–30 March 2000; pp. 46–53.

35. Lucey, P.; Cohn, J.F.; Kanade, T.; Saragih, J.; Ambadar, Z. The Extended Cohn-Kanade Dataset (CK+): A Complete Dataset for Action Unit and Emotion-Specified Expression. In Proceedings of the 3rd IEEE

Workshop on CVPR for Human Communication Behavior Analysis, San Francisco, CA, USA, 13–18 June 2010; pp. 94–101.

36. Lyons, M.; Akamatsu, S.; Kamachi, M.; Gyoba, J. Coding Facial Expression with Gabor Wavelets. In Proceedings of the 3rd IEEE International Conference on Automatic Face and Gesture Recognition, Nara, Japan, 14–16 April 1988; pp. 200–205.

37. SPIDER.   Available online: http://www.people.kyb.tuebinggen.mpg.de/spider/index.html (acc essed on 20 October 2011).

Index